SPACE SAFETY AND RESCUE
1979 - 1981

including

Worldwide Disaster Response, Rescue and Safety Employing Space-Borne Systems

Organized under the aegis of
The International Academy of Astronautics

Proceedings of the Twelfth - Fourteenth International Symposia

PUBLICATIONS OF THE AMERICAN ASTRONAUTICAL SOCIETY

Following are the principal publications of the American Astronautical Society:

JOURNAL OF THE ASTRONAUTICAL SCIENCES (1954-)

Published quarterly and distributed by AAS Business Office, 6060 Duke Street, Alexandria, VA 22304

AAS NEWSLETTER (1962-)

Published bimonthly and distributed by AAS Business Office, 6060 Duke Street, Alexandria, VA 22304

ADVANCES IN THE ASTRONAUTICAL SCIENCES (1957-)

Proceedings of major AAS technical meetings. Published and distributed for the American Astronautical Society by Univelt, Inc., P.O. Box 28130, San Diego, CA 92128

SCIENCE AND TECHNOLOGY SERIES (1964-)

Supplement to Advances in the Astronautical Sciences. Proceedings and monographs, most of them based on AAS technical meetings. Published and distributed for the American Astronautical Society by Univelt, Inc., P.O. Box 28130, San Diego, CA 92128

AAS HISTORY SERIES (1977-)

Supplement to Advances in the Astronautical Sciences. Selected works in the field of aerospace history under the editorship of Dr. Eugene M. Emme, NASA Historian. Published and distributed for the American Astronautical Society by Univelt, Inc., P.O. Box 28130, San Diego, CA 92128

AAS MICROFICHE SERIES (1968-)

Supplement to Advances in the Astronautical Sciences. Consists principally of technical papers not included in the hard-copy volumes. Published and distributed for the American Astronautical Society by Univelt, Inc., P.O. Box 28130, San Diego, CA 92128

Subscriptions to the Journal and the AAS Newsletter should be ordered from the AAS Business Office. Back issues of the Journal and all books and microfiche should be ordered from Univelt, Inc.

Thanks are due Teresa Wilson, Univelt, Inc., for preparing copy for publication.

Front Cover Illustration:

Satellite service operations (see paper IAA 81-257).

Frontispiece:

Full disc image from METEOSAT 2 in the IR II micro "window" channel used to monitor surface conditions and weather systems (see paper IAA 81-263).

AN AMERICAN ASTRONAUTICAL SOCIETY PUBLICATION

SPACE SAFETY AND RESCUE 1979 - 1981

Edited by Jeri W. Brown

Volume 54, Science and Technology

A Supplement to Advances in the Astronautical Sciences

Proceedings of Symposia of the International Academy of Astronautics held in conjunction with the 30th, 31st, and 32nd International Astronautical Congresses. Organizing Chairman of the Symposia: Gloria W. Heath, Chairman, Space Safety and Rescue Studies Committee

Published by permission of
The International Academy of Astronautics

AMERICAN ASTRONAUTICAL SOCIETY

AAS Publications Office
P. O. Box 28130
San Diego, California 92128

Affiliated with the American Association for the Advancement of Science
Member of the International Astronautical Federation

ISBN 0-87703-177-0 (Hard Cover)
ISBN 0-87703-178-9 (Soft Cover)

Published by Univelt, Inc., P.O. Box 28130, San Diego, California 92128

Printed and Bound in the U.S.A.

FOREWORD

Among the papers selected for this Volume in the SCIENCE and TECHNOLOGY SERIES, several illuminate safety challenges that, if not addressed in the near future by the entire international space community, hold the potential for dramatically constraining future space operations. Other papers record the state of the art of safety standards and safety investigation modeling. Others point to policy questions that have yet to receive effective international attention - questions fundamental, from a safety viewpoint, to the orderly use of space. Still others present innovative proposals that would yield significant improvements in the alleviation of the human losses suffered from the effects of severe natural forces on earth.

The papers presented in this Volume were selected from those comprising the Symposia organized annually by the Space Safety & Rescue Studies Commitee of the International Academy of Astronautics. Each year a two-session Symposium is held as part of the International Astronautical Congress, hosted, in turn, by a National Space-Scientific Academy of member nations of the International Astronautical Federation. One session of the Symposium addresses activities relating to operations that take place in space itself; the other, space-segment systems and techniques that have applications to earth-surface disaster and distress response and prevention.

In the three-year period covered (1979-1981), the Symposia have provided leadership in enunciating the challenges presented by the accumulation of man-made space debris and those presented by reentering radioactive elements of nuclear-powered spacecraft. It has coalesced understanding of earth-surface distress response requirements and the application of appropriate space-system capabilities, drawing to its Studies Committee representatives of the preeminent international disaster and distress response agencies and associations. At the request of the Second United Nations Conference on the Exploration and Peaceful Uses of Outer Space (UNISPACE '82), it prepared a background paper on disaster applications (CHECKLIST: DISASTER WARNING & PREVENTION, EMERGENCY MEDICAL CARE, EMERGENCY COMMUNICATIONS - Paper No. A/CONF.101/BP/NGO/3(E)).

In the spirit that wide acquaintance with the Symposia purviews will benefit space investigations and operations universally, proposals for papers to be presented at future annual Symposia are welcomed, particularly the thoughtful analyses that direct attention to safety aspects insufficently grasped in the pursuit of the full range of space opportunities. The purviews of the Studies Committee and its Symposia appear on the following page.

Gloria W. Heath

Gloria W. Heath, Chairman
SPACE SAFETY & RESCUE Studies Committee
International Academy of Astronautics

SPACE SAFETY & RESCUE SYMPOSIUM
Organized by
International Academy of Astronautics

TWO SESSIONS: SPACE SAFETY & RESCUE (activities in space itself)

WORLDWIDE DISASTER RESPONSE, SAFETY, & RESCUE EMPLOYING SPACE-BORNE SYSTEMS

Session I: Space Safety & Rescue

Forward-looking analyses of safety design and operating principles underlying selection of future space programs. Models for risk assessment, accident investigation and safety data collection. Early warning detection schemes. Rescue techniques/equipment. Safety design precepts for equipment and work techniques. Safety standards.

Session II: Worldwide Disaster Response, Safety & Rescue Employing Space-borne Systems

Demonstrable space-system capabilities in aiding conduct of actual or simulated distress response functions. Feasibility of incorporation in global systems. Assessment of unified global system requirements, costs, performance. Evaluation of international institutional arrangements for distress-response task conduct.

Submit abstracts of paper proposals to the Organizing Chairman for both sessions:

G.W. Heath,
SAR-ASSIST
One Island Lane
Greenwich, CT 06830 USA

Telephone: (203) 869-1322

PREFACE

An international Space Safety and Rescue Symposium is held annually under the aegis of the International Academy of Astronautics. The Studies Committee for these symposia considers that a record of all technical papers presented at the symposia should be available from a single source. This volume along with earlier volumes in the Science and Technology Series and in the AAS Microfiche Series is a response to this requirement.

Be it understood, however, that the editors of Acta Astronautica, the official journal of the International Academy of Astronautics, select certain papers for publication. In these cases, only an abstract appears in the Science and Technology Series, with reference, if known, to the Acta or other source for the complete article.

It is planned henceforth to publish a proceedings volume every second year covering two Space Safety and Rescue Symposia.

The current volume includes selected papers from Space Safety and Rescue Symposia held 1979-1981. However, those papers omitted are covered by abstract or summary and abstracts of papers from 1976-1978 are also included. In all cases reference to sources, if known, for the complete paper is given.

A complete list of Space Safety and Rescue Symposia and available proceedings appears on page *x*.

The Editors

SPACE SAFETY AND RESCUE SYMPOSIA

These symposia held annually since 1968 by the Space Safety and Rescue Committee of the International Academy of Astronautics in conjunction with International Astronautical Congresses are now available for purchase through the American Astronautical Society. They are identified as follows:

Proceedings No.	Location of Symposium & Date	Publication
1 2 3	New York City, USA (1968) Mar del Plata, Argentina (1969) Constance, West Germany (1970)	Vol. 23, AAS Microfiche Series
4 5 6	Brussels, Belgium (1971) Vienna, Austria (1972) Baku, USSR (1973)	Vol. 24, AAS Microfiche Series
7	Amsterdam, Holland (1974)	Vol. 37, Science and Technology
8	Lisbon, Portugal (1975)	Vol. 41, Science and Technology
9 10 11	Los Angeles, USA (1976) Prague, Czechoslovakia (1977) Dubrovnik, Yugoslavia (1978)	In full in Vol. 40, AAS Microfiche Series Abstracts in Vol. 54, Science & Technology
12	Munich, Germany (1979)	Vol. 54, Science and Technology and Vol. 39, AAS Microfiche Series
13 14	Tokyo, Japan (1980) Rome, Italy (1981)	Vol. 54, Science and Technology and Vol. 41, AAS Microfiche Series
15	Paris, France (1982)	

Published by Univelt, Inc., for the American Astronautical Society
P.O. Box 28130, San Diego, California 92128

CONTENTS

CONTENTS (Cont'd)

CONTENTS (Cont'd)

CONTENTS (Cont'd)

CONTENTS (Cont'd)

This volume was produced by the photo-offset process using, to a considerable degree, manuscripts supplied by the authors. These manuscripts were typed or typeset using various fonts and there are some variations in format. It was not possible to review all copy editorially.

SPACE SAFETY AND RESCUE
1981

The following papers also appear in full in Volume 41, AAS Microfiche Series, supplement to this volume:

IAA 81-250, -251, -252, -255, -256, and -257

IAA 81-256

MAN-MADE SPACE DEBRIS - DOES IT RESTRICT FREE ACCESS TO SPACE?

M. Wolfe*, V. Chobotov*, D. Kessler**,
R. Reynolds***

The presence of man-made debris in orbit about the earth presents a hazard to spacecraft operating in that environment. Analysis has identified three areas of concern: (1) The hazard from the tracked population; (2) the hazard from the untracked population; and (3) the hazard from the future population.

At the present time, the collision hazard is real but not severe. However, continuation of present design and operational practices and procedures ensures that the probability of collision will increase and eventually reach unacceptable levels, perhaps within a decade. The space debris issue should be faced by all space users, and coordinated action taken if the future use of space is not to be seriously restricted; there is a critical need for an international dialogue to be initiated on the issue.

The space debris problem can be forestalled by immediate action in five major areas: (1) Education, (2) Technology, (3) Space Vehicle design, (4) Operational procedures and practices, (5) National and international space policies and treaties.

Keywords: Collision; debris; space; hazard; safety; explosion; operations; space policy.

SUMMARY

The presence of man-made debris in orbit about the earth presents a hazard to spacecraft operating in that environment, the level of hazard to a given spacecraft depending on the size of the spacecraft, the number and size of debris fragments in its operating environment, and the length of time the spacecraft remains on orbit. Efforts to provide a definitive assessment of this problem have been directed toward analyzing the hazard level presented by particular debris populations and predicting how this hazard level will change with time. Much less effort has been directed toward satellite design and strategies to minimize the short-term and long-term effects of debris deposition.

* *The Aerospace Corporation, El Segundo, California 90245*
** *NASA Lyndon B. Johnson Space Center, Houston, Texas 77058*
****Battelle Columbus Laboratories, Columbus, Ohio 43201*

A key aspect of the on-orbit debris hazard is that it is self-perpetuating. This arises from three factors: (1) a single spacecraft launch can be responsible for a multitude of hazardous objects in space; (2) orbital debris tends to disperse randomly, producing high intersection velocities and making avoidance extremely difficult; and (3) objects accumulate in earth orbit rather than passing through the near earth space in the manner of meteoroids. Impact protection may not be feasible in most cases because of the likelihood of very high approach velocities and the fact that certain protuberances, especially those of relatively large area such as solar arrays and antennae, cannot easily be permanently shielded. Evasive maneuvering techniques may reduce the present probability of collision for specific satellites in certain circumstances, but do not provide a practical long-term solution.

The only natural mechanism to oppose debris buildup is removal by atmospheric drag. However, this process can take a very long time, especially from high altitudes, and causes debris to migrate from higher to lower altitudes. Another mechanism, collection by a spacecraft (orbital garbage truck), would be extremely difficult and expensive. Prevention of debris formation is the most effective approach.

At the present time, the collision hazard is real but not severe. However, continuation of present policies and practices ensures that the probability of collision will eventually reach unacceptable levels, perhaps within a decade. Future problems can be forestalled by initiating studies and implementing their results in five major areas: (1) education, (2) technology, (3) satellite and vehicle design, (4) operational procedures and practices, and (5) national and international space policies and treaties.

There is presently no coordinated national or international concern for space use management. All countries and organizations using the near-earth space environment should perform a combined and detailed evaluation of debris control design and operations practices and policies. A dialog should be initiated as soon as possible to involve the United Nations and the member states in developing practical design and operating standards and regulatory policies. The impact of program and policy decisions, relevant to the deposition of objects in orbit, should be understood by all space users for both immediate and future space operations.

INTRODUCTION

Although the natural meteoroid environment has been considered in the design of past and existing spacecraft, future satellite design may have to take account of space debris in addition to the natural environment.

Man-made space debris differs from natural meteoroids because it is in permanent earth orbit during its lifetime and is not transient through the regions of interest. As a consequence, a given mass of material presents a greater problem in the design and operation of spacecraft because of the extended time period over which there is a collision risk.

Past design practices and deliberate and inadvertent explosions in space have created a significant debris population in operationally significant orbits. The debris consists of spent spacecraft and rocket stages, separation devices, and explosion products. Much of this debris is resident at altitudes of considerable operational interest. Products larger than 10 cm^2 in low orbits can be observed directly. The existence of a substantially larger population of small fragments can be inferred from terrestrial tests in which the particle distributions from explosions have been assayed. From these tests it is reasonable to infer small particle numbers of the order of 10 thousand for each low intensity explosion and ten million for high intensity explosions.

Two types of space debris are of interest: (1) large objects whose population, while small in absolute terms, is large relative to the population of similar masses in the natural flux (by a factor of about 1000); (2) a larger number of smaller objects, whose size distribution approximates natural meteoroids and which are additive to the flux in those size ranges. The interaction of these two classes of objects, combined with their long residual times in orbit, leads to the further concern that inevitably there will be collisions producing additional fragments and causing the total population to grow rapidly.

PROBLEM DEFINITION

The space debris population consists mostly of explosion fragments, but also includes: operating and expended satellites; mission-related objects, such as shrouds, clamps, and separation components; and spent propulsive elements. The larger of these can be tracked by the North American Air Defense (NORAD), but the smaller ones cannot. In addition to the problem of future orbit prediction, those objects that can be tracked are difficult to track with the precision required to prevent collisions.

Analysis has identified three areas of concern: (1) the hazard from the tracked population, (2) the hazard from the untracked population, and (3) the hazard from the future population. These are manifested in both low earth orbit (less than 2000 km) and geostationary orbit.

Tracked Population

NORAD is presently tracking more than 5000 objects in space. Most appear to be larger than about 10 cm in diameter, are in low earth orbit, and originate from explosions -- either spent propulsive stages or anti-satellite tests. (Over the last 10 yrs this population has grown at a rate of about 10% per year.) Figure 1 shows how the average number density of objects varies with altitude for the observed (tracked) population of objects. The greatest concentration is between altitudes of 500 km and 1100 km. Maximum density occurs near 850 km. Figure 2 provides a graphical representation of the hazard levels presented by tracked debris for several altitudes. This hazard is characterized by the expected time between collisions (ETBC), which is inversely proportional to the spatial density given in Fig. 1, and depends on the size of the object and the altitude at which it is operating. The sizes of typical spacecraft, actual as well as projected, are also indicated in Fig. 2.

The probability of collision for the Space Shuttle Orbiter is only 4×10^{-6} for a typical 7 day flight. However, the probability of collision for a large 50 m diameter spacecraft in one year is about 0.001 at 500 km, and 0.003 at 850 km. For geostationary satellites, the probabilities of collision with other tracked objects in one year are between 10^{-6} and 10^{-5}. However, only objects larger than approximately 1 m diameter are detected at this altitude and the numbers may actually be greater.

Planned spacecraft cover many orders of magnitude in size; debris densities vary by several orders of magnitude. The results in ETBCs will vary from days to millions of years. Although, at present, the Shuttle Orbiter and smaller spacecraft at 850 km and below do not face a significant collision problem, future very large area spacecraft at these altitudes could be seriously endangered.

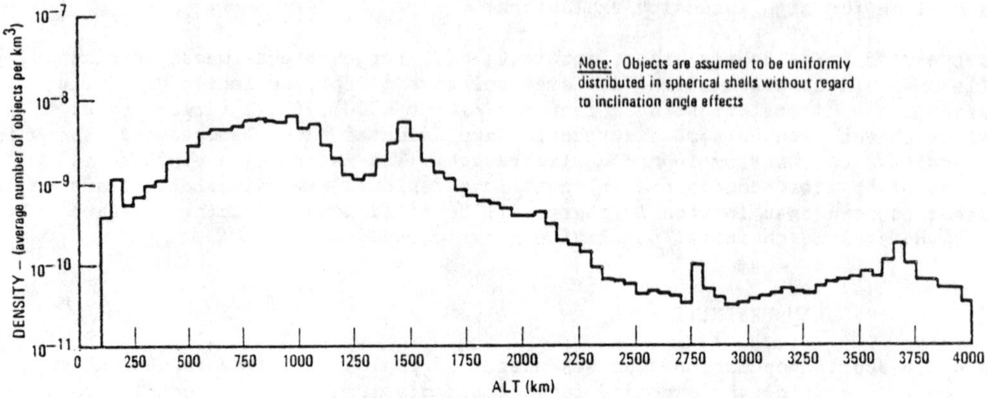

Source: Adapted From: Chobotov, V. A. (1981). Collision Hazard in Space. The Aerospace Corporation Report TR-0081(6790)-1.

Fig. 1. Observed space object density vs altitude.

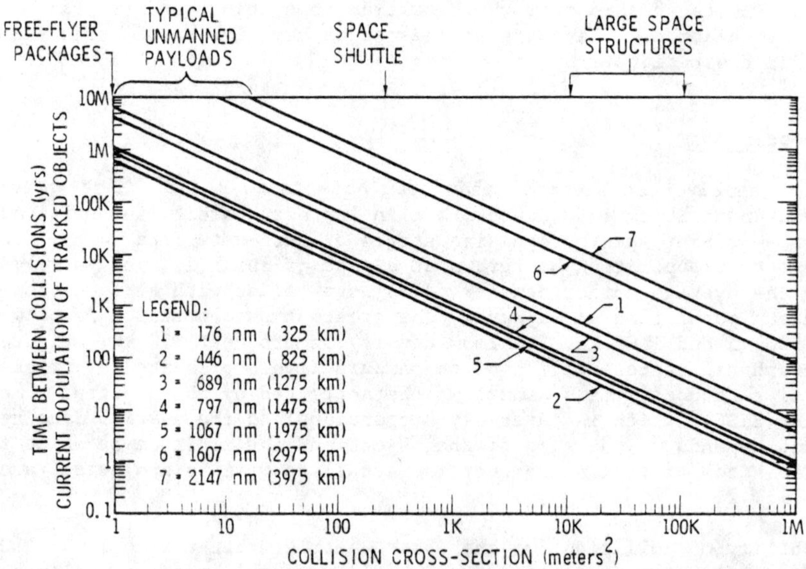

Source: Adapted From: Reynolds, R. C., and D. S. Edgecombe (1981). Status of the On-Orbit Debris Problem for Spacecraft in Low Earth Orbit. Battelle Briefing Notes.

Fig. 2. Time between collisions as a function of collision cross-section and altitude for the current tracked debris population.

Untracked Population

An untracked population of objects is known to exist because: (1) more than 50 explosions have occurred in space, undoubtedly producing many objects too small to be tracked; (2) objects apparently as small as 4 cm in diameter are frequently detected just prior to reentry. Objects of this size, not detectable at higher altitudes, can be detected when their orbits decay to about 400 km, just before reentry. Based on the observed number of these objects at 400 km and the size distribution characteristics of particles generated in low intensity explosions on the ground, collision probabilities can be estimated. The collision probability of a 50 m diameter spacecraft with the 4 cm and larger population in one year is estimated to be 0.007 at 500 km and 0.016 at 850 km. These rates are characterized in Fig. 3 by showing the ETBC as a function of spacecraft cross-section. The probabilities of collision are surely higher from fragments smaller than 4 cm but the data which exist are inadequate to make definitive estimates.

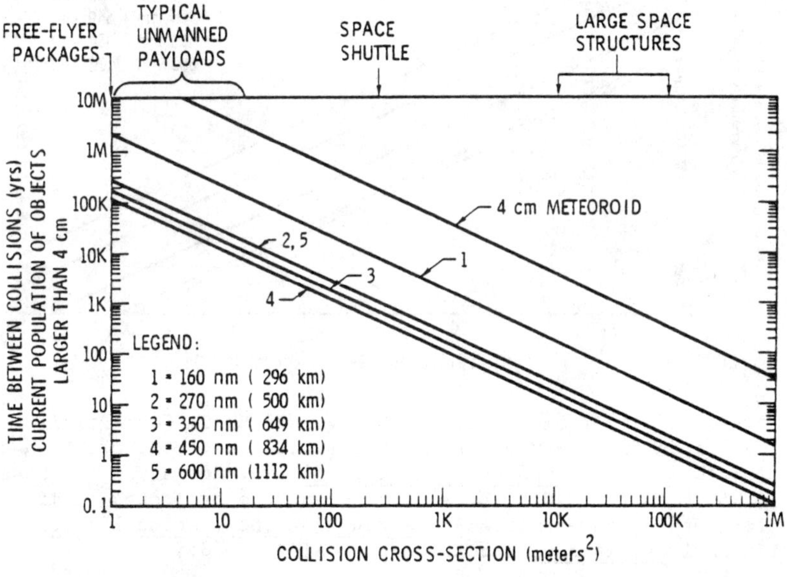

Source: Kessler, D. J. (1980). Sources of orbital debris and the projected environment for future spacecraft. Presented AIAA International Meeting and Technology Display, 6-8 May 1980. AIAA-80-0855.

Fig. 3. Time between collisions as a function of collision cross-section and altitude for the current background debris population larger than 4 cm.

Future Population

If past trends continue, the number of tracked objects in space is predicted to increase by a factor of 2 to 8 (depending on the rate of future explosions) within the next 20 years, significantly increasing the probabilities of collision. Furthermore, there may be sufficient objects in space that collisions could be expected between non-operational objects in low earth orbit. Just one of these collisions could produce 1.4×10^4 particles larger than 1 cm, and 3.5×10^6 particles larger than 1 mm. A 1 cm object in low earth orbit could penetrate a 5 cm thickness of solid aluminum. The collision frequency between a 50 m diameter spacecraft and the 1 cm population in the year 1995 could be about 0.033/yr at 500 km and 0.250/yr at 1100 km. The increased number of small fragments could significantly reduce the reliability of even small structures. In addition, this high frequency of collision will propagate an irreversible growth in fragment population. For illustration, the ETBCs from debris equal to or larger than 1 cm and 1 mm in diameter are shown in Fig. 4 and 5, respectively, for the year 1995.

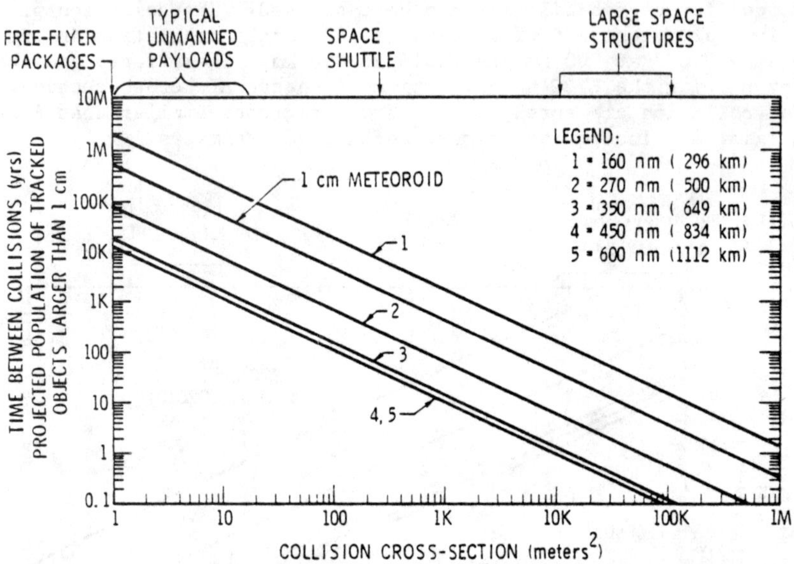

Source: Based on Data From: Kessler, D. J., and B. G. Cour-Palais (1980). Collision frequency of artificial satellites: the creation of a debris belt. J. Geophysical Res., 83, No. A6, and D. J. Kessler (1980). Sources of orbital debris and the projected environment for future spacecraft. Presented AIAA International Meeting and Technology Display, 6-8 May 1980. AIAA-80-0855.

Fig. 4. Time between collisions as a function of collision cross-section and altitude for the projected debris populations larger than 1 cm in the year 1995.

Periodically, over the course of the solar sun spot eleven year cycle, the earth's atmosphere is excited and rises significantly above its median altitude. During these periods, drag effects cause particles to migrate to lower altitudes. Eventually, such a process is cleansing, but is extremely slow. It is inadequate to deal with past or present rates of debris generation.

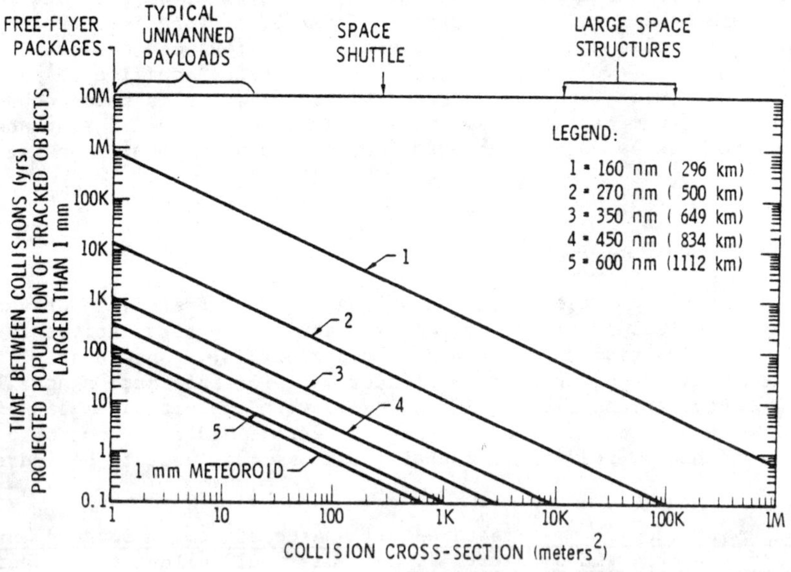

Source: Based on Data From: Kessler, D. J., and B. G. Cour-Palais (1980). Collision frequency of artificial satellites: the creation of a debris belt. J. Geophysical Res., 83, No. A6, and D. J. Kessler (1980). Sources of orbital debris and the projected environment for future spacecraft. Presented AIAA International Meeting and Technology Display, 6-8 May 1980. AIAA-80-0855.

Fig. 5. Time between collisions as a function of collision cross-section and altitude for the projected debris populations larger than 1 mm in the year 1995.

WHAT IS BEING DONE

There is presently no strong national or international concern for space debris management, although individual investigators are studying the problem, and operational satellite program offices have expressed concern about the safety of their own satellites. Questions on the subject have been posed by the U.S. State Department, the Office of the Under Secretary of Defense for Research and Engineering, the United Nations, the National Aeronautics and Space Administration (NASA), and representatives of a number of other entities. However, there is no central organization to coordinate and implement the actions which may be required. In general, reclamation or disposal of space objects is not perceived by the user community as being economical or even important. For this reason, there is a tendency to view the collision problem only as it impacts the economic loss of a specific satellite or in terms of the legal liability of the responsible agency. The issue of space debris control needs to be dealt with, not in this parochial fashion, but as a common problem shared by all space users.

Funded work is being carried out by the NASA Johnson Space Center (NASA/JSC), the U.S. Air Force Space Division, The Aerospace Corporation, Battelle Columbus Laboratories, and the European Space Agency. Studies are continuing to better define and predict the threat of collision. Devices to track smaller objects and to improve the accuracy of tracking are being tested. NASA/JSC is attempting to evaluate the magnitude of the problem and the effectiveness of various methods to minimize the propagation of debris. Operational procedures are being modified to reduce the probability of collision, and the probability of collision is being considered in selecting orbital parameters for new satellites. The problem of space debris has already been shown to be controllable to some degree when sound planning decisions are made.

WHAT SHOULD BE DONE

Before any potential "solutions" become policy, it is extremely important to fully understand the problems generated by orbital debris, the alternatives for debris control, and the effectiveness of the proposed controls. Consequently, it is imperative that the formal and informal interchange of information at all levels (national and international) begin immediately, with the ultimate goal of forming a central group to coordinate and encourage research, and to recommend policy. Some of the fundamental policy/procedural questions that need to be addressed are:

Should a policy be adopted that requires all spacecraft to be boosted out of geostationary orbit at the end of useful life? This would alleviate a specific, very important problem.

Should a policy be adopted to regulate which objects may be left in long-life orbits? The reduction in hazard implied by the adoption of this type of policy will have to be traded off against the penalty in deliverable payload. On-orbit debris modeling will provide a quantitative assessment of such a policy. Unless this question is addressed, certain regions in the near-earth environment may become unusable.

If the on-orbit debris hazard becomes significant, will the use of collision avoidance systems relieve the problem? The velocity of a spacecraft relative to the debris objects will typically be on the order of 10 km/sec. Hence, objects which are quite small (and which will be the most numerous) may cause considerable damage if a collision occurs. The small size of these objects will probably require sophisticated detection equipment on the spacecraft. The added weight for the detection system and propellant needed to perform the evasive maneuvers will reduce useful payload.

If the on-orbit debris hazard becomes significant, will the employment of impact protection (bumpers) relieve the problem? The major considerations in answering this question come from the field of fracture mechanics under hypervelocity impact conditions. Bumpers will have to be able to stop particles moving at velocities on the order of 10 km/sec, while producing a minimum number of debris particles.

Although bumpers of any size will reduce the useful payload that can be carried, decisions on the maximum particle size to be protected against and the parts of the spacecraft which are critical and must be protected (and, therefore, on the amount of bumper material required) will be guided by a definitive knowledge of the debris population and its kinematic properties.

What are the implications of anti-satellite operations? There are several military space activities which leave debris in orbit. The consequences of such actions must be understood before they are undertaken and even if such actions are essential to national security, they should be carried out with a clear understanding of the consequences. An analysis of the debris production will not only provide information on the short and long-term effects, but also suggest strategies and procedures for minimizing their impact.

Comprehensive documents should be produced which specify operational and procedural techniques for minimizing debris in critical areas. In the long term, the issue must be faced cooperatively by all space users and international agreements should be drawn up to ban, or restrict to low orbit, the explosion of satellites, and to institute efficient procedures for the disposal of expended satellites and space debris. In the interim, each separate space user community should immediately institute programs for evaluating actions to minimize the collision problem for all users. Consideration should be given to:

Education

Education of the total space-user community in the critical nature of the problem

Education of space system designers in the need for litter-free systems

Technology

Advanced ground and space-based detection techniques

Comprehensive models of the earth-space environment

Design and operation of a debris monitoring system

Design and operation of a debris collection system

Space Vehicle Design

Litter-free systems

Space systems disposal by: retrieval, earth reentry, earth escape, transfer to selected "dump" regions or less-frequented orbits

Addition of shielding to survive collisions from smaller debris

Rocket designs which limit the likelihood of accidental explosions

Designs that decentralize or minimize critically sensitive areas

Operational Procedures and Practices

Safe spacing of satellites

Avoidance of crowding specific orbit locations

Limitation of explosions to low altitudes so that fragments reenter quickly

Planning of launch trajectories for early reentry of spent rocket stages and dead payloads

Space Policies and Treaties

Identification within each agency of responsible groups for recommending and implementing required actions

Initiation of conferences at the international level to develop policy and agreements

Identification of national and international level groups responsible for coordinating research, establishing standards and policy, and enforcing agreements.

CONCLUSIONS AND RECOMMENDATIONS

It is concluded that the space debris issue is real, that action to resolve it is imperative, and that no obvious, simplistic resolution is evident. The nature of the issue can be summarized as follows:

At the present time, the collision hazard posed by space debris is real but not severe. However, continuation of present design and operational practices and procedures ensures that the probability of collision will increase and eventually reach unacceptable levels, perhaps within a decade.

The space debris issue should be faced by all space users, and coordinated action taken immediately if the future use of space is not to be seriously restricted.

Design to tolerate debris impact (bumpers), or to provide evasive capability, may provide a measure of protection to particular satellites but the most effective approach is to eliminate the need for such action by constraining the generation of further debris.

The space debris problem can be forestalled by immediate action in five major areas:

- Education
- Technology
- Space vehicle design
- Operational procedures and practices
- National and international space policies and treaties

There is an immediate need for an international dialogue to be initiated on the space debris issue, with the goal of forming responsible groups to coordinate research and recommend policy.

ACKNOWLEDGMENTS

This paper was prepared under the auspices of the American Institute of Aeronautics and Astronautics (AIAA) Space Systems Technical Committee and is representative of an official AIAA position on space debris. The AIAA has developed a position in recognition of its responsibility to further the understanding of the importance of the space debris problem, so that all space users are encouraged to take necessary and appropriate action.

SAFETY ASPECTS OF NUCLEAR WASTE DISPOSAL IN SPACE

E.E. Rice, D.S. Edgecombe*

P.R. Compton**

The future desirability and acceptance by society of nuclear waste disposal in space depends upon safety, cost, and perceived risk issues. This paper addresses only the safety issues. As a part of the U.S. Program for assessing nuclear waste disposal in space as a complement to mined geologic repositories, we have developed, over the years, a general safety philosophy for the disposal of nuclear waste in space. The philosophy has been developed by considering the safety requirements used for radioisotope thermal generators and other nuclear payloads, as well as the specific problems and needs for disposing of nuclear waste. This paper outlines for space disposal: (1) the general safety guidelines for waste disposal in space; (2) the current Reference Concept (August 21, 1981); and (3) the safety aspects that are currently built into the current Reference Concept.

Keywords: Safety; safety requirments; safety guidelines; risk; accident environments; Space Shuttle; STS; nuclear; nuclear waste; radioactivity; radiation exposure; nuclear accidents; space rescue; international; treaties; U.S. DOE: DOE; NASA.

INTRODUCTION

The importance of nuclear waste management has made it necessary that studies of reasonable disposal alternatives be carried to the point that determinations can be made regarding their relative merit. One of the disposal alternatives that have been studied is a space disposal option, as a complement to the mined geologic repository. Various concepts for the space disposal of nuclear waste have been studied in the U.S.A. Most recent studies guided by NASA have concluded that space disposal is feasible and that it can be carried out in a safe manner (Rice and co-workers, 1980a). The unique feature of the space disposal of nuclear waste is the expectation of total and permanent separation of certain wastes from the human environment. It should be emphasized again, that nuclear waste disposal in space would be a complement, not a replacement for mined repositories.

* *Space Systems and Applications Section, Battelle's Columbus Laboratories, Columbus, Ohio, U.S.A.*

** *Space Utilization Systems, NASA/Headquarters Energy Systems Division, Washington, D.C., U.S.A.*

During 1981, NASA and DOE are performing a sustaining level assessment of the concept of disposing of certain high-level nuclear wastes in space as a complement to the currently planned U.S. nuclear waste management program (Rice and co-workers, 1981). One of the most important factors in the ultimate decision making process for this concept is public health safety. For space disposal to be an acceptable approach, it is likely that the total long-term health risk of a space disposal concept coupled with terrestrial disposal must be at a comparable or preferably at a much lower level than that of terrestrial disposal of all the waste. Short-term health risk must be at an acceptable level.

Over the years of studying space disposal, a "safety concept" has been developing. The various aspects of this safety concept are presented in this paper. The concept has taken the form of a document "Draft System Safety Guidelines for Space Disposal of Nuclear Waste" (August 26, 1981, draft-unpublished); previous drafts can be found in Rice's work (1980b) and in Priest's work (1980). Work done on safety specifications for radioisotope thermal generators (U.S. DOE, 1977), was included in the development of safety guidelines for space disposal. As a result of 1981 study, the safety guidelines have been modified and a preliminary risk assessment will soon be made to determine if the current Reference Concept for space disposal, as a complement, has anything to offer over terrestrial disposal of all high-level nuclear waste material. The current study (Rice and co-workers, 1981) is assessing possible accidents and events which contribute most to radiological risk to people. Both the short-term accidents and long-term events are being assessed.

Some of important issues facing the total nuclear waste management system are: What level of health risk needs to be achieved, technically? What is the actual level of health risk? What is the perceived risk to the public and how important is the "not in my back yard syndrome"? Can other options, such as space disposal, sub-seabed, or very deep hole help satisfy public concerns such that mined geologic disposal can be implemented without major problems and nuclear power development can continue?

Major issues that relate to the safety/long- and short-term risk of a space disposal complement are: the short-term risk due to accidents during booster ascent and transfer out of low Earth orbit, or the possible return to Earth of a payload after improper placement in solar orbit. What degree of safety is required to achieve an acceptable level of risk at a reasonable cost increment to the total waste management system? For nuclear waste disposal missions, Fig. 1 shows how safety and cost can be traded. The level of public and regulatory concerns will ultimately determine safety and cost for space disposal. The aspects discussed in this paper attempt to balance reasonable systems with reasonable costs to what is believed an acceptable risk level for space disposal.

The three major sections that follow include: (1) the general safety guidelines for space disposal; (2) a brief overview of the current (August 21, 1981) Reference Concept; and (3) the safety aspects built in to the Reference Concept.

GENERAL SAFETY GUIDELINES

The general safety objectives for the nuclear waste disposal in space mission are: (1) to contain the solid radioactive waste materials and (2) to limit the exposure of humans and the environment to the radioactive waste materials. For normal operations, complete containment and minimum radiological exposure are required. For potential accident situations, the degree of containment and

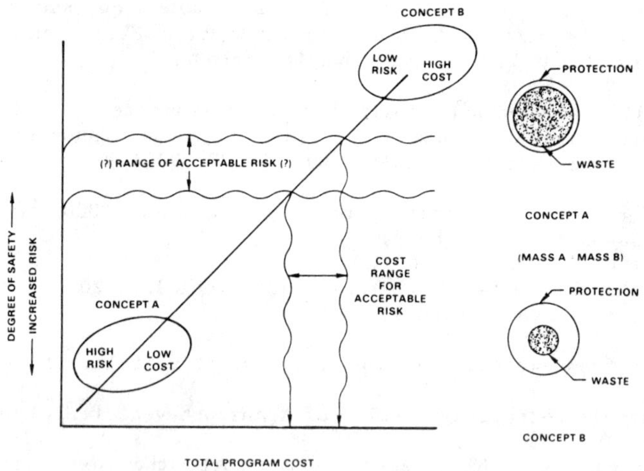

Fig. 1. Safety versus cost for space disposal of nuclear waste.

degree of interaction shall result in an acceptable risk to humans and the environment and be as low as reasonably achievable (ALARA). Many of the general safety guidelines have been selected using our best judgement and do not have the benefit of detailed analysis.

Future work is necessary to improve these guidelines and to insure confidence that the guidelines will provide for assured safety.

The general system safety guidelines for the nuclear waste disposal in space mission involve the following safety aspects:

(1) Radiation Exposure and Shielding
(2) Containment
(3) Accident Environments
(4) Criticality
(5) Postaccident Recovery
(6) Monitoring Systems
(7) Isolation.

The information given below defines the guidelines (as of August 26, 1981) that should be followed for the development of any nuclear waste disposal in space mission, employing conventional space technology approaches, such as the Space Shuttle.

Radiation Exposure and Shielding

Radiation exposure guidelines for normal operations for the public and ground crews are those contained in ERDA-MC-0524 (U.S. DOE, 1975) and shown in Table 1. Radiation exposure limits for astronaut crew members during normal operations are those contained in the Space Shuttle Flight and Ground Specification, JSC 07700 (NASA, 1979) and shown in Table 2.

The normal radiation exposure limits for the current terrestrial transportation of nuclear waste materials would also apply to ground transportation of nuclear waste payloads. The radiation limits [49 CFR 173.393(j)] are given as:

- 1 m from external container surface...1000 mrem/hour (closed transport vehicle only)
- External surface of transport vehicle...200 mrem/hour (closed transport vehicle only)
- 2 m from external surface of transport vehicle...10 mrem/hour
- Normally occupied position of transport vehicle...2 mrem/hour.

For accident conditions of terrestrial transport, the post accident dose rates are limited to 1000 mrem/hour at 1 meter from the external surface of the waste package.

A general guideline for the waste package shipped to space is that the radiation dose at 1 meter from the flight shield surface is not greater than 1000 mrem/hour. This value can be obtained by including shielding contributions from outer layers of the payload. In the absence of these layers, no more than 2000 mrem/hr at 1 m should be allowed. The shield should be carried all the way to the destination.

Containment

Containment must be defined for the various portions of the disposal mission. Five general mission phases include fabrication/assembly of the waste form, terrestrial transport, launch site handling, launch to Earth orbit, and orbit transfer to destination. For all normal operations, the systems should be designed so that no release of radioactive material occurs. For accident environments the system should be designed so that the risk to the public is acceptable. Current U.S. federal regulations cover little beyond transportation and general handling aspects. The discussion below outlines the general guidelines for containment of the high-level waste form during each phase of the space disposal mission.

Containment/Philosophy

The ideal goal for containment of high-level waste material is to (1) provide an absolute barrier between the waste and the environment, and (2) minimize the radiation exposure to humans under all normal and accident conditions. Various government regulations have been developed and applied to current terrestrial transportation activities involving radioactive materials, including irradiated nuclear fuel. No regulation applies to the space transport of high-level waste. Consequently, the containment guidelines developed here are based on considering

TABLE 1 Normal Operations Exposure Limits for Individuals in Controlled and Uncontrolled Areas

INDIVIDUALS IN CONTROLLED AREAS:

Type of Exposure	Exposure Period	Dose Equivalent (Dose or Dose Commitment[(a)], rem)
Whole body, head and trunk, gonads, lens of the eye[(b)], red bone marrow, active blood forming organs.	Year Calendar Quarter	5[(c)] 3
Unlimited areas of the skin except hands and forearms). Other organs, tissues, and organ systems (except bone).	Year Calendar Quarter	15 5
Bone.	Year Calendar Quarter	30 10
Forearms.[(d)]	Year Calendar Quarter	30 10
Hands[d] and feet.	Year Calendar Quarter	75 25

INDIVIDUALS IN UNCONTROLLED AREAS:

Annual Dose Equivalent or Dose Commitment (rem)[(e)]

Type of Exposure	Based on dose to individuals at points of maximum probable exposure	Based on an average dose to a suitable sample of exposed population
Whole body, gonads, or bone marrow	0.5	0.17
Other organs	1.5	0.5

(a) To meet the above dose commitment standards, operations must be conducted in such a manner that it would be unlikely that an individual would assimilate in a critical organ, by inhalation, ingestion, or absorption, a quantity of a radionuclide(s) that would commit the individual to an organ dose which exceeds the limits specified in the above table.

(b) A beta exposure below an average energy of 700 Kev will not penetrate the lens of the eye; therefore, the applicable limit for these energies would be that for the skin (15 rem/year).

(c) In special cases with the approval of the Director, Division of Operational Safety, a worker may exceed 5 rem/year provided his average exposure per year since age 18 will not exceed 5 rem per year.

(d) All reasonable efforts shall be made to keep exposures of forearms and hands to the general limit for the skin.

(e) In keeping with ERDA policy on lowest practicable exposure, exposures to the public shall be limited to as small a fraction of the respective annual dose limits as is practicable.

Source: U.S. DOE, 1975.

TABLE 2 Radiation Exposure Limits for Space Shuttle Flight Crews[(a)]

Constraints, rem	Bone Marrow, 5 cm	Skin, 0.1 mm	Eye, 3 mm	Testes[(c)]
1 year average daily rate	0.2	0.6	0.3	0.1
30-day maximum	25	75	37	13
Quarterly maximum[(b)]	35	105	52	18
Yearly maximum	75	225	112	38
Career limit	400	1200	600	200

Notes:

(a) These exposure limits and exposure rate constraints apply to all sources of radiation exposure. In making trade-offs between man-made and natural sources of radiation, adequate allowance must be made for the contingency of unexpected exposure. These data are from Space Shuttle Flight and Ground Specification, JSC 07700, Volume X, Revision F, Chapter 7.4 (NASA, 1979). Estimated doses for Shuttle crew members, as based upon STS-1 launch are $\sim$0.01 rem per day.

(b) May be allowed for two consecutive quarters followed by six months of restriction from further exposure to maintain yearly limit.

(c) These dose and dose rate limits are applicable only where the possibility of oligospermia and temporary infertility are to be avoided. For most manned space flights, the allowable exposure accumulation to the Germinal Epithelium (3 cm) will be the subject of a risk/gain decision for particular program, mission, and individuals concerned.

existing regulations for other radioactive material handling, storage, and transport activities. The containment philosophy is applied first to meet current regulations and, second, to minimize human exposure to as low as reasonably achievable (ALARA) where regulations do not exist.

The acceptable amount of radioactive material that may be released should be a function of the mission phase, accident severity and frequency of the loading condition, keeping in mind (ALARA) criteria. Allowable releases from the primary container are expected to range from zero, for normal conditions, to minimal values (ALARA) for extremely severe, low probability accidents.

Containment guidelines are presented in terms of three independent categories: (1) specific parameters indicative of the response of various containment systems; (2) specific systems for containing the waste (waste form, container, etc.); and (3) various mission phases during which specific levels of containment conditions are required. Table 3 lists the components of these categories. The use of these three levels of containment categories can be used to define any aspect of containment.

TABLE 3 Specific Components of Containment

Parameters	Components	Mission Phases
• Thermal	• Waste Form	• Fabrication/Assembly
• Mechanical	• Primary Container	• Terrestrial Transport
• Chemical	• Radiation Shield	• Launch Site Handling
	• Impact Absorber	• Launch to Earth Orbit
	• Ablation Shield	• Orbit Transfer to Destination
	• Shipping Cask	

Parameters

Containment guidelines take the form of specific limits for key parameters. For the space disposal mission, the significant parameters can be grouped into three major categories: (1) thermal; (2) mechanical; and (3) chemical. Within each category, many specific parameters can be included if the design is known; only the generic parameters are included in this discussion. Once a conceptual design is known/postulated specific technical data (temperatures, etc.) may be substituted to obtain specific working guidelines.

Thermal. Thermal guidelines consider only limiting conditions which, irrespective of critical parameters in other areas, serve as upper bounds to determine permissible designs and response. In interrelations with other major parameters, a single parameter is considered as the limiting guideline and, through its dependence upon other parameters, in effect produces corresponding limits. Thus, parameters such as melting temperature will be independent limits, while yield strength will be a function of temperature. The limiting thermal conditions for all forms of containment and mission phases should require that the containment barrier must not be altered in physical or chemical phase during operations that are not remote from the human environment. For normal conditions, 40% of the material melt absolute temperature is the guideline (material creep not expected to be significant below this). If the melt absolute temperature is not known, then 90% of the fabrication absolute temperature is the guideline. For accident conditions, the temperature limit is the 90% of the melting absolute temperature. (See Table 4.)

Mechanical Strength. For all normal mission phases, and containment barriers, the mechanical strength must maintain stress and strain limits within 90% of the normal yield limits for given temperature conditions (standard 0.2% offset). For accident conditions, where stress/strain limits are provided, one should not exceed 90% of ultimate strength requirements at the temperature anticipated. Mechanical strength limits are assumed to be dependent on temperature and loading condition. In addition, it is assumed that all containment barriers must also

TABLE 4 Thermal Guidelines for Containment of High-level Waste for Space Disposal

	Mission Phase				
Component	Fabrication/ Assembly	Terrestrial Transport	Launch Site Handling	Launch to Earth Orbit	Orbit Transfer to Destination
Waste Form	40% Melt/ 90% Melt*	40% Melt/ 90% Melt	40% Melt/ 90% Melt	40% Melt/ 90% Melt	40% Melt/ 90% Melt
Primary Container	40% Melt/ 90% Melt	40% Melt/ 90% Melt	40% Melt/ 90% Melt	40% Melt/ 90% Melt	40% Melt/ 90% Melt
Flight Radiation Shield	40% Melt/ 90% Melt	40% Melt/ 90% Melt	40% Melt/ 90% Melt	40% Melt/ 90% Melt	40% Melt/ 90% Melt
Impact Absorber	--	--	40% Melt/ 90% Melt	40% Melt/ 90% Melt	40% Melt/ 90% Melt
Ablation Shield	--	--	40% Melt/--	40% Melt/--	40% Melt/--
Shipping Cask	--	DOT, NRC Reg.	--	--	--

*Note: The normal absolute temperature limit is given first: the accident absolute temperature limit is given second. If the melt absolute temperatures are not appropriate for the material in question, then the 90% of the fabrication absolute temperature should apply.

have sufficient fracture toughness, fatigue endurance, and buckling stability to withstand normal and accident conditions. (See Table 5.) For accidents associated with reentry, total mechanical integrity for all components is not feasible. Rather, a guideline to be used would be to allow deformation but not allow major breach of containment.

Chemical. Containment materials shall be compatible with adjacent media to the extent that no significant detrimental chemical reactions occur and the material is non-pyrophoric in an air environment at sea-level pressure. For conditions not covered by existing U.S. federal regulations, guidelines are provided for the various containment barriers and mission phases. (See Table 6.)

Containment Components

Containment components constitute the barrier between the payload and the external environment. Depending on the mission phase, the containment barrier may be a single item (e.g., waste primary container) or multiple items (e.g., primary container, radiation shield, impact absorber, etc.). Consequently, a particular component may not be a part of containment in all mission phases. If it is not a

TABLE 5 Mechanical Guidelines for Containment of High-level Waste for Space Disposal

Component	Mission Phase				
	Fabrication/ Assembly	Terrestrial Transport	Launch Site Handling	Launch to Earth Orbit	Orbit Transfer to Destination
Waste Form	90% Yield/ 90% Ultimate*	90% Yield/ 90% Ultimate	90% Yield/ 90% Ultimate	90% Yield/ No Breach at 95% Confidence	90% Yield/ No Breach at 95% Confidence
Primary Container	90% Yield/ 90% Ultimate	DOT, NRC Reg.	90% Yield/ 90% Ultimate	90% Yield/--	90% Yield/--
Flight Radiation Shield	90% Yield/ 90% Ultimate	90% Yield/ 90% Ultimate	90% Yield/ 90% Ultimate	90% Yield/--	90% Yield/--
Impact Absorber	--	--	90% Yield/--	90% Yield/--	90% Yield/--
Ablation Shield	--	--	90% Yield/--	90% Yield/--	90% Yield/--
Shipping Cask	--	DOT, NRC Reg.	--	--	--

*Note: The normal mechanical limit is given first: the accident mechanical limit is given second.

TABLE 6 Chemical Guidelines for Containment of High-level Waste for Space Disposal

Component	Mission Phase				
	Fabrication/ Assembly	Terrestrial Transport	Launch Site Handling	Launch to Earth Orbit	Orbit Transfer to Destination
Waste Form	Container Compatible, Non-pyrophoric in air at SL	Container Compatible, Non-pyrophoric in air at SL	Container Compatible, Non-pyrophoric in air at SL	Container Compatible, Non-pyrophoric in air at SL	Container Compatible, Non-pyrophoric in air at SL
Primary Container	Similar to DOT, NRC Reg.	DOT, NRC Reg.	Similar to DOT, NRC Reg.	Similar to DOT, NRC Reg.	Waste form Compatible Non-pyrophoric in air at SL
Flight Radiation Shield	Similar to DOT, NRC Reg.	Similar to DOT, NRC Reg.	Similar to DOT, NRC Reg.	Similar to DOT, NRC Reg.	Container Compatible, Non-pyrophoric in air at SL
Impact Absorber	Similar to DOT, NRC Reg.	Similar to DOT, NRC Reg.	Similar to DOT, NRC Reg.	Similar to DOT, NRC Reg.	Shield Compatible Non-pyrophoric in air at SL
Ablation Shield	--	--	Similar to DOT, NRC Reg.	Similar to DOT, NRC Reg.	Impact Absorber Compatible Non-pyrophoric in air
Shipping Cask	--	DOT, NRC Reg.	--	--	--

part of the containment barrier, then the specific limits for containment, as applied to a particular subsystem, do not apply.

Waste Form. The principal containment barrier for the space disposal option is the primary container. A strong, non-dispersible, waste form is required to minimize the possibility and quantity of nuclide release. To meet these requirements, the waste form will still have limits for thermal, mechanical and chemical parameters. The nuclear waste mix and form will be designed to meet the criticality limit specified in later paragraphs.

A meeting was held at Battelle Columbus Laboratories on July 19, 1979 to evaluate waste forms for the space disposal of commercial and defense high-level waste (HLW). Participants included ONWI, NASA, BCL, and DOE-Richland Operations personnel and waste form experts from Battelle Northwest Laboratories, Oak Ridge National Laboratories, Idaho Chemical Processing Plant, and Sandia Laboratories. During that meeting, the following set of parameters (listed in order of priority) was determined to be applicable and important to the selection of a space disposal waste form.

1. **High waste loading** - This is an important parameter when considering the disposal of the large mass of commercial or defense HLW. Waste forms having high waste loadings will require fewer launches. Fewer launches not only lower operational costs but also yield lower construction costs for dedicated launch facilities. Rating: primary importance.

2. **High thermal conductivity** - Commercial HLW, especially in waste forms having high waste loadings, generates a significant quantity of heat. To prevent central regions from becoming excessively hot, the waste form should possess a relatively high heat transfer coefficient. Similarly, upon unplanned reentry of a waste package, the waste form should be capable of rapidly conducting heat away from the container surface. Low heat waste payloads imply that this is not an important consideration. Rating: primary importance.

3. **Toughness** - An aspect of dispersion resistance is material toughness. The waste form should be shatter and abrasion resistant upon impact, and should deform to absorb impact. Retrieval of the waste form as a single piece rather than many fragmented parts is desirable. Powdered forms are not desirable; liquid forms are unacceptable. Rating: primary importance.

4. **Thermochemical stability** - In launch pad or reentry accidents, the waste form should remain chemically stable. It should not degrade, decompose, or otherwise be altered in its chemical form in such a way that a significant release of radionuclides occurs during such postulated accidents. Rating: primary importance.

5. **Resistance to thermal shock** - The waste form should be resistant to shattering which may be caused by thermal shock. This is a property that will help in achieving low dispersibility of the waste form. Rating: secondary importance.

6. **Resistance to leaching** - A low leach rate may be important if the waste form package impacts into water after an unplanned reentry. While leach rate may be important, it is not as important as in terrestrial disposal where radionuclide transport by groundwater is considered the most probable mechanism for loss of isolation. Rating: secondary importance.

7. **Resistance to oxidation** - Another aspect of dispersion resistance is waste form resistance to oxidation. If an unplanned reentry of the damaged waste package occurred, the surface of the waste form should not rapidly oxidize and break away from the main body of the waste package. Rating: secondary importance.

8. **Economics and resource utilization** - Waste form materials and fabrication processes should not be prohibitively expensive. Also, waste form materials should not severely deplete valuable raw materials. Since the cost of a booster launch is expected to be the major part of the total cost, waste form material and process costs will not be overly important. Rating: secondary importance.

For the waste form, the maximum temperature limit for normal conditions is 90% of the fabrication or creep absolute temperature; for accident conditions the limit is 90% of the melting absolute temperature. Mechanical limits, when applicable for containment, are yield (normal) and ultimate strengths or low dispersion (accident). Chemical limits are that the waste form be compatible with container materials, exhibit low reactivity, and be non-pyrophoric. It is also required that a subcritical condition be maintained at all times (see discussion on criticality).

Primary Container. The primary container, designed to enclose the waste form throughout all mission phases beyond waste form fabrication, is also the primary containment boundary. The thermal limit for normal conditions is 40% of the melt absolute temperature. For accident conditions, 90% of the absolute melt temperature is the guideline. Mechanical limits are yield (normal) and ultimate strengths or low dispersion (accident). Chemical limits are covered by existing federal regulations (U.S. NRC, 1978).

Radiation Shield. The radiation shield for flight should be designed to function during all mission phases through transfer to the final destination. The radiation shield should be supplemented with auxiliary shielding materials, as needed during various mission phases, such that radiation exposure limits are not reached. For mechanical strength, 90% of the yield (normal) and 90% of the ultimate (accident) stress limits apply (ultimate does not apply for launch and orbit transfer operations). Thermal limits are 40% of the melt absolute temperature (normal conditions) and 90% of the absolute melt temperature (accident conditions). Chemical requirements will be similar to those in existing federal regulations (U.S. NRC, 1978).

Radiation shielding limits for the payload package (1000 mrem/hr at 1 m) have been assumed for conditions not covered by existing regulations. Conservative limits (such as those for transportation) have not been selected due to the sensitivity of the overall system design (payload/shield mass ratio) to the dose limits. Rather, the guideline limits chosen reflect the fact that the waste payload package will be isolated from the general public throughout all but a small fraction of its lifetime.

Impact Absorber and Ablation Shield. The impact absorber and ablation shield, have similar containment limits. For thermal guidelines, 40% of the melt (normal) and 90% of the melt (accident) absolute temperature apply to the impact absorber; 40% of absolute melt temperature applies as the upper limit for the ablation shield under normal conditions. For mechanical strength, yield (normal) limits exist. The absorber and ablation shield will be chemically non-reactive with other containment layers (similar to other DOT/NRC regulations). They will be non-pyrophoric. The impact absorber is designed to absorb mechanical energy during accidents. The yield strength of the absorber material is expected to be exceeded. Therefore, the ablation shield which is designed to reduce heating effects during possible reentry phases, is not expected to survive ground or water impact.

Shipping Cask. During ground-based Earth transport, the high-level waste package will be enclosed within a shipping cask. Current U.S. federal regulations [10 CFR 71 (U.S. NRC, 1978), 49 CFR 173 (U.S. DOT, 1979) and IAEA Safety Series Number 6 (IAEA, 1973)] define the requirements for this component, which is expected to be similar to conventional shipping cask designs.

Mission Phases

As described previously, the containment guidelines are also a function of mission phases, and, more specifically, the conditions within each phase. The definition of mission phase, as used for containment guidelines, is chronological.

Fabrication/Assembly. This phase begins with the insertion of the waste form into the primary container, and ends with the beginning of transport of the flight-shielded primary container out of the fabrication facility. During this phase, auxiliary cooling and additional shielding may be required. The primary container is the principal containment barrier during this phase.

Terrestrial Transport. This phase begins at the time of loading of the waste container and shield into the shipping cask and ends with the unloading at the launch site. Throughout this phase, active auxiliary cooling may be required to maintain thermal limits. At both ends of the phase, additional shielding may be required. The requirements are defined for irradiated nuclear materials in existing U.S. regulations. They are expected to be similar to regulations governing transport of processed waste. While in the shipping cask, the cask vessel will be the principal containment barrier during transport.

Launch Site Handling. This phase, similar to the initial one, begins with the arrival of the shipping cask at the launch site and ends with the completion of transfer into the launch system. Auxiliary cooling, and additional shielding, may be required during this phase depending on the waste form characteristics. The principal containment barrier remains the primary container.

Launch to Earth Orbit. This phase begins with the loading of the waste payload into the launch system, and ends after Earth orbit has been achieved. Auxiliary cooling may be required for most of this phase depending on the waste form characteristics. Containment will rely principally on the primary container. Accident conditions that might occur during this phase are among the most severe. The guidelines during this phase allow for: no melting, and no significant failure of the waste container.

Orbit Transfer to Destination. This mission phase commences with the removal of the waste payload from the launch system upon arrival at Earth orbit, and concludes with the payload arrival at the final destination. No active auxiliary cooling will be required during this phase. Containment guidelines for this phase (and the long-term) are expected to be less restrictive than those for near-term phases involving greater chances of public exposure. The radiation shield, primary container and the waste form provide the principal containment barrier for the waste.

Accident Environments

The accident environments that need to be considered in the design of containment and other auxiliary systems are as follows:

- Shipping accidents
- Ground handling accidents
- On-pad or near-pad booster vehicle failures
- Reentry accidents.

Shipping Accident Environments

DOT and NRC regulations, as defined in 49 CFR 170 to 179 (U.S. DOT, 1979) and 10 CFR 71 (U.S. NRC 1978), will be assumed for the ground shipment of nuclear waste payloads from the waste payload fabrication facility to the launch site. The following sequential test environments for shipping cask accidents are given below. Initial conditions are to be assumed the same as the normal condition.

- A 9-m drop in worst orientation onto an unyielding surface

- A 1-m drop in the worst orientation onto the end of 15-cm-diameter, 20-cm-high bar (mild steel)

- A 30-min ground fire at 800 C followed by 3 hours of no artificial cooling; with a cask emissivity of 0.9 and cask absorbtivity of 0.8

- An 8-hour emersion in 0.9 m of water.

At the end of this test, the shipment will meet the conditions specified in 10 CFR 71.36 including:

(1) An external radiation dose rate not exceeding 1000 mrem/hr at 1 meter from the external surface of the waste package

(2) No release of radioactive material from the package, except for gases and contaminated coolant containing total radioactivity exceeding neither:

(a) 0.1 percent of the total radioactivity of the package contents; nor

(b) 0.01 Curie of Group I radionuclides
0.5 Curie of Group II radionuclides
10 Curies of Group III radionuclides
10 Curies of Group IV radionuclides, and
1000 Curies of inert gases irrespective of transport group.

Ground Handling Environments

The payload systems, auxiliary support equipment and facilities must be designed to minimize the occupational radiation exposure to workers (see Table 1). Care must also be taken to ensure that if certain subsystem failures occur during ground handling, radiation exposure and contamination is kept to as low as reasonably achievable. The waste payload preparation facility at the launch site should be designed to be a total containment facility.

On- or Near-Pad or Ascent Booster Accident Environments

The payload package must be designed to survive the predicted accident environments for a given time in the flight for a given vehicle (see Rice and co-workers, 1980b) in expected <u>sequence</u> without a major breach of primary containment. Initial conditions are assumed to be the normal condition. An example of predicted environments are given below for the on- or near-pad for liquid boosted Space Shuttle (from Rice and co-workers, 1980b). This example would <u>generally</u> provide the worst-case accident potential, assuming proper consideration of ground impact.

- A blast side-on overpressure of 250 N/cm^2, a reflected overpressure of 1700 N/cm^2 and side-on and reflected impulses of 2.0 and 15.0 $N\text{-}s/cm^2$, respectively, in worst orientation. (Based upon a 10 percent explosive yield of the ET propellants)

- A potential edge-on penetration of 1 per m^2 of impacting fragments, assumed to be discs 100 cm in diameter and 0.56 cm thick, having a mass of 12 kg, and moving at 500 meters per second. The worst orientation is assumed.

- A heat flux of 3500 kW/m^2 for 15 seconds from a liquid propellant fireball.

- A 60-min ground fire at 1100 C followed by 2 hours of no artificial cooling.

- An impact in the worst orientation onto an unyielding surface at 10 percent higher than the predicted terminal velocity or an impact on to land such that the payload is buried, in low conductivity soil (k = TBD), but does not reach 90% of the melt absolute temperature.

- An impact in the worst orientation into 25 C water at a velocity 10 higher than the predicted terminal velocity, followed by a descent into the ocean to a depth corresponding to a hydrostatic pressure of 12,000 N/cm^2.

Reentry Accidents

The payload package shipped to its space destination must be able to withstand inadvertent reentry into the Earth's atmosphere and impact onto the Earth's surface without the dispersion of significant quantities of radioactive material. The reentry environments that must be considered for each type of disposal mission are defined as follows:

- A decaying reentry trajectory (shallow angle Skylab type) to provide maximum heating energy possible

- A reentry trajectory (steep angle) which provides the maximum heating flux possible

- An impact in the worst orientation onto an unyielding surface (western granite) at a velocity 10 percent higher than the predicted terminal velocity or an impact onto land such that the reentering waste payload is buried in low conductivity soil (k = TBD), but the waste form does not reach 90% of the melt absolute temperature.

- An impact in the worst orientation into 25 C water at a velocity 10 percent higher than the predicted terminal velocity, followed by a descent into the ocean to a depth corresponding to a hydrostatic pressure of 12,000 N/cm^2.

The response of the payload package to the reentry environments mentioned above should be calculated after the possible reentry conditions have been determined by analysis for a specific disposal mission type.

Criticality

The radioactive waste payload package must be subcritical (calculated K-effective $+3\sigma \leq 0.95$) for normal operations or any possible credible accident during processing, fabrication, handling, storage, or transport to the space destination. Calculations should show that any credible change in waste form geometry and any credible grouping of packages will not cause K-effective $+3\sigma$ to exceed 0.95.

Postaccident Recovery

Postaccident recovery teams should be made part of the operational disposal system. They should be responsible for all accident recovery operations, including accidents involving processing, payload fabrication and railroad shipment, payload preparation at the launch site, the launch and possible reentry. Special recovery equipment should be developed and be provided for possible use in postaccident recovery activities. Every credible eventuality should be considered in developing recovery plans. Every effort should be made to recover as much waste material as possible.

Monitoring Systems

Monitoring systems should be developed for the overall system such that overall mission safety can be assured. Examples of such systems include: devices for measuring radiation; temperature and pressure in the waste payload package;

instruments to provide data for tracking the payload during its transit to its desired destination; and permanent labeling specifying the product, contents, history and radiation projection of the waste contents.

Isolation

The nominal space destination should insure, at a minimum, an expected isolation time from the Earth's biosphere in excess of one million years, and should not adversely interfere with normal space operations projected to be carried out by future generations. Careless contamination of celestial bodies should be avoided.

REFERENCE CONCEPT

The current Reference Concept, as defined in a program-level, working meeting on August 21, 1981, represents a concept that basically evolved over the years, is believed to include adequate safety, and is expected to provide benefits to the mined geologic repository. The ground operations are shown in Fig. 2; the space operations are shown in Fig. 3. Aspects of the overall disposal mission are discussed below.

Waste Source and Mix

The primary waste source would be high-level nuclear waste generated by the operation of commercial nuclear power plants and recovered by reprocessing. The waste mix to be disposed of in space is reprocessed high-level waste (containing 0.5% of the Pu and U that is present in the fuel rods at the time of reprocessing) that has been out of the reactor for 50 years. Also, at the time of reprocessing, 95% of the Sr and Cs is removed. Gases and trans-uranic (TRU) wastes, plus 95% of Sr and Cs, would go to disposal in the mined repository. Plutonium would be processed out of the TRU wastes; this fraction would be added to the mix and go into space for disposal. The combination of Cs and Sr removal and the 50 year aging of the waste is needed to avoid post-burial meltdown for the "reference sized" sphere packages flown on a given mission. (Smaller spheres or dilution of the waste form would allow the transport of 10 year old waste.)

Waste Form

The reference waste form for space disposal is the Oak Ridge National Laboratory (ORNL) iron/nickel based cermet (ceramic/metal matrix), a dispersion of ceramic particles in a continuous metallic phase. This waste form has been chosen over other waste forms because of the expected responses to possible accident environments. The cermet is expected to have a waste loading of the order of 67% where 100% is defined as high-level waste in oxide form. The thermal conductivity is expected to be about 9.5 Watts/m-C and the density is about 6.5 g/cc.

Waste Processing and Payload Fabrication

The cermet waste form would be made into cylindrical billets approximately 5 cm long and 5 cm in diameter. They would be placed into a solid spherical steel container system where holes have been milled into the solid sphere accommodate

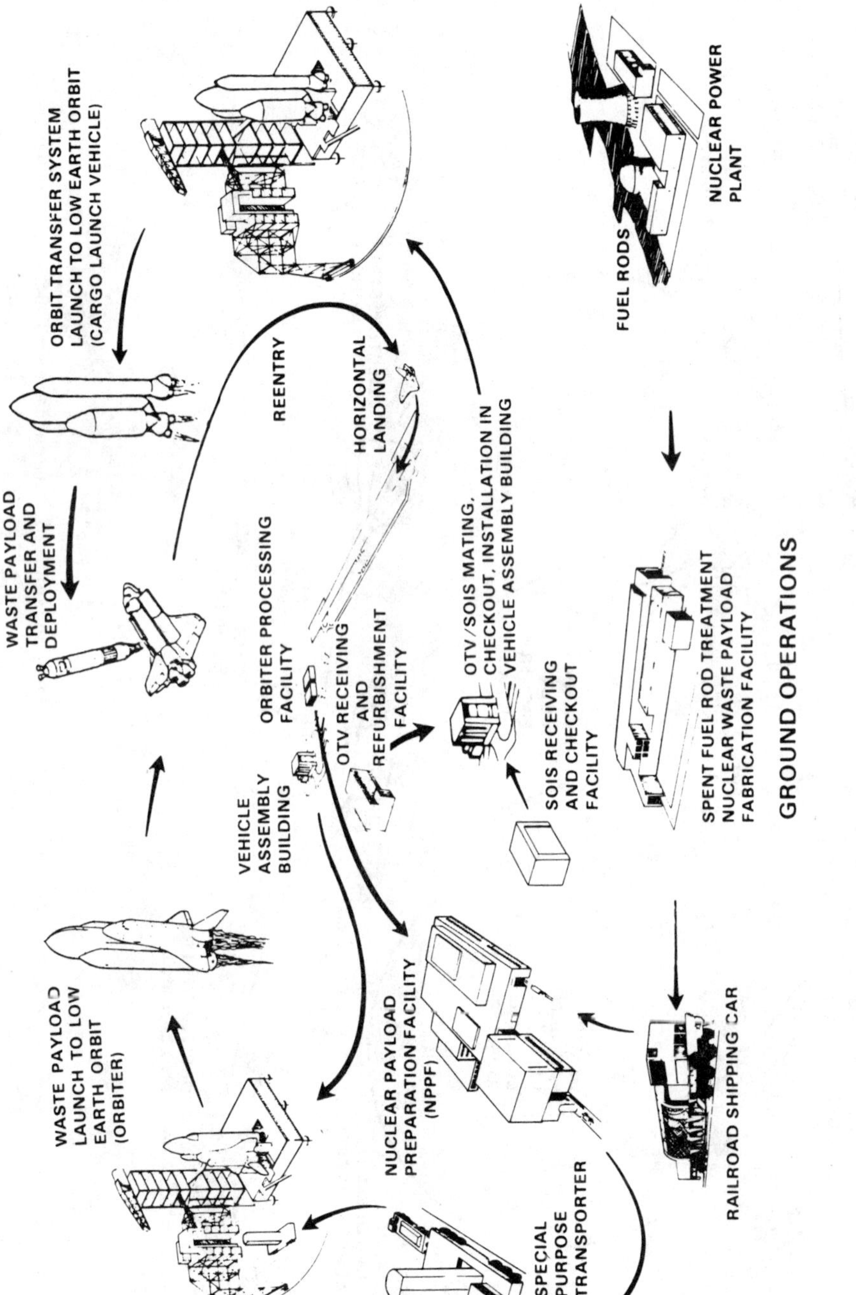

Fig. 2. Ground operations for the current reference concept for space disposal.

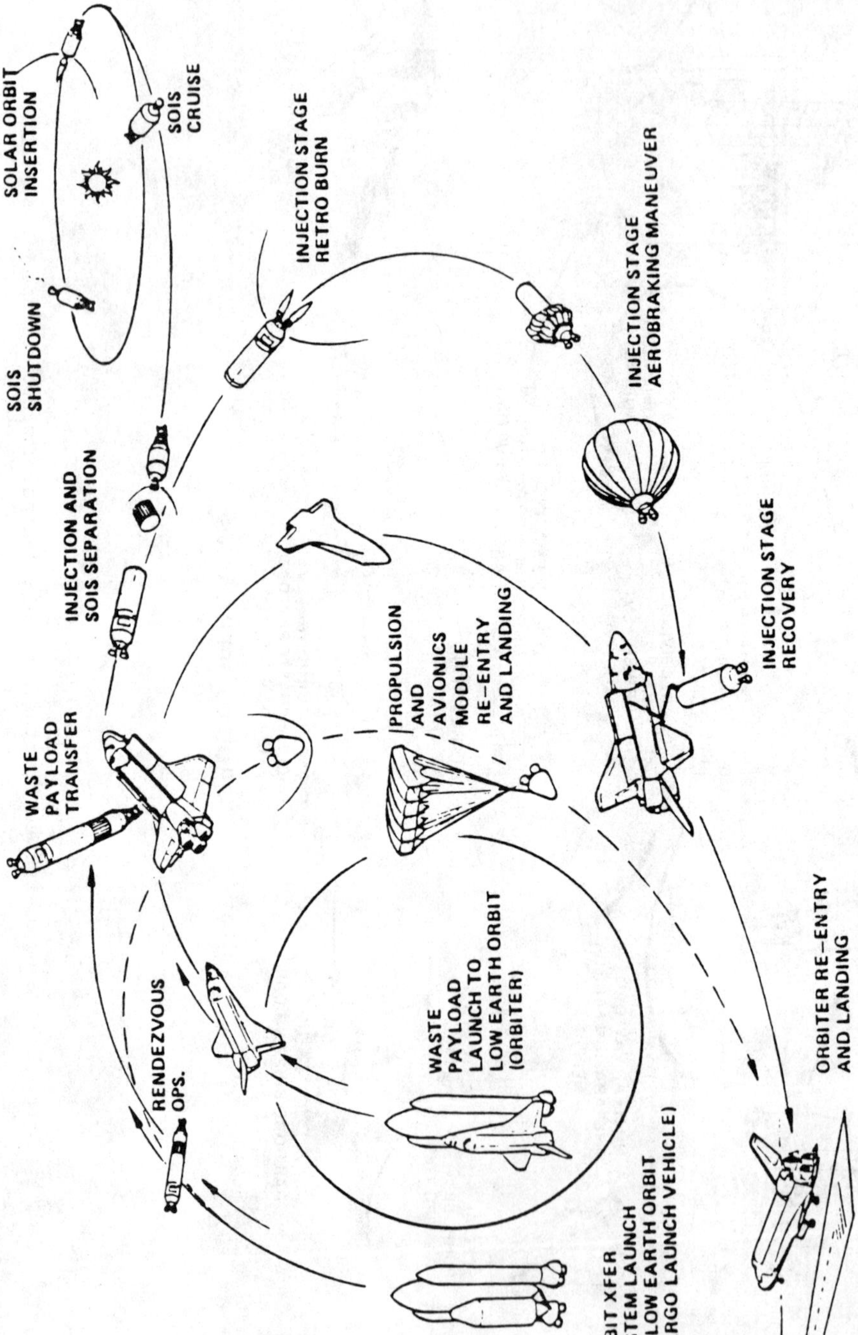

Fig. 3. Space operations for the current reference concept for space disposal of nuclear wastes.

the waste form billets. Fig. 4 shows the waste container/integral shield concept. At the payload fabrication facility, after the billets have been installed into the container system, the "lid" on the container/integral shield would be placed into position and then welded with an electric beam. The thick steel container wall serves as a radiation shield, as well as an impact structure to protect against breaches of containment. Before further handling, the container outer wall would be decontaminated. Individual graphite reentry tiles would be placed on the container by means of bolts. Then two half shells of steel would be placed around the graphite covered sphere (see Fig. 5) and seal welded. Then the system is placed into a shipping cask for shipment to the launch site.

Shipping Casks and Ground Transport Vehicles

For transport from the waste processing and fabrication facilities, the waste package would be housed in a shipping cask. The cask would be licensed by the U.S. NRC and would be transported by rail to the launch site. The cask system would require an active cooling system.

Launch Site Facilities and Operations

Upon arrival at the launch site the waste package would be removed from the cask and placed into its flight support structure system and stored for launch.

Uprated Space Shuttle Vehicle

The Uprated Space Shuttle vehicle is defined as having oxygen/RP-1 liquid rocket boosters (LRB's) replacing the Solid Rocket Boosters (SRB's). This not only provides for a 45,000 kg payload, but allows increased safety for the launch ascent phase and a lower launch cost.

Shuttle Derived Vehicle (SDV)

The Shuttle Derived Vehicle (SDV) would be the basic Uprated Space Shuttle, where the Orbiter is replaced by the SSME propulsion pod and a large aerodynamic payload shroud (see Fig. 3). This vehicle would be used to launch orbit transfer vehicles (OTV) and the Solar Orbit Insertion Stage (SOIS) into low Earth orbit. The vehicle would also use liquid rocket boosters.

Orbit Transfer Vehicle and Solar Orbit Insertion Stage

The Orbit Transfer Vehicle (OTV) would be a large hydrogen/oxygen cryogenic stage, employing existing technology. The Solar Orbit Insertion Stage (SOIS) would be a cryogenic propellant vehicle with long-life subsystems.

Orbital Operations

The SDV would be launched first to place the OTV/SOIS on orbit (38°, 300 km). Two waste packages would then be launched by the Uprated Space Shuttle to rendevous with the OTV/SOIS. The OTV/SOIS would then dock with the payload in the cargo bay of the Shuttle Orbiter. The payload package would be removed and the

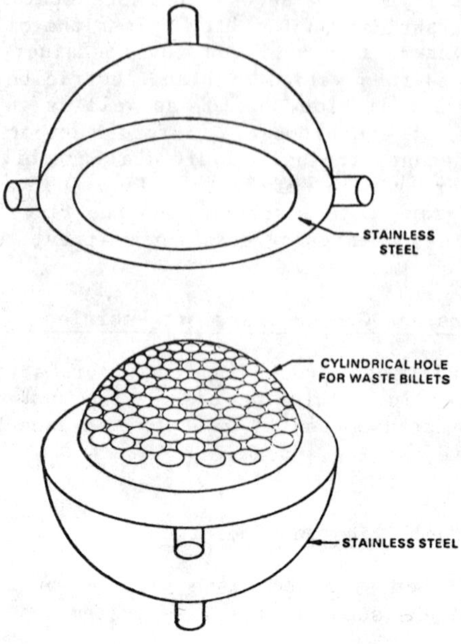

Fig. 4. Waste container and integral radiation shield.

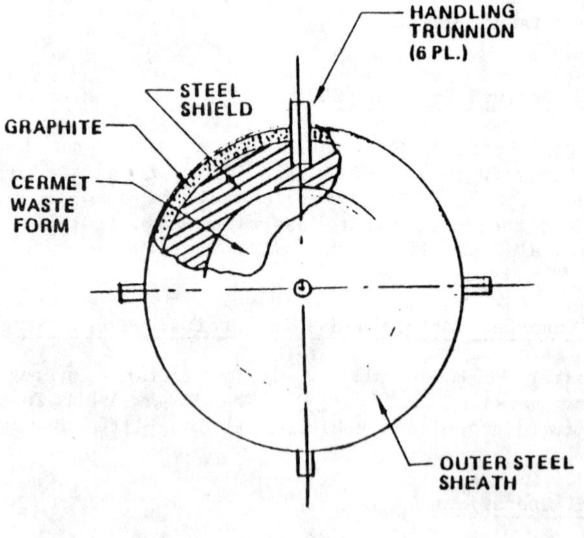

Fig. 5. Waste package with container, integral shield, graphite thermal protection and steel protective sheath.

OTV/ SOIS would carry out proper maneuvers to deliver the payload package to its destination. The OTV would be recoverable. The expendable SOIS would provide the velocity increment at perihelion (0.85 AU) needed to circularize the solar orbit.

Space Destination

The space destination for the Reference Concept is the orbital region between the orbits of Earth and Venus. The nominal circular orbit is defined as $0.85 \pm .01$ AU, with a one degree inclination from the ecliptic plane.

SYSTEM SAFETY ASPECTS FOR REFERENCE CONCEPT

The following paragraphs define specific safety design aspects established for the Reference Concept for space disposal of nuclear waste (as defined above). Safety aspects not stated here are inferred from the general safety guidelines. As the Reference Concept changes, these aspects may also change. Safety aspects are discussed in terms of Reference Concept elements and mission phase.

Safety Aspects of Elements

Safety aspects for the various major elements of the Reference Concept (August 21, 1981) for space disposal are described below.

Waste Form

For normal conditions, a cermet temperature of 1050 C (90% of fabrication absolute temperature) will not be exceeded. For accident conditions, a cermet temperature of 1280 C (90% of melt absolute temperature) shall not be exceeded. Criticality requirements will also be met.

Waste Processing and Payload Fabrication Facilities

The design and operation of these facilities will follow current proposed regulations, as specified for reprocessing plants.

Shipping Casks and Ground Transport Vehicles

Shipping casks and ground transport vehicles will comply with DOT and NRC regulations. The maximum outside diameter of the shipping cask will be 3.05 meters (10 feet). When required for heat rejection, a redundant cooling system for the shipping cask will be required.

Payload Primary Container

For normal conditions, the primary stainless steel container shall not exceed a temperature of 416 C (40% of melt absolute temperature). No chemical and physical interaction will occur between the cermet waste form and the container. For accident conditions, the primary container must not exceed a temperature of 1280 C (90% of melt absolute temperature).

Flight Radiation Shielding

Radiation shielding including outer layer shielding contributions for flight systems, will be designed to limit radiation to no more than 1000 mrem per hour at 1 meter from the package surface under normal conditions. The shield itself, when stripped of all outer "non-shielding" layers of the payload package, will not exceed 2000 mrem per hour at 1 meter from the shield. Auxiliary shielding will be designed such that radiation exposure limits (see Tables 1 and 2) for ground personnel and flight crews are not exceeded during handling or flight operations.

For normal conditions, the temperature limit for the stainless steel flight radiation shield is 416 C (40% of melt absolute temperature). For accident conditions, the stainless steel radiation shield must not exceed the temperature of 1280 C (90% of melt absolute temperature).

Reentry Systems

The reentry system for the Reference Concept includes two basic systems: The booster vehicle reentry system and the payload package reentry system.

The booster vehicle reentry system is the Space Shuttle Orbiter. It has the capability to detach from the ET and perform a controlled maneuver to a proper safe landing site (return-to-launch site, abort-to-contingency landing strip, abort-to-orbit, abort-to-sea or abort-to-land) at almost any time in the flight. The Orbiter has sophisticated and redundant guidance and control systems, an elaborate thermal protection system, as well as a manned crew, which will all aid in the safe return to Earth of the payload package as a result of a critical ascent booster system failure. In addition, the Orbiter will carry a structural pallet (supports the waste during launch) that will reduce the Orbiter crash landing loads placed on the payload package. Also, the Orbiter will provide systems which will allow for Orbiter flotation in the event of a ditching at sea.

The reentry system for the waste payload package must include provisions to survive expected on-pad and reentry accident environments. The system must include: (1) provisions for absorbing the expected external impact loads; (2) a fire and reentry thermal protection system; (3) a transmitter for recovery. The thermal protection system will not ablate more than 50% of its initial thickness during postulated worst-case reentry environments. The outer side of the package will have proper labeling.

Launch Site Facilities

The launch pad used for launching nuclear waste into space should be a dedicated pad. The Nuclear Payload Preparation Facility (NPPF) should be designed to be a total containment facility.

Safety Aspects of Nuclear Waste Disposal in Space

Uprated Space Shuttle Launch Vehicle

The Uprated Space Shuttle launch vehicle design will reflect considerations of keeping on-pad accident environments as low as possible. The overall minimum vehicle launch reliability will be greater than 0.999 (meaning one chance in 1000 that the Orbiter vehicle will be catastrophically lost during liftoff and ascent). The External Tank and Liquid Rocket Booster design will include destruct systems, properly located, which provide for a minimum explosive yield of the liquid propellants when a catastrophic event is in process or when life or property on the ground is at risk. Every effort will be made to save the payload and crew from adverse accident environments.

Earth Parking Orbits

Intermediate Earth parking orbits shall be incorporated into the flight profiles of space transportation systems to allow a minimum of 6 months before orbital decay of the nuclear waste payload package could occur.

Upper Stages

All upper stages will have a combined mission delivery reliability greater than 0.99. Achievement of delivery is defined as starting in the proper Earth parking orbit and ending within the bounds of the following: $0.85 \pm .01$ AU and 1.00 ± 0.20 degrees inclination.

Space Destination

The nominal space destination solar orbit at 0.85 AU, 1° from the Earth's orbital plane, will be verified by proper analysis to provide an expected isolation time of at least one million years.

Safety Aspects of Mission Phases

Safety aspects for the Reference Concept during the various mission phases have been developed over several years of study. The philosophies presented here are important to the future safe development of the space disposal concepts. Safety aspects peculiar to accidents and malfunction contingency plans for the general phases of the space disposal mission are listed and addressed below:

- Ground transportation from the payload fabrication sites to the launch site
- Preflight operations prior to ignition of the Shuttle's engines
- Launch operations from the launch pad to achieving parking orbit
- Orbital operations.

Ground Transportation

Ground transport (via rail) of the shipping cask would be assigned to the U.S. Department of Energy (DOE), which would supply the necessary accident recovery plans and systems, as needed. Two types of incidents that must be considered are: loss of auxiliary cooling to the waste container, and possible breach of the waste container with a loss of radioactive material. In case of cooling loss, adequate provisions should be made to have self-contained, auxiliary cooling units available within reasonable time. Monitoring equipment for both container temperature and radiation will be required during all ground transport operations. A continuous capability to cope with a container breach will be necessary. A specially trained accident recovery crew will always be ready to act, if necessary.

Preflight Operations

Contingency plans should be provided for potential malfunctions and accidents that could occur while waste payload packages are in the Nuclear Payload Preparation Facility (NPPF), being transported to the launch pad, being transferred from the pad Payload Changeout Room (PCR) to the Uprated Space Shuttle cargo bay, and awaiting liftoff in the Shuttle. Accidents and contingency plans would be similar to those discussed in the paragraph above.

Launch Operations

Contingency plans, procedures and systems envisioned to minimize the hazard caused by on- or near-pad failures are given below. These would minimize the effects of severe blast wave, high velocity fragments, fire, and possible high velocity impact.

- **A two Shuttle launch option** will allow the two waste payload packages to be launched in a cargo bay completely dedicated to the support of the payload package. This allows additional safety in that (1) no propellant-loaded orbit transfer vehicle is available to contribute to the close in accident environment (explosion, fire, high velocity fragments), and (2) additional structure can be used to protect the payload packages in the event of a crash land or catastrophic failure.

- **A destruct system** on the External Tank and LRB's properly designed and located (on the far sides and on the fuel tanks only) to minimize the blast wave and fragment environments caused by the exploding hydrogen/oxygen and RP-1/oxygen propellants. This would aid in minimizing the effect on the containment systems, as well as the crew and the Shuttle Orbiter.

- **Engine shutdown** is an important factor in vehicle survivability as a result of subsystem failures. For the Uprated Space Shuttle, all booster engines are liquid fueled, and as such, they can be shut down if a failure occurs during the engine start-up process, and prior to actual liftoff. This capability is expected to greatly reduce the probability of an on-pad catastrophic vehicle failure. Also, the abort modes mentioned below, become possible.

- **Flotation gear and locator beacons** in the Orbiter to assist in the recovery operation after ditching at sea.

- **Conservatively designed containment systems** (e.g., container, shielding impact and thermal protection and systems) to maximize the probability of surviving the possible hostile environments.

- **The use of a tough, solid waste form** not easily dispersed under adverse conditions.

- **The application of appropriate launch constraints** (e.g., wind direction) to reduce human radiological exposure resulting from a potential containment breach from a catastrophic launch accident.

- **The use of a recovery team** ready to rescue the payload at sea or on land.

Systems and procedures, in addition to some of those mentioned above, which would minimize the hazard caused by subsystem failures during the boost phase are:

- **Intact aborts** can be implemented after a few seconds into the flight. Three types of intact aborts are possible for the Uprated Space Shuttle. These are: the return-to-launch-site (RTLS), abort-once-around (AOA) and abort-to-orbit (ATO).

- **Contingency aborts** could lead to either a return-to-land (runway or crash land) or to a ditching at sea.

- **Design of the boost trajectory** to avoid land overflight, for example the 38° inclination orbit, should help in reducing overall risk for the early portion of the flight.

Orbital Operations

The orbit transfer vehicle (OTV) propulsion phase provides for transportation from low Earth orbit to the intermediate destination. In the initial years* of the disposal mission the OTV would be a high-thrust, chemical propulsion (liquid hydrogen/liquid oxygen) stage. To minimize possible failures the following systems, procedures and design guidelines are envisioned:

- **The use of command OTV engine shutdown** in the event of a grossly inaccurate propulsive burn

- **The capability to separate** the Solar Orbit Insertion Stage (SOIS) and attached payload from the OTV and the use of the SOIS to place the payload in a safe orbit for eventual recovery by a rescue vehicle or Shuttle Orbiter

- **A jettison system** incorporated into the SOIS payload adapter to separate the waste payload from the OTV/SOIS configuration when necessary to preclude a possible reentry

---

*Later on, low-thrust technology (e.g., solar electric propulsion using argon propellant) might be used. With low-thrust systems, both the probability of reentry and magnitude of an explosion are decreased. In addition, there is a much longer decision and response time available in case of a malfunction of the low-thrust propulsion systems while in low Earth orbit. However, because of the large solar arrays needed, the probability of solar array damage caused by an impact with on-orbit, man-made debris could become significant in the future.

- **The use of a rescue vehicle** to retrieve a waste payload stranded in any given orbit

- **The use of redundant systems** where feasible to ensure high reliability

- **On-orbit OTV launch crew** to obtain instantaneous visual and telemetric status of the OTV propulsive burn (from the Orbiter)

- **The proper design of trajectories** and propulsive burns of the OTV to reduce the probability for reentry, if a failure occurs

- **A waste form** which helps insure intact reentry and recovery of the payload, should an unplanned reentry occur and the requirement that the waste payload will not melt after self-burial in low conductivity soil

- **The use of thermal protection** material on the outside of the package to reduce the risk of atmospheric dispersal on the ground and in the air, as well as an outer steel shield to protect the reentry material in the case of explosion

- **The use of a relatively high melting point container and shield material** (stainless steel) to reduce the risk of atmospheric disposal of waste.

The Solar Orbit Insertion Stage (SOIS) provides for transportation from an intermediate to the final destination. For the Reference Concept, the SOIS is used to reduce the aphelion from 1.0 to 0.85 AU. Systems, procedures and design requirements envisioned to minimize hazards due to SOIS failures are:

- **The use of a rescue vehicle** to retrieve a cooperative or noncooperative payload stranded in any orbit in heliocentric or Earth orbital space

- **The use of redundant SOIS systems** where feasible to ensure high reliability

- **The proper use of trajectories** and orbits inclined to the Earth's orbital plane that exhibit long-term orbital stability

- **The use of tracking systems** on board the SOIS to aid in deep space rescue operations.

Rescue Operations

Provisions must be made to rescue the SOIS and the nuclear waste payload in Earth orbit in the event of a failure of the OTV during the Earth escape burn. The approach is to rendezvous and dock the rescue OTV with the SOIS and continue the mission from the failed orbit. The rescue mission is based on the premise that, with proper control of the OTV launch, any failure of the OTV will result in an elliptic orbit about Earth. The mission profile for payload rescue is to deliver a rescue OTV to low Earth orbit, transfer by a burn of the OTV to a phase-adjust orbit, and transfer from the phase-adjust orbit at the proper time for rendezvous and docking with the failed system. The lifetime of the rescue OTV, considering the coast time in the phase-adjust orbit, must be as much as 300 hours, compared to the 50 hours for OTV lifetime on the nominal reference mission.

After injection into deep space, the nuclear waste payload could fail to achieve its stable destination orbit, because of a premature shutdown of the OTV engine beyond Earth-escape conditions or a failure of the SOIS to ignite at solar orbit conditions. Studies that address the probability of Earth reentry under these failure conditions have recommended the use of a deep space rescue mission capability as a way of further reducing the overall risk during this phase of the mission (Rice, E. E., and coworkers, 1980b). A deep space rescue mission capability is defined as the ability to send another propulsion system (e.g., OTV and SOIS) to rendezvous with the failed payload in solar orbit and to place it into the desired stable orbit (circular 0.85 AU solar orbit). A capability for uncooperative payload rescue will also need to be provided.

SUMMARY REMARKS

General safety guidelines have been developed and are being used to guide the development of potentially acceptable concepts for the disposal of nuclear waste in space. In some cases, these safety guidelines were selected based upon judgement and not on the basis of detailed analysis. Future work is necessary to develop an improved basis for the guidelines and additional definition and/or change in the guidelines.

To exercise the general safety guidelines, as well as to define a system that can be used to study the risk of space disposal, a Reference Concept, described briefly herein, has been developed using near-term space and nuclear technology. As a result, various safety aspects have been identified in response to the need of achieving what is believed to produce an acceptable health risk for space disposal.

The major questions that remain to be answered are: (1) can space disposal, complementing mined repository, provide a significant benefit, and (2) if so, what is the cost?

REFERENCES

IAEA (1973). Requirements for the safe transport of radioactive materials. Revised code, Safety Series Number 6, Vienna, Austria.

NASA (1979). Space Shuttle flight and ground specification. JSC 07700, Vol. X, Rev. F, Chapter 7.4, NASA's Johnson Space Center, Houston, Texas. USA.

Priest, C. C. (1980). Concept definition document for nuclear waste disposal in space. NASA/Marshall Space Flight Center, Huntsville, Alabama. USA.

Reinert, R. P., et al. (1981). Analysis of space systems - study for the space disposal of nuclear waste. Contract NAS8-33847, Boeing Aerospace Company, Seattle, Washington. USA.

Rice, E. E., C. C. Priest, and A. L. Friedlander (1980a). U.S. program assessing nuclear waste disposal in space: a status report. International Astronautical Federation, XXXI Congress, Tokyo, Japan, Paper No. 80-IAA-50, September 21-28, 1980.

Rice, E. E., N. E. Miller, K. R. Yates, W. E. Martin, and A. L. Friedlander. Analysis of nuclear waste disposal in space - Phase III - Final Report. (1980b). NASA CR-161419, Battelle's Columbus Laboratories, Columbus, Ohio. USA.

Rice, E. E., D. S. Edgecombe, P. R. Compton, and R. E. Best (1981). U.S. program assessing nucler waste disposal in space: a 1981 status report. International Astronautical Federation, XXXII Congress, Rome, Italy, Paper No. IAF-81-188, September 6-12, 1981.

U.S. DOE (1975). U.S. Energy Research and Development Administration (now U.S. DOE)--ERDA manual. Chapter 0524, Standards for Radiation Protection, Washington, D.C. USA.

U.S. DOE (1977). Safety specification for plutonium-238 developmental heat sources, NRA-3. Division Nuclear Research and Applications, Energy Research and Development Administration (now U.S. DOE), Washington, D.C. USA.

U.S. DOT (1979). Requirements for transportation of radioactive materials. Title 49, Parts 171-177, Code of Federal Regulations, U.S. Department of Transportation, Government Printing Office, Washington, D.C. USA.

U.S. NRC (1978). Packaging radioactive materials for transport. Title 10, Part 71, Code of Federal Regulations, U.S. Nuclear Regulatory Commission, U.S. Government Printing Office, Washington, D.C. USA.

IAA 81-257

SPACE STATION SAFETY
DESIGN AND OPERATIONAL CONSIDERATIONS

H.R. Griswold, A.J. Hoffman*

Establishment of a permanently manned space station in earth orbit that would serve as a base for space operations represents a USA space program goal. An orbital operations base would provide international capability for commercial space ventures which could serve to improve the quality of life on earth. Studies are currently under way to define such a facility which expand upon past space station studies and utilize experience and technology from Apollo, Skylab, and STS programs. With the long-duration space station mission, complexity of new space hardware, and expanded crew activity that is projected, a fresh view of space mission safety is required.

This paper reviews projected space station requirements and space operations. Safety issues are discussed that will continue to guide evolving space station designs and operational plans. Specific safety approaches, defined in past space programs and in space station studies, which would support operational safety, are described.

Keywords: Space Station Safety, Crew Safety, Emergency and Rescue Operations, Spacecraft Design, Long-Duration Missions, Extravehicular Activity (EVA).

INTRODUCTION

As the Space Shuttle Columbia lifted from the earth to begin its historic maiden voyage into space, concrete meaning was added to the words of space pioneer, R.H. Goddard, "There can be no thought of finishing, for aiming at the stars, both literally and figuratively, is the work of generations, but no matter how much progress one makes there is always the thrill of just beginning." At that moment, past space achievements of man could not eclipse the fact that this flight marked the beginning of a new and exciting era in space. The advent of the Space Shuttle will unfold a broad new spectrum of space activity designed to exploit the resources of space. In the USA space program, this flight demonstrated a fundamental capability requisite in the future establishment and logistics support of a permanently manned space station operations base for the commercialization of space.

* *Hamilton Standard, Division of United Technologies Corporation, Windsor Locks, Connecticut, 06096 U.S.A.*

The concept of a space operations center has evolved in space station studies conducted over nearly 20 years and from experience and technology gained from previous space programs including Skylab, the USA's first operational space station. In contrast to Skylab emphasis on space research, current space station studies emphasize a permanent space labor force to industrialize space through new technology advancement, construction of large communication platforms, satellite servicing, orbital transfer vehicle (OTV) facilities, manufacturing and space processing of materials, and science and applications missions. The long-duration missions of the future will require regenerative life support systems to reduce logistics resupply penalty for consumables and new systems to support space work operations. An expanded work role is seen for crewmembers, in support of mission objectives, utilizing man's ability to mentally process large amounts of data, observe and analyze nondeterministic processes, and repair hardware which represent only a few processes difficult to accomplish with automated machines. With the projected space station long-duration mission requirement, complexity of space hardware and expanded crew work activity, a fresh view of space mission safety is required.

SPACE STATION REQUIREMENTS

Numerous space station studies have been conducted that define space facility concepts to support activities ranging from scientific research to full-scale construction and manufacturing operations in space. The more recent studies in the USA assume use of Space Shuttle transport capability for initial space station construction and operational logistics support and a zero-gravity facility design. This latter design characteristic stems from Skylab and Salyut mission experience which indicates that while zero-gravity environment does affect the human body, as evidenced by fluid shifts, some cardiovascular deconditioning, loss of some bone calcium and muscle mass, and modest changes in the blood, the changes appear reversible. It is generally conceded that longer zero-gravity exposure times than the 185-day Salyut mission are safe for man and that a high level of human performance can be expected in space for extended periods of time.

The most recent space station study[7] has defined a concept, depicted in Fig. 1, which could support a broad range of space operations. This space station would be initially constructed on-orbit from multiple modules transported as Shuttle payloads. With Shuttle resupply of consumables and spares and on-orbit maintenance, it would operate continuously with limited self-sufficiency for an indefinite period. Two pressurized habitation modules, shown in Fig. 1, are interconnected at one end by the pressurized docking module and at the other end by two joined service modules. This configuration would provide two exits from a habitation module in the event of an emergency. Each habitation module would support a nominal crew of four and contain a command center capable of controlling all space station functions, sleeping quarters, and areas for food preparation, hygiene, health maintenance, waste management, exercise, and recreation. The life support system in each habitat module would be regenerative where all metabolic water and oxygen requirements would be generated by the closed-cycle system to minimize the need for resupply of consumables. The life support system would provide for cabin air revitalization, heat rejection, ventilation, temperature control, water processing, carbon dioxide reduction, and atmosphere makeup supply. The service modules would provide electrical power generated by two large solar arrays supported on the long booms. Subsystems that would be contained in the pressurized service modules would be: power conditioning and distribution, control moment gyros for attitude stabilization and control, communications, some of the environmental control and life support system equipment, and airlocks for EVA. The logistics module, shown in Fig. 1 berthed to the upper service module, would store consumables and spare parts. The Shuttle would exchange the logistics module during periodic resupply visits.

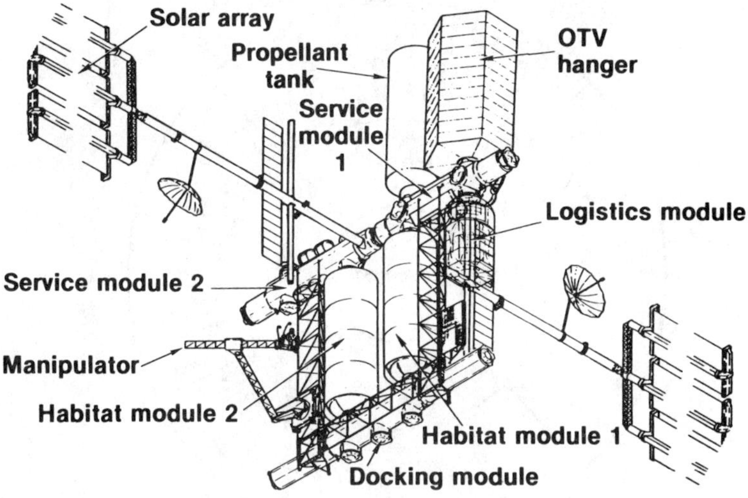

Fig. 1. A space station concept

The space station of Fig. 1 would include work facilities for on-orbit assembly, launch, recovery, and servicing of satellites. A manipulator and tools associated with the holding fixture would aid the berthing of satellites to the station. EVA would be required to provide needed dexterity since most operations would not be repeated often enough for development of automated maintenance systems. The space station would include facilities to support EVA. Transport of distant satellites to the space station or the deployment of satellites would require the use of orbital transfer vehicles (OTV). Before launch, the OTV would be mated to the satellite and fueled from a propellant tank, shown in Fig. 1, using propellant transfer equipment. After completion of a mission, the OTV would be returned to the space station and stored by the manipulator in an OTV storage hangar. The space station would support construction of large space structures and systems and their checkout and transfer to operational orbit. The use of manipulators and external work platforms are included in the concept. Services such as electrical power, active cooling, computer data link, and illumination would be provided at each work platform. The concept provides an on-board computer capability for subsystems monitoring, fault isolation and failure analysis, and storage and call-up of extensive sets of data for on-board control of operations.

A space station can be characterized in terms of its functional requirements needed for projected space operations. The requirements of a space station, such as that of Fig. 1, designed to support a broad spectrum of space operations, are indicated in Fig. 2. Since the space station concept of Fig. 1 embodies most of the operational capabilities and complexities defined in past space station studies, it has been used as a model in this paper in discussing space station safety. While it is recognized that more effort will be required before the next generation space station concept has been definitively specified and optimized, the safety issues discussed here are generic in nature and independent of specific space station configuration.

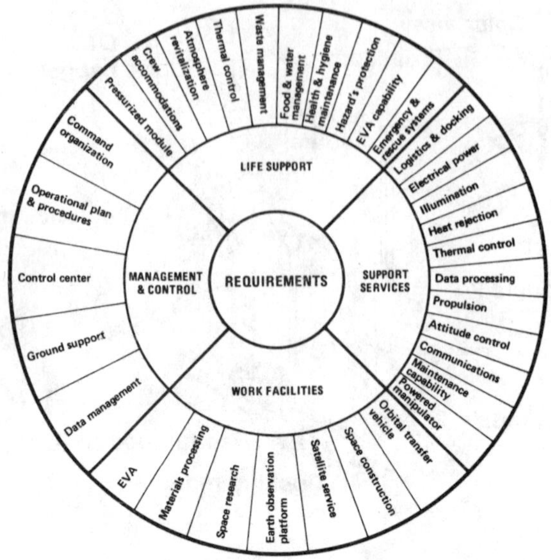

Fig. 2. Space station requirements

SAFETY CONSIDERATIONS

The first step in establishing a high level of space mission safety is the identification of hazards that could lead to operational contingencies. Once identified, steps can be taken to eliminate hazards through appropriate designs, prevent hazards through use of safety features and devices, and control hazards through use of warning devices, special procedures, and emergency and rescue protection systems. In viewing the space station requirements of Fig. 2, it is apparent that the future space station mission differs from those of earlier manned space flights in a number of important respects. The mission will be of long duration, involve new systems of added complexity, and involve greater crew participation in a broad spectrum of work operations. Following is a discussion of space station safety issues, listed in Fig. 3.

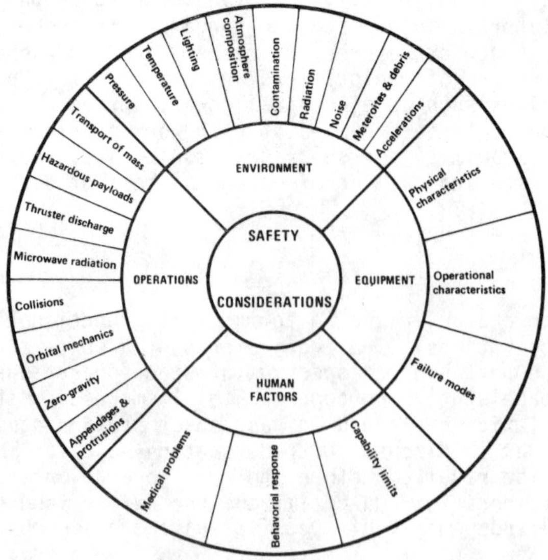

Fig. 3. Space station safety considerations

ENVIRONMENT - A limiting characteristic of space environment is its vacuum which forces man into a pressurized enclosure. In earth orbit, beyond the protection of the earth's atmosphere, radiation increases markedly due primarily to electrically charged particles trapped in the earth's magnetic field. Meteoroids, traveling through interplanetary space, can reach the earth traveling at high velocity. Since orbital missions expose crewmembers to these hazards of near-earth space, special wall and spacesuit constructions are required. Illumination in near-earth space is substantially different from that on earth. When viewed from outside the earth's atmosphere, the sun appears nearly twice as bright as from the earth's surface. In addition, space structures can become a source of reflective illumination and, during portions of most earth orbits, periods of total darkness will be encountered. Therefore, visors for work in the sun and artificial lighting for use in shaded work activities are required. Surfaces in space can reach temperatures in excess of 200°F in the sun or, in full eclipse, fall below -200°F. Consequently, spacesuit and spacecraft wall constructions must be designed to provide thermal protection. The known population of debris in earth orbit is currently in excess of 10,000, where many more small objects may exist, having resulted from collisions or disintegration of larger bodies. Because of the dominant use of preferred orbital inclinations, debris is a residual safety concern.

EQUIPMENT - Physical characteristics of space equipment, such as shape, size, mass, and surface configuration must be considered with regard to crew safety and may require the use of special handling procedures and work aids. Operational characteristics of equipment such as temperature, dynamics, voltage, and pressure may require crew safeguards. Identifying equipment failure modes is the important first step in designing reliable and safe spacecraft equipment. There are three basic design approaches which are used in designing space systems. The first, redundancy, is the prime design approach which involves placement of a backup component in parallel with the critical component. The design safety margin approach, different from redundancy, provides performance margin in the critical component, beyond that required at the nominal operating condition. Finally, the maintenance design approach involves design of components or systems with provision for on-orbit maintenance. This approach involves modular designs employing mechanical fasteners and electrical and fluid maintenance disconnects where replacement spares must be provided.

The operational design criterion generally adopted for critical space systems is that of fail-operational/fail-safe. The fail-operational/fail-safe criterion requires that the function of the system continue to be provided and available to the crew should a first failure in the system occur. This criterion is generally met through use of redundant components. Should a second failure occur, the fail-safe criterion requires that adequate functional services continue to be provided.

HUMAN FACTORS - Most crew work safety concerns are removed through providing proper crew protective equipment and work aids, planning work compatible with known astronaut capability limits, and extensive training. Astronaut training[1] using mockups and simulators promotes safe space operations by helping to define crew capability limits, identify potential work hazards, evaluate and select safety measures, and provide familiarity with work tasks and zero-gravity effects. Behavioral response concerns in the form of crew judgment errors or faulty execution of procedures are minimized through crew training and the use of fail-safe backup protection devices. Medical problems are minimized through proper crew selection and safety planning designed to prevent crew injury and conditions that could induce illness.

OPERATIONS - In planning space operations involving transport of mass, the use of proper restraints and aids are required. Safe maximum limits must be set on transport rates to avoid collisions which could result in injury and equipment damage. High temperature payloads, toxic chemical payloads, microbe experiments, and propellants represent payloads that may require development of special handling techniques and crew protective equipment. The zero-gravity and orbital mechanics of earth orbit require special work aids, restraints, and procedures to ensure proper crewmember stability and orientation during work activity. Work activity must be scheduled and situated so that crewmembers are not exposed to any harmful microwave radiation from communication or radar equipment or hot gas discharge from propulsion or attitude control system thrusters. Appendages such as antenna, solar arrays, and radiators, as well as other external protrusions, need to be accounted for in planning work activity outside of the space station.

SAFETY APPROACH

The general approach to space mission safety is indicated in Fig. 4. Following is a discussion of specific approaches to space station safety in both design and operational areas. It is assumed in the discussion that specific safety approaches cited, which have evolved from space program and industrial experience and in past space station studies, would be taken in designing the future space station.

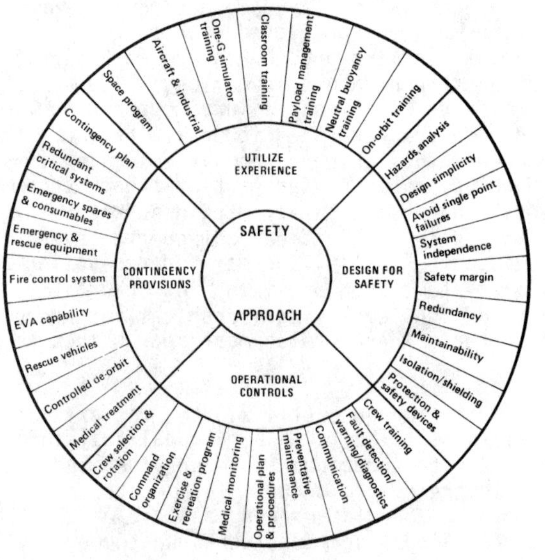

Fig. 4. Space station safety approach

UTILIZE EXPERIENCE - Prior space program, aircraft, and earth-based industrial experience provides a valuable source of information for future space station safety planning. In particular, the experiences and lessons of Skylab have pointed the way to design, develop, and manage a large space station operations base. Skylab was established as a first generation space station with objectives of proving the value of a permanent space station, exploring physiological limits of man's endurance in space flight, and conducting a series of important scientific space experiments. A review of written accounts by Skylab program participants, including the astronauts, reveals important lessons that will be helpful in planning future space stations. First, it was learned that man can work and operate for extended periods in zero-gravity with no apparent serious physical or mental side effects; that astronaut health may be maintained by proper diet, exercise, recreational activities, and psychological stability may be maintained by a proper habitat environment and by suitable communication with the ground. It was concluded, based on contingencies resolved through crew activity, that large space facilities should be designed to make use of man as a trained observer and maintenance mechanic and that EVA capability should be included; that man in space can do essentially anything a person on earth can do if given a place to stand and the proper restraints. It was learned that zero-gravity can be an aid in work activities rather than a handicap and safety hazard as long as systems and components have been properly designed. In-flight maintenance and hardware replacement was found generally to be easier and safer than was originally anticipated using tools similar to those required in earth applications. Other lessons for future space station designers and safety planners, learned in the Skylab program, were:

- Provide spares and maintenance capability for critical systems.
- Provide operational flexibility in designs.
- Provide for excess consumables.
- Investigate all failure modes and develop contingency plans.

DESIGN FOR SAFETY - A design objective would be to make the space station accident-proof and free of debilitating system failures. This would require that the basic structure, configuration layout, work tools, and support equipment be designed to eliminate and control hazards to crewmembers. As indicated in Fig. 4, design of space hardware proceeds with a comprehensive hazards analysis to define failure modes and safety hazards. Once identified, the design can proceed using the design approaches indicated in Fig. 4. Following are examples of specific design approaches that would promote space station mission safety.

Satellite service and space construction operations will require extensive use of EVA. It is therefore desirable that a cabin pressure be selected, compatible with existing spacesuit technology, that eliminates the requirement for crew prebreathe of pure oxygen prior to EVA to remove nitrogen from the body. Removing requirement for prebreathe would eliminate a residual hazard, namely, improper crewmember denitrogenization prior to EVA. Recent studies suggest that a cabin atmosphere of 24% oxygen and 76% nitrogen at a pressure of 11.8 psia would remove EVA prebreathe requirement using a spacesuit pressure of 5.75 psia. This choice of atmosphere composition, based on flammability study findings, together with proper space station materials selection, would minimize the flammability hazard.

Since space in earth orbit is finite, collisions between man-launched objects and the space station represents a residual operational hazard. In addition, meteorites from deep space can reach the earth traveling at high velocity. Therefore, the space station design must include provision for debris and meteoroid protection. Maximum use of on-board mass should be used as collision shielding. The habitat modules can be located toward the center of the configuration where crewmembers spend the majority of their time. In addition, thermal control system radiators for rejecting heat loads to space can be located outside of the habitat modules to add mass between the outside environment and the crew. OTV's parked at the station can be provided meteoroid and debris collision and passive thermal protection by storage hangars. Each space station window would employ redundant panes and be designed for high pressure retention with a safety factor of at least 3. Pressurized structure design would be consistent with prior spacecraft designs employing a factor of safety of 2. As a goal, fail-safe design concepts would be applied to all critical structure so that failure of a single structural member would not seriously degrade the strength or stiffness of the remaining space station structure. All pressure vessels, including habitat walls, would be designed to satisfy a leak-before-rupture criteria.

In Fig. 5, EVA is shown involved in cleanup and replacement of a communications antenna which has suffered collision damage. A replacement antenna is shown being erected in the Shuttle payload bay having been transported from earth. EVA capability would be included as a space station capability, not only to support scheduled work operations, but for handling contingency maintenance of equipment. EVA support equipment, shown in Fig. 5, which includes the Extravehicular Mobility Unit (EMU), the spacesuit system, and the Manned Maneuvering Unit (MMU), the propulsion unit, includes many design safety features[1] which provide the EVA crewmember wide margin of safety. EVA crew translation corridors and work site areas would be designed compatible with the size and mobility characteristics of an EMU. The space station equipment would be configured to provide EVA crew protection from high voltage electrical sources, corrosive fluids, microwave radiation, and without sharp edges, corners, and protrusions which might damage the EMU spacesuit. A minimum of two EVA airlocks would be provided where redundant pressurization capability will be provided to ensure safe reentry of crewmembers from EVA. A window placed close to the airlocks would allow an inside observer to view the EVA astronaut immediately after he has left the airlock and during work activity. All airlock hatches would be designed to close in the direction of positive pressure differential, and employ hinge linkages to control hatch motion, and a window to permit visual contact across the pressure hatch when closed. Areas into which hatches open would be designed so that the full open position of the hatch would not block crew passage.

Fig. 5. EVA supports mission safety

A major space station technical challenge will be achievement of long-life subsystems. The long-duration mission results in a high statistical probability that certain components will wear out or fail if not maintained or periodically replaced. While redundancy of critical systems would be included in designs, primary reliance on redundant components would lead to an impractical design for even a ten-year mission. Adequate system reliability and therefore mission safety can only be practically achieved through use of maintainable hardware with a provision of in-flight spares. The use of common components in different space station subsystems would promote safety by minimizing crew training requirements and providing added operational flexibility where spares could be borrowed in a contingency to repair a failed critical system. Space station subsystem reliability would be designed high enough so that maintenance activities would not place excessive demands on crew time. Components would be designed to be replaceable by a single man within four hours. The possibility of incorrect assembly, installation, and connections could be eliminated by using keyed connectors and fittings. Captive fastners for mounting of components would avoid the loss of parts, and the use of coarse threads would limit the possibility of cross-threading damage to fastners and connectors. All electrical cable and flexible hose installations would be designed with sufficient flexibility, length and protection to permit easy removal without damage to connector, wiring, or hose. Electrical subsystems would include safety interlock protection devices. To provide ease of maintenance, ceiling, floor, and wall panels in the pressurized modules could be removeable for equipment access. Equipment would be packaged so that critical components would not be more than one layer deep with adequate perimeter access. Adequate hand space would be provided to allow complete movement and full utization of maintenance tools. Fig. 6 depicts a life support system pump and accumulator module changeout within the space station. A spare replacement module is shown in the foreground being readied for installation. Planned redundancy in critical space station subsystems would permit safe scheduled maintenance on limited-life items avoiding unnecessary breakdowns.

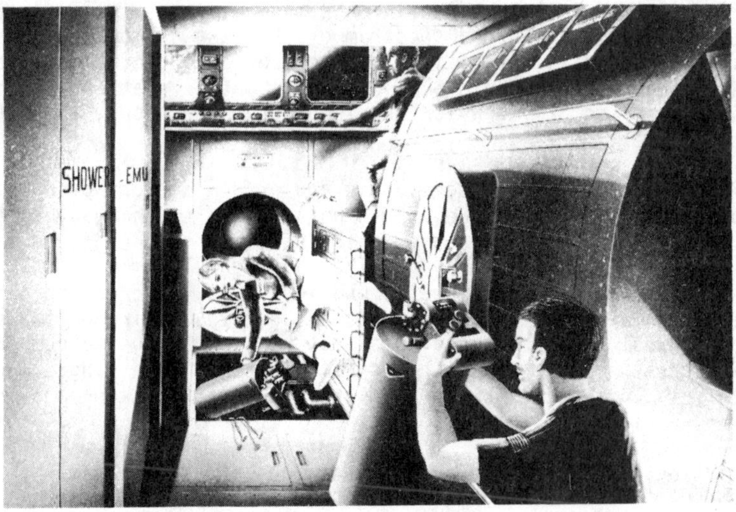

Fig. 6. Subsystem maintenance

Isolation and shielding represent design safety approaches that would be used extensively in future space station designs. For example, new life support equipment that uses or generates hydrogen would be located away from the main habitat. Water would be used as the coolant in habitat modules since it is non-toxic. Water lines would be adequately protected by insulation and isolated from exposure to space to avoid freezing and possible rupture. Tubing, ducting, and electric cables would be routed away from component and module mounting points where damage could occur from maintenance tools. Redundant equipment, fluid lines, and cables, critical to crew safety and space station operation, would be routed in separate compartments or separated by a structural wall and protected against fire, smoke, contamination, and overpressurization to preserve system independence and avoid sequential propagation of critical component failures. Fluid pressure vessels would be placed outside of habitat modules. Rotating machinery would be isolated from the crew to avoid the possibility of personal injury or damage to other equipment. Inherently hazardous activities, materials, or equipment that may be present in the space station would be located in areas other than habitat modules. Designs would include provisions for containment of metabolic solid and liquid wastes and for biological experiment microbe cultures. While communicable disease in the space station is unlikely, as a result of medical screening, use of redundant habitat modules would provide for isolation of suspected or communicable disease cases from the rest of the crew for observation and treatment or to await evacuation.

OPERATIONAL CONTROLS - An integrated space station system of controls could enhance operational safety. An operations plan would be adopted to be executed by a carefully selected, highly motivated crew responsive to a dictated command organization. Familiarity with contingency plans would be necessary for safety. Periodic crew training would be required on orbit where drills for all personnel would be conducted to practice response to simulated emergencies. Crews would be rotated at periodic intervals and be selected to be in good physical and mental health at the start of their oribital stay. At least one crewmember would have extensive medical training or be a physician. The central space station computer could provide storage capability for crew biomedical data and receive input by the medical crewmember of all significant crew physiological data obtained on a periodic basis. A computer medical diagnosis scheme could cover a broad spectrum of the most common medical conditions. A store of drugs and medications, similar to those of Space Shuttle medical systems but with increased quantities, would be included. Dental treatment capability would be similar to that used in Skylab which provided for extractions, temporary fillings, and dental pain suppression.

Effective communication between space station and ground support pesonnel would be essential to operational safety. The communications and tracking subsystem would be designed to provide communications and tracking services to the ground station by way of relay satellites, the Shuttle, EVA and OTV's. Internal communications including duplex voice, caution and warning signals, public address, and closed circuit video would be available in all habitable areas and all active docking port areas. Internal communications would be configured so as not to be interrupted nor degraded within remaining pressurized modules should a malfunction in a single module occur. Closed circuit TV would be available to support spacecraft docking and monitoring of work activity.

Space-flight mission control has been characterized in past programs by large numbers of ground personnel and support facilities monitoring and controlling flight-crew activities. In future space station missions, dependence on earth for control and monitoring will be reduced by providing an on-board data management system. The on-board system would provide for systems checkout, monitoring, warning, and fault isolation to a level consistent with safety and on-orbit maintenance capability. The computer would be programmed to access subsystem parameters and generate required control outputs through switching interfaces. A control center would be provided in the habitat module, as shown in Fig. 7, where the data management system could be accessed and control and monitoring of all subsystems could occur. The command center would be located so as to provide direct visiblity of high activity areas such as where space construction and satellite service, involving substantial EVA, would be performed. Computer programs featuring collision avoidance software to set maximum excursion limits for manipulators and other supporting equipment could be provided. Critical subsystems of docked OTV's could be continuously monitored with appropriate audio and visual warning provisions at the control center to indicate out-of-tolerance conditions. Caution and warning alarms associated with any hazardous events would be displayed in all inhabited areas so the crew could initiate necessary emergency or remedial action. A program for monitoring consumables usage would be used to detect excessive consumption rates and low remaining supply levels. As indicated in Fig. 7, the control center would feature multifunction, programmable CRT's and audio, and flashing display techniques to ensure crew attraction. CRT displays employing color, alpha-numerics, graphics, and vector symbology would enhance presentation of information and crew interaction.

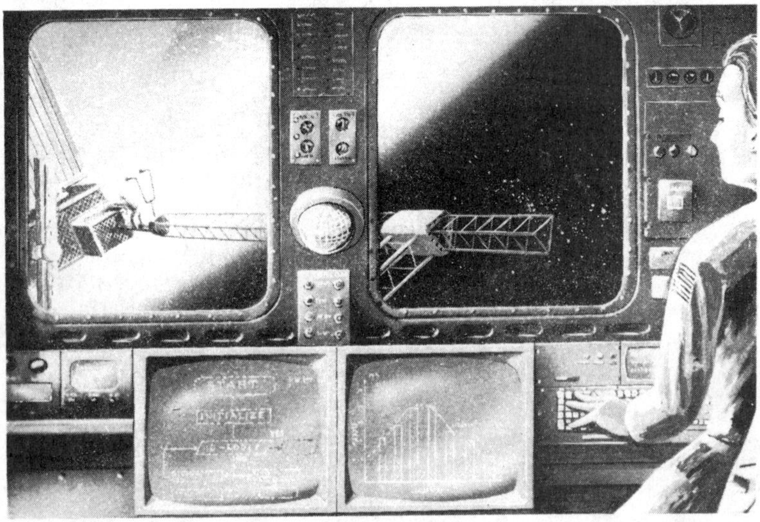

Fig. 7. Space station control center

Projected future space station operations indicate a high level of EVA usage. For example, EVA would be required in the service of satellites as depicted in Fig. 8. Satellite service operations could involve handling of coolant system fluids or propellants, or chargeout of high voltage, high temperature, or microwave components. Special safety features will be required on satellites[4] designed for service and

Fig. 8. Satellite service operations

the development of safe operational procedures. A crewmember within the space station at the control center would monitor EVA operations as shown in Fig. 7. A cumulative radiation exposure record would be maintained for each crewmember and reviewed at periodic intervals in order to prevent overexposure to radiation. Any personnel approaching the limit of safe radiation exposure would be returned to earth on the next Shuttle visit. Radar or communications equipment on-board the space station, Shuttle, or nearby satellite payloads, with radiated power sufficent to be a hazard, would be shut down during EVA or EVA would be restricted to work areas free from exposure. Operational procedures and space station systems would be developed to ensure that thrusters would not endanger EVA personnel.

CONTINGENCY PROVISIONS - The space station design would include two separate pressurized habitation modules with independent life support capability and contingency provisions for the crew for a period of time adequate to allow for repair of critical systems or rescue by the Space Shuttle. Should habitat atmosphere become contaminated, as a result of a fire, or release of biological or chemical toxins, a portable oxygen system (POS), developed for the Shuttle program, could be donned by the crew, as shown in Fig. 9, and worn while correcting the problem or during transfer to the other habitat module. A response to a fire might involve shut-down of ventilating equipment and oxygen supply, partial depressurization of the module, crew donning the POS, and repressurizing of the module with an inert gas to extinguish or retard fire propagation.

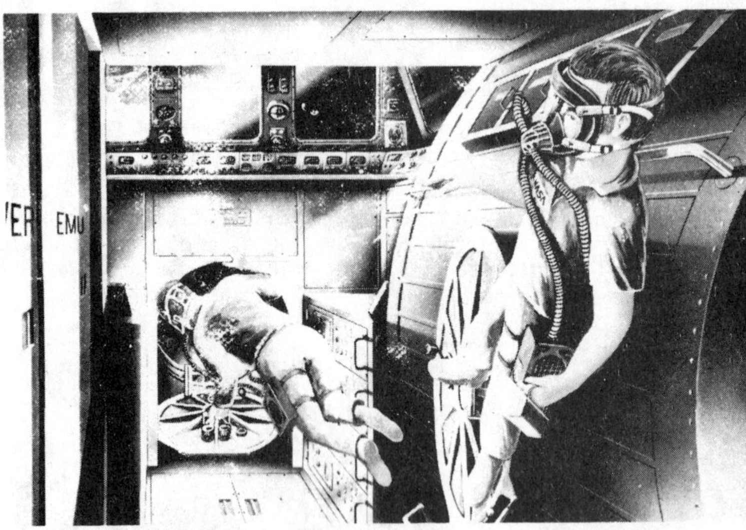

Fig. 9. Portable oxygen system

Should a module become uninhabitable to the extent that it cannot be pressurized or cooled adequately, then EVA equipment could provide the capability to conduct unpressurized IVA to resolve the problem as shown in Fig. 10. The EMU spacesuit system, with its independent life support and cooling capability, would protect the crewmember while conducting repair activities. Portable lights, protective helmet cage, portable TV camera, radio communication, and repair tools would be available for effecting the repair.

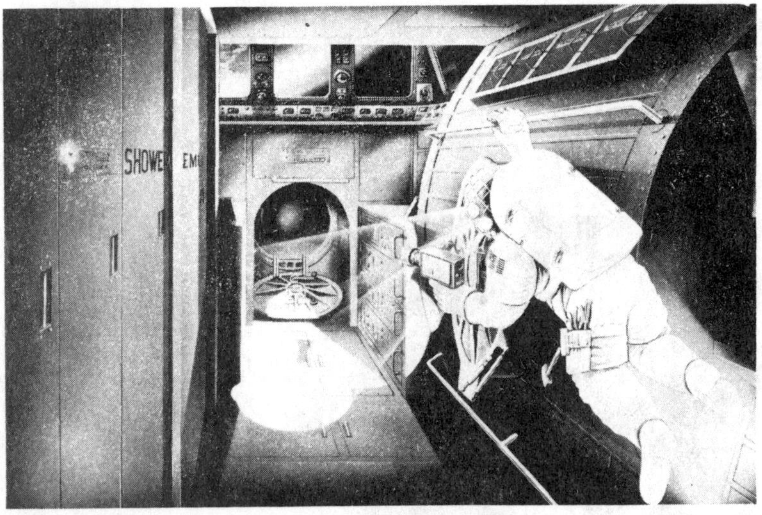

Fig. 10. Unpressurized IVA

While the probability of debilitating crew illness or injury would be low as a result of careful crew selection and rotation, safety provisions, and a preventive medicine program, medical emergencies could occur. Facilities for carrying out minor surgical and dental procedures would be provided including capability to treat fractures, lacerations, and minor trauma. Provisions would be implemented to contain and control body fluids, including blood, urine, feces, and vomitus to avoid atmosphere contamination. Equipment would be provided to monitor vital signs, with capability to transmit cardiac rhythm to the ground. A video system of communications would include image transmission so that the on-board medical crewmember could show images of an injured crewmember, microscope slides, or X-ray films while in consultation with the ground physician. The Space Shuttle could be used for medical evacuation of an incapacitated crewmember. Another approach, as shown in Fig. 11, would make use of a two-man reentry evacuation vehicle where the vehicle would be capable of carrying a pilot and the incapacitated crewmember to be returned to earth. In Fig. 11 one such vehicle is shown docked to the space station and one is shown initiating a deorbit burn. These rescue vehicles would be delivered to the space station by the Shuttle to remain on orbit as a backup emergency system. Modified Gemini or Apollo command module spacecraft represent candidate vehicle choices or more advanced reentry vehicles that could maneuver during reentry to maintain low-g loads on the crew. The evacuation vehicle would reenter the atmosphere and deploy a parachute as shown for sea recovery by a waiting ship with required medical facilities.

In the improbable event that a series of failures required abandonment of the space station and crew rescue, a rescue Space Shuttle would be launched. Should the Space Shuttle be unable to effect a safe dock to the space station, then the rescue of personnel would be effected through the use of EVA as shown in Fig. 12. EVA equipment shown includes the EMU, MMU, Personnel Rescue System (PRS) and POS. The PRS developed for the Space Shuttle is a 34-inch diameter, inflatable enclosure that provides pressure retention and thermal and micro-meteoroid protection. The crewmember to be transferred to the Space Shuttle would don the POS and enter the PRS through a zipper opening. The sphere would then be pressurized using the POS. An EMU equipped crewmember would then effect the transfer of the PRS ball out of the space station airlock to a standby EMU-MMU equipped astronaut for transfer to the Shuttle. During EVA transfer, the POS would provide the PRS occupant oxygen for breathing and maintain PRS pressure. Should permanent abandonment of the space station be required, then to assure terrestrial safety, the space station design would provide sufficient reserve capacity in the propulsion system to accomplish a controlled de-orbit.

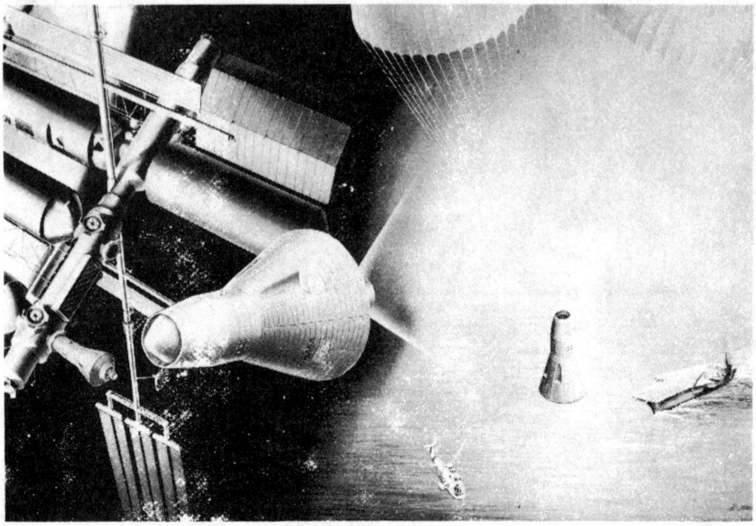

Fig. 11. Medical evacuation

Fig. 12. Shuttle rescue operation

SUMMARY

Space program planning in the USA can be expected to exploit Space Shuttle capabilities, where a major goal is the establishment of a permanent manned space station in earth orbit, to serve as a base for space operations with limited self-sufficiency. The Space Shuttle would provide the capability required for space station construction and operational logistics support. A space operations base could be used for earth observation, materials processing and manufacture, satellite service, construction of space structures and orbital transfer vehicle support. The long-duration mission and operations planned for this type of facility would require new space systems and expanded use of crew in both IVA and EVA work. A fresh view of mission and crew safety will be required to insure adequate safety.

As in past space programs, safety will be both preventive and corrective in nature, with primary emphasis on preventing the occurrence of emergency situations and the need for rescue. The response to a space station contingency will not be mission abort and return to earth, but rather reconfiguration of systems followed by maintenance and return to normal operational status. The first step in establishing safety will be identification of hazards that could lead to operational contingencies. Once identified, steps can be taken to eliminate hazards through appropriate designs, prevent hazards through use of safety features and devices, and control hazards through use of warning devices, special operational procedures, and emergency and rescue protection systems. Safeguards successfully employed in terrestrial industrial operations, similar to those of projected space station operations, will serve as useful guides in planning for mission safety.

ACKNOWLEDGEMENTS

The Authors wish to acknowledge use of valuable information from NASA and Boeing Aerospace Company reports. Recognition is also due Paul Tremblay for his helpful critique of this paper.

REFERENCES

1. Breeding, R. E., and H. R. Griswold (1980). Safer way with EVA. Proceedings XXXI Congress IAF, Tokyo, Japan.
2. DaRos, C. J. (1977). Space station systems analysis study. McDonnell Douglas Astronautics Company Report, NAS9-14958.
3. Griswold, H. R., and R. B. Trusch (1980). Emergency and rescue considerations for manned space missions. 6th IAA Symposium on Basic Environmental Problems of Man in Space, Bonn, Germany.
4. Griswold, H. R., and R. C. Wilde (1980). EVA equipment for satellite service. ASME Intersociety Environmental Systems Conference, San Diego, California.
5. Johnston, R. S., and L. F. Dietlein (1977). Biomedical Results From Skylab. NASA Scientific and Technical Information Office, Washington, D.C.
6. McCabe, E. L., and E. P. Goodrich (1970). Space station safety study. Boeing Aerospace Company Report, NASA-CR-108286.
7. Woodcock, G., and W. Nassiff (1981). Space operations center. Boeing Aerospace Company Report, NAS9-16151.

IAA 81-250

ITALY'S CONTRIBUTION FROM A MEDICAL STANDPOINT TO THE SPACE SAFETY OF PAYLOAD SCIENTISTS AND PERSPECTIVES FOR THE FUTURE

G. Rotondo*, C.A. Ramacci**, G. Meineri***
G.C. Modugno+, F. Monesi++

In Italy, the selection of Italian payload scientists has been performed according to the Spacelab Program of ESA. Twenty-four subjects underwent a screening performed by the Health Service of the Italian Air Force, following the procedures suggested by NASA. A preliminary physical examination was carried out at the Psycho-physiological Institute for the Italian Air Force in Rome. All who did well underwent another, more exacting, specific examination. They were requested to pass an exercise test on a treadmill and another ten-minute test in a centrifuge, subject to the effect of $+3G_Z$. The authors briefly describe the results of the tests and the EKG records for the candidates. Noteworthy is the determination of Central Flicker Fusion Frequency. This parameter makes it possible to assess the endurance level of the subject, much earlier than other techniques (EKG or EEG). The authors emphasize the importance of an accurate preliminary screening as well as of successive training periods. These are highly advisable to enhance tolerance to prolonged accelerations $+G_Z$ and $+G_X$ in the scope of an adequate training for space flight of subjects who are expected to perform as payload specialists on board spacecraft. Future studies will be undertaken to compare evoked cortical potentials with behavior parameters of space safety, with a view to setting up a subtle tool of evaluation for both future candidates and payload scientists.

Space safety is of preeminent interest. Therefore it has attracted the attention of physiologists, psychologists and physicians for many years before the actual beginning of manned space flights. Both from a physiological and a psychological standpoint, the first and most important basis of safety is represented by an accurate preliminary selection, other than by a thorough training of probable candidates for space flight.

* *Major General, IAF (MC), Chief, IAF Health Service*
** *Brig. General, IAF (MC), Director, IAF Aeromedical School, Rome, Italy*
*** *Colonel, IAF (MC), Director, IAF Aerospace Medical Centre, Rome, Italy*
\+ *Professor, Ophthalmologist, State University, Rome, Italy*
\++ *Colonel, IAF (MC), Chief, Department of Neurology, Aerospace Medical Centre, Rome, Italy*

Such a need is to be felt more acutely when selection is concerned with people such as payload scientists on board a space laboratory. Of course, they are not supposed to possess the exceptional qualities one is accustumed to expect from space pilots. Anyway their physical and psychical fitness will have to be exceptional at least within the limits fixed by ESA Programs.

Although this will be an absolute necessity at any time, it is foreseeable that, in the near future, the requirements of endurance and performance of payload scientists will be comparatively less demanding if compared with present requirements.

Our aim, in this paper, is to expose some aspects of the physiological and functional selection to which our Italian candidates were subjected for participation in the ESA Spacelab Program as payload scientists.

First of all, we would like to summarize a few contributions made by a group of medical investigators of the Italian Air Force on the problems related to man in space. We ask the reader to keep in mind that the experiments we will describe were carried out in the 1950s and shortly after 1960 (before the first manned space flight) and, therefore, should be regarded as a simple preliminary approach to actual investigation in space.

At that time a special apparatus was devised, the so-called "Subgravity Tower". Its structure was a bit different from certain devices, also called subgravity towers, used for similar purposes by foreign investigators. Properly, it consisted of a vertical accelerator, capable of producing an acceleration of 2 to 3 G, followed by a very short period (not exceeding 2 or 3 sec) of free fall, and consequently by weightlessness.

In this way, some elementary experiments were conducted concerning the execution of simple tasks (such as tapping or tracking). Their results, although modest, were confirmed by experiments conducted in foreign countries by means of parabolic flights (ref. 1,4,5,8,9,10).

Later on, a new and different apparatus ("Subgravity Axis") was built in order to investigate man's motility following a reduction of variable amounts of body weight so as to simulate partial or total weightlessness.

Subjects were suspended, by means of a special harness, to a cable provided with variable counterpoise, or to elastic ropes attached to a light trolley running along a horizontal metal beam, 23 m long. The energy expenditure was determined, and the peculiar type of deambulation, adopted spontaneously by the subjects, was studied. This was a contribution to anticipation of possible ways of locomotion both on space vehicles and on the lunar surface. In particular, some aspects of the experiments of locomotion on different types of artificial soils provided a type of motion not too dissimilar from actual walking on the Moon. Similar results were obtained from like experiments carried on separately and independently in the USA (ref. 3,6,16).

And now we will deal with the selection proper of candidate payload scientists.

They were subjected (in 1977) to an accurate medical screening in the Psycho-physiological Institute of the Italian Air Force in Rome. Candidates who had passed this test underwent a specific psychological and and physiological selection review. We will deal only with the latter. This was represented by an exercise test (on a treadmill) and in two more tests aimed to ascertain tolerance to acceleration (positive G or $+G_z$). During a preliminary run in the centrifuge, candidates were submitted to the effects of $+1.5G_z$, while, in the subsequent run, they were required to stand $+3G_z$ for ten minutes.

The exercise test presented no particular features related to spatial activities, being represented by a few minutes of walking and running on a treadmill, performed in accordance to the specific technique adopted by the Italian Air Force for the selection of candidate pilots and the control of flying personnel (ref. 2,7,14,15).

On the other hand, it is important to consider in detail the acceleration test, conducted in accordance to the methods currently adopted for training and control of aircraft pilots and other flight personnel (ref. 13), and following the procedures suggested by NASA. Now we will describe methods employed and the results of the test (ref. 11, 12).

After the exercise test and a preliminary run in the centrifuge at $+1.5G_z$, 24 Italian candidate payload scientists (two of them women), were subjected, on the following day, to the second and principal test (10 min at $+3G_z$, onset 0.05G/sec). Subjects were continuously monitored by means of electrocardiographic recording (EKG).

It is well known that EKG is generally used as an index of tolerance to G. Although one needs to record it in any case, EKG is by no means the best suited system.

Electroencephalographic recording is, in some investigators' opinion, a more desirable technique; notwithstanding, its recording is not as feasible as EKG, and its interpretation not so immediate.

That is why we felt the need for supporting EKG monitoring of the candidates (i.e. a conventional technique) with the recording of Central Flicker Fusion Frequency (CFF). Nowadays this technique is very easily employable, thanks to Modugno's and Rispoli's apparatus (ref. 11, 12). Such a device is essentially represented by a series of LEDs, each one flickering at a different frequency.

It is known that, above 15 cps, repeated luminous stimuli are perceived as an uninterrupted, although flickering, light. This holds true from 15 till about 40 cps. But, around or above 40 cps, luminous stimuli are perceived as fixed, continuous light. The lowest frequency that a subject perceives as a continuous, non-flickering illumination is defined as "critical flicker fusion" (CFF). This phenomenon is dependent upon the conditions of blood circulation and of oxygenation of retinal cells.

The apparatus for CFF recording was fixed in the gondola of the centrifuge at the same level as the subject's eyes and about 45 cm apart. Inside the gondola, lighting was dimmed, although we must remember that CFF may be adequately recorded in quite different conditions of illumination with no serious impairment. On the contrary, the distance of the apparatus from the eyes of the observer should be standardized (40 to 45 cm). Subjects are allowed to wear glasses.

CFF was recorded one minute prior to the beginning of the run, at one-minute intervals during the test, and again one minute after the end of run.

As previously stated, the programmed duration of the test was ten minutes. Its interruption (either on subject's request or decided by the operator) was a cause for rejection. Causes of interruption were: subject malaise, discomfort, dizziness or nausea; blackout, bradycardia, etc.

Thus, only one half of the candidates (12 out of 24) passed the test. Relationship between CFF and Heart Rate (HR) were studied. Synthetic results are summarized in two diagrams, in which CFF and HR (expressed as percent variations from initial values) are plotted against time.

Diagram A concerns the behavior of CFF, while diagram B represent HR. In both diagrams a continuous line indicates those subjects who could withstand acceleration till the end of the test, while a dotted line refers to those subjects who had to give up before its termination.

The following facts are interesting:

1. In the first group (diag. A and B continuous line) CFF undergoes a slow, quite moderate and uniform decrease. In practice it remains almost constant till the end. Complete blackout is never recorded. HR shows an abrupt increase during the two or three minutes from the beginning and, thereafter, a remarkable stabilization around the highest values.

2. In the second group of subjects (poor tolerance, diagrams A and B dotted line) CFF shows an early and constant decrease toward levels averaging around 50 percent of initial values. HR, on the other hand, presents, at first, an increase (therefore one cannot forecast, a supervening failure) on this basis, followed after a while by an abrupt decrease.

For these reasons CFF should be regarded as a simple means for an early assessment of a subject's inability to withstand acceleration without undergoing heart failure. It is therefore a valuable tolerance index.

As far as EKG recording is concerned, we must remember that the following changes were recorded: tachycardia, bradycardia, extrasystoles, problems of repolarization, etc. Problems were consistently more frequent and severe in unfit people.

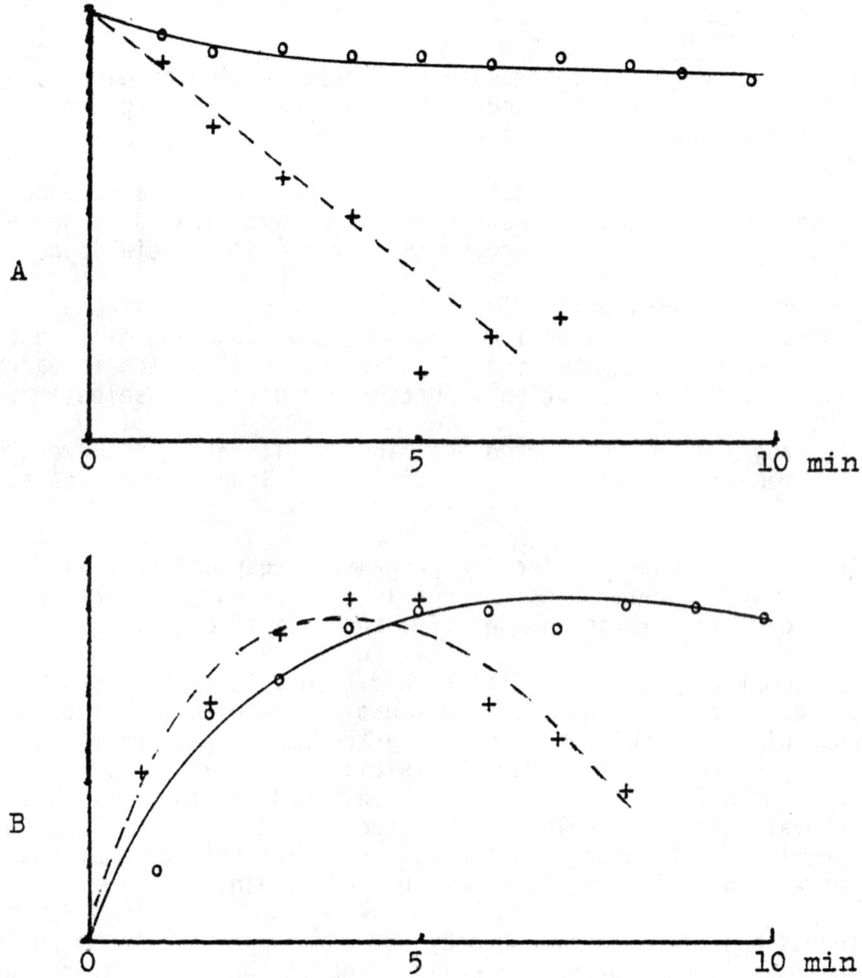

Diagrams A and B: Complex percent changes of CFF (diagram A) and HR (diagram B) in fit (° and continuous line) and unfit (+ and dotted line) subjects, during exposition to acceleration.

Note precocious changes of CFF in comparison with HR changes.

(From ref. 12, modified).

We emphasize the fact that, thanks to CFF recording, unfitness to tolerate acceleration can be detected at an early moment, much in anticipation of cardiac problems even when they are limited to heart rate decrease, so that the test can be interrupted with no risk at all for the subject himself. Indeed, CFF changes appear even two or three minutes before EKG signs.

Later on, after the favorable result of the test, some candidates were subjected to one or more training runs in the centrifuge (positive G and $+5G_x$, i.e., chest to back, for ten minutes).

In conclusion, looking to the future, it may be stated that changes in requirements for astronaut selection have been anticipated as scientific and related space tasks have become more diverse and specialized.

Although there are many basic similarities between both astronaut-pilot and scientist-astronaut selection programs, understanding of scientific mission requirements and the type of individual best suited to carry out the mission are factors of prime importance in astronaut selection.

Selection programs must be worked out closely within the framework of a unique mission task requiring scientific professional competence and qualification.

Thereafter, more frequent selection programs, accepting relatively few astronauts, may be preferable rather than less frequent selection programs, bringing in a larger number of astronauts at one time.

Since the complexity of space flight conditions, testing of space equipment, and conduct of scientific investigations require that individuals other than pilots be selected, such as astronomers, biologists, geophysicists, engineers, navigators, physicians, etc, it has become necessary to modify certain criteria established for the first astronauts in evaluating the health of selected candidates as well as to verify problems pertaining to selection by taking into account characteristics of age and deficient level of physical training.

A continuous search for active prophylactic measures has to be performed prior to spaceflight, during spaceflight and in the period of restoration of functions disturbed in spaceflight, so that selection becomes a continuous process.

Thus, a development of new neurophysiologic and psychologic selection methods and improvement in the system as a whole will be required. In particular, the selection of the most suitable neurophysiologic measurements has become of paramount importance (1) to make appropriate modifications to the chosen methodologies, (2) to follow changes in the underlying biochemical activity of CNS, and (3) to monitor behavior during spaceflight.

REFERENCES

1. Fabris L., Lomonaco T.: Congegno per creare e studiare le condizioni di subgravità nell'uomo, *Atti II Congr. Mond. Med. Aeron.*, Roma, 1959.

2. Lomonaco T., Scano A.: Esplorazione degli apparati respiratorio e cardiovascolare ai fini della selezione e del controllo fisico del personale aeronavigante, *V Congr. Med. Aeron.*, Napoli, Sept. 1953.

3. Lomonaco T., Scano A., Meineri G.: Osservazioni fisiologiche sulla motilità dell'uomo con parziale e totale alleggerimento del peso corporeo, 1) Meccanica della deambulazione e suo dispendio energetico, *Riv. Med. Aeron. Sp.*, 25, 1962, 623.

4. Lomonaco T., Scano A., Rossanigo F.: Comportement de quelques données physiopsychologiques chez l'homme soumis à des variations d'accélération comprises entre 3 et 0 G., *Congr. Intern. Med. Aeron.*, Louvain, 1958.

5. Lomonaco T., Scano A., Rossanigo F.: Comportamento di alcune funzioni percettivo-motorie durante il passaggio da circa 2 a 0 G ed influenza dell'allenamento. Esperimenti eseguiti con la torre di subgravità, *Riv. Med. Aeron. Sp.*, 23, 1960, 439.

6. Lomonaco T., Scano A., Rossanigo F., Caporale R., Meineri G.: Osservazioni fisiologiche sulla motilità dell'uomo con parziale e totale alleggerimento del peso corporeo. II) L'attività elettrica muscolare e il dispendio energetico dopo allenamento. *Atti XII Congr. Int. Med. Aeron. Sp.*, Roma, 1963.

7. Lomonaco T., Scano A., Rossanigo F., Strollo M.: Metodi e dati per una moderna selezione psicofisica del personale aeronavigante e previsioni per il futuro, *II Congr. Mond. e IV Europ. Med. Aeron. Sp.*, Roma, 1959.

8. Lomonaco T., Strollo M., Fabris L.: Sulla fisiopatologia durante il volo nello spazio. Comportamento della coordinazione motoria in soggetti sottoposti a valori di accelerazione variante da 3 a 0 G, *Riv. Med. Aeron.*, 1957, suppl. 1, 76.

9. Lomonaco T., Scano A., Strollo M., Rossanigo F.: Alcuni dati sperimentali fisiopsichici sugli effetti delle accelerazioni e della subgravità previsti nell'uomo lanciato nello spazio, *Riv. Med. Aeron.*, 20, 1957, 363.

10. Meineri G.: Gli effetti della subgravità e i metodi per riprodurla a terra e in volo, *Riv. Med. Aeron. Sp.*, 26, 1963, 80.

11. Meineri G., Modugno G.C., Ramacci C.A., Rispoli E., Pietronero E.: Analisi del comportamento della frequenza critica di fusione centrale in soggetti sottoposti ad accelerazioni $+G_z$ in centrifuga umana, *Riv. Med. Aeron. Sp.*, 40, 1977, 259.

12. Modugno G.C., Meineri G., Ramacci C.A., Rispoli E., Romanin A: Impiego del rilevamento della frequenza critica di fusione centrale retinica (FCFCR) quale indice di sopportazione delle accelerazioni e sue correlazioni con le modificazioni circolatorie, *Congr. Naz. Med. Aeron. Sp.*, Roma, 1980.

13. Ramacci C.A., Meineri G.: Impiego della centrifuga umana per l'addestramento dei piloti militari all'esecuzione di manovre anti-g, *Riv. Med. Aeron. Sp.*, 39, 1976, 71.

14. Rotondo G.: Lo sport nell'età evolutiva e la formazione sportiva dei giovani con particolare riferimento all'attitudine dei giovani al pilotaggio aereo ed al contributo dello sport quale mezzo di potenziamento fisiopsichico dell'aviatore, *Riv. Med. Aeron. Sp.*, 36, 1973, 107.

15. Rotondo G., Meineri G.: Metodi e tecniche per la valutazione fisiologica funzionale degli sportivi nell'Aeronautica militare, *Rel. al VII Incontro tra Specialista e Medico Pratico*, Castrocaro Terme, 1979,

16. Scano A., Meineri G.: Aspetti fisiologici della deambulazione con parziale alleggerimento del peso corporeo e vario attrito con il suolo, *Giornate Mediche delle Forze Armate*, 1965, *Min. Med.* 56, 1965, 3960.

IAA 81-255

INCREASING RELIABILITY OF COSMONAUT-OPERATOR'S ACTIVITY

N.V. Krylova, A.K. Bokovikov*

The security of a spaceflight is provided not only by reliability of a spacecraft, its systems and equipment, but also by working reliability of the man-machine system as a whole. Reliable human performance in more and more complex tasks on board a spacecraft is guaranteed by preliminary professional and psychological training.

Cosmonaut-operator performance is carried out on sensory, sensomotor and verbal-intellectual levels. The probability of errors increases in situations when the human operator performs the function of an automatic reserve unit (passive control) or executes monotonous rhythmically repeated operations (sensory and sensomotor levels) for a long period of time.

Lomov (1979) stressed that "in situtations when an individual functions as a reserve link of a control system, he remains a human being with all his characteristics and these characteristics manifest themselves one way or other in the process of his informational interaction with technology" [7].

Among the situations of this type, emergencies are the most complex. They differ from normal conditions not only by heightened emotional strain, time limit for decision making and following complex performance, but also in the necessity of action reprogramming and formation of a new image to regulate the action [1].

In order to eliminate the probability of errors it is necessary to maintain "an active operator" status. Operator reporting (verbal-intellectual level) can be used as a possible method for activity maintenance because speech stimulates the condition and organization of intellectual activity [2, 3, 4].

Under conditions of complex integrated performance, the operator's speech carries out two functions: (1) communication and (2) intellectual, i.e., serves as an intellectual unit in the case when an operator uses the communicative function of speech to himself.

---

* *Institute of Psychology, Academy of Sciences, Moscow, USSR*

Operator reporting (OR) is a specially organized speech with quite definite functions: self-control, control after sequences of actions, improvement of information interchange etc.

Introducing OR in the operator performance pattern contributes to fuller utilization of human abilities for increasing system reliability. Moreover operator performance can be evaluated according to his speech; i.e., an additional control unit can be incorporated into the system.

In radio traffic conditions OR serves as an effective feedback channel. It helps the operator to receive information about more distant consequences of the action performed. Feedback realized through OR can be varied in amount and presentation frequency; besides, information can be addressed in the form of definite instructions to an operator. OR gives us an opportunity to get information not only about instrumental signals but also noninstrumental signals perceived by a cosmonaut-operator. OR must include not only regulating statements (e.g., reading out apparatus values, delivering information about different systems and equipment) but also contain expository statements made while performing different tasks (e.g., during flight control, in decision-making situations, while evaluating and analyzing unexpected situations).

Differences in patterns and methods of OR evaluation are connected with cosmonaut-operator performance characteristics under various regimes.

In standard situations under little load it is necessary to maintain operator activity by means of rational intensification of his performance. It can be achieved by OR introduced into the general scheme of complex combined operator performance.

We can distinguish a number of tasks when speech (radio communications) is the main element of operator performance, but there is also a set of tasks in which it is just a secondary task.

Visual information processing and delivery during manual and semiautomatic control of a moving vehicle is an example of the latter set of tasks.

In tasks like this, OR must not interfere with the main task of performance. From a psychological point of view, the introduction of OR into the complex performance scheme gives us an opportunity to investigate thoroughly the psychological aspects of operator performance. On the one hand OR must not affect the performance level of the main operator task; on the other hand OR must be informative enough in order to produce expanded characteristics of the performance.

Analysis of the studies in this direction enables us to form the following principles of OR organization:

1. Providing semantic relationship between OR and task performed
2. Availability of instructions directing uniform OR use
3. Conducting preliminary training in use of OR simultaneously with control performance (until stability has been obtained).

Possibility and effectiveness of OR use have been tested and collaborated in experiments.

Two-dimensional tracking was used as an operator main task. The target was presented by a spot of light moving along various tracks. Operators had to compensate for deviations of the target from its initial position in the center of the screen with the help of the display level or to match the marker with the target. During each of three-hour performance blocks, operators executed up to 40 tasks with different types of control and operating conditions and this performance was partly accompanied by OR.

Analysis of results in terms of fineness of operator performance and OR shows that the introduction of OR improves performance in the most complex control tasks. It is suggested that attempts to "verbalize" the structure of performance executed helps operators to analyze more objectively their decision-making strategies and thus choose the most optimal among them.

It is important to stress that OR content is a source of additional information for analysis of individual psychological characteristics of operators.

A number of studies examining operator speech as a component of complex combined performance enables us to establish general requirements for OR [5, 6].

The main significant "unit" of OR is a statement that has its own informational meaning and can be treated as a single OR unit. Every statement must be understood by the operator and contain as much information as possible while expressed in brief form. All statements can be divided into groups according to certain properties. The number of groups depends upon the properties chosen for classification.

For OR evaluation in standard treatments it is necessary to set up a list of possible OR groups. For this purpose we must emphasize the most important stages of operator performance that require control. Also we should set up a list of the most frequent errors committed by the operator within the system taking into account their occurrence and possible consequences. Then it is necessary to examine the significance of statement groups. Using results of the study we can establish a structural OR scheme for standard treatment. For the same purpose statements can be ranged according to the times of their occurrence in standard situations. In this case the parameters that characterize OR shall be the following:

1. Informativeness, defined as the relationship between the whole number of states taking into account group significance and the whole duration of OR use
2. Correlation between the structural scheme of particular OR and the same scheme of optimum OR.

For unexpected working conditions we can form only the general principles of the OR structural scheme. Such an OR must include the following information: description and evaluation of emergency situations; decision making regarding necessary actions; statements reflecting particular actions in order to resolve an emergency situation. OR in emergency conditions must contain statements that reflect actions and that are transmitted simultaneously with the execution of these actions. OR informativeness can be evaluated by the same method as in the case of the standard treatment but taking into account statements related only to the emergency situation.

SUMMARY

An OR structural scheme that can be used in operator's learning may be formed on the basis of expert evaluation.

The introduction of OR into the structure of complex combined performance enables the establishment of an additional control unit in the man-machine-environment system, namely, "cosmonaut-spacecraft-control center".

OR contributes to more successful execution of the most complex control tasks and aids in the more complete use of human activities in terms of increasing operational reliability of the whole system.

REFERENCES

1. Beregovoy, G.T., N.D. Zavalova, B.F. Lomov, V.A. Ponomarenko. *Experimental Psychological Studies in Aviation and Cosmonautics.* Moscow, Nauka, 1978.
2. Vitt, N.V. *Papers on Experimental Studies in Psycholinguistics.* Moscow, Nauka, 1974.
3. Galunov, V.J. *Studies in Variability of Human Speech Behavior.* Doctorial dissertations abstracts, Leningrad, 1975.
4. Gurevich, K.M. *Psychological Aspects of Professional Fitness of Power System Operators.* Moscow, Prosveshenije, 1965, vol. 4.
5. Zamaletdinov, I.S., Krylova, N.V. *Psychological Aspects of Space Flights.* Moscow, Nauka, 1979.
6. Krylova, N.V. *Psychological Journal,* 1981, vol. 2, No. 3.
7. Lomov, B.F. *Cybernetics Questions.* Moscow, 1979, vol. 50.

IAA 81-251

ANNUAL SURVEY OF SPACEFLIGHT SAFETY SYSTEMS: 11th SUPPLEMENT SURVEY PERIOD: JULY 1980 -- JUNE 1981

Nelson E. Brown, Jeri W. Brown*

The beginning of a new era in manned spaceflight--the Operational Spaceflight Era--is marked by the first launch of the United States' Space Shuttle in April 1981, 175- and 185-day record-breaking habitation of the USSR Salyut space station, the frequent-flight schedule for the reusable Shuttle and European Spacelab, and the reboost of the Salyut 6 to a higher orbit to prevent unwanted reentry. During past programs, missions were widely spaced, and man's reactions to space and his ability to conduct scientific experiments and perform meaningful work were learned on a step-by step basis and applied to subsequent flights. The operational era is a significant change, with a vast variety of scientific payloads and routine, frequent launches. Throughout the U.S., European, and other manned space program evolution, safety has been maintained in the forefront.

During Shuttle and Spacelab design, development, and testing phases, significant emphasis has been and continues to be placed on assuring that all safety concerns are satisfied. Extensive testing, implementation of safety programs, and crew and ground controller training contribute to overall program safety. Since 1971, international literature concerning spaceflight safety and crew rescue has been surveyed and reported in yearly survey supplements. Specifically, in 10 annual supplements, 281 referenced sources have addressed safety projects, systems design and testing to ensure the safety of crewmembers onboard spacecraft, the space vehicles during all flight phases, and payloads and payload operations. Each year, results of safety programs and testing, descriptions of new safety provisions, developments of innovative safety equipment and techniques, and safety training and trainers are reported by U.S., European, and other space agencies, industry and academia. This paper, the eleventh supplement, surveys safety reports published between July 1980 and June 1981. Several categories of safety-related publications are examined: program safety guidelines and requirements, ground support and inflight crew operations, spacecraft hardware and components, and crew equipment and procedures. The yearly number of publications that address some aspect of spaceflight safety is too large for complete coverage in this survey. Therefore, the spaceflight safety information from the survey considered most applicable to current and future space activities is reported in this paper.

Keywords: Safety; spaceflight; hazards; safety requirements; extravehicular activity; spacecraft testing; maintenance; repair; servicing.

* *NASA, Lyndon B. Johnson Space Center, Houston, Texas 77058*

INTRODUCTION

The well-being, health and survival of crewmembers in space are of utmost importance in the overall, long-term success of the new operational era of spaceflight. Advances in technology and basic experience in human response to the weightless environment expand on-orbit capabilities for crew operations and enhance total crewmember safety. Much literature is published yearly discussing continued assurance for space safety and rescue during future missions. This paper, the 11th annual supplement to a 1971 basic literature survey, reviews international spaceflight safety information published between July 1980 and June 1981. To reflect the increase in the utilization of space for operations rather than strictly for flight testing and to accommodate the larger proportion of operationally-oriented literature, the format of the survey has changed accordingly. This new index/categorization is shown in Fig. 1.

FLIGHT RESCUE AND SAFETY--LAUNCH, ON-ORBIT, REENTRY, LANDING

The flight of the U. S. Space Shuttle Columbia beginning on April 12, 1981, met or exceeded all flight goals from launch through orbital operation, reenty, and landing. The integrity of the vehicle's propulsion, avionics, and structural systems through the most challenging ascent profile flown by a space vehicle was affirmed. This successful launch cleared the way for completion of the Shuttle flight test program and rapid entry into the U. S. Space Transportation System (STS) operational era. As the preparation date of this paper closely follows the inaugural Orbiter flight, little official documentation is as of yet available.

- I. FLIGHT RESCUE AND SAFETY--LAUNCH, ON-ORBIT, REENTRY, LANDING
  - A. Vehicles
  - B. Systems and Support Equipment
  - C. Operations and Techniques
  - D. Space Hazards
  - E. International Rescue Cooperation
- II. HUMAN CAPABILITIES AND CREWMEMBER ROLES
  - A. Human Performance
  - B. Crew Operations
  - C. Crew Training
  - D. Man-In-Loop Testing and Simulations
- III. EQUIPMENT REQUIREMENTS AND DESIGN FOR SAFETY--SPACECRAFT, PAYLOADS, EXPERIMENTS
  - A. Specifications
  - B. Design and Analysis
  - C. Development
  - D. Testing and Qualification
- IV. LARGE SPACE STRUCTURES
  - A. Space Construction
  - B. Space Habitats
  - C. Unmanned Structure Servicing
- V. LONG-RANGE SPACE PLANNING
  - A. Earth-Orbit
  - B. Lunar
  - C. Interplanetary
  - D. Extraterrestrial Contact

Fig. 1. Spaceflight safety systems--text and bibliography index.

The *STS-1 Mission Report*, Reference 282, summarizes the major activities and findings of the first Shuttle flight test (STS-1). The primary purpose of STS-1 was to demonstrate a safe launch, ascent, and return of the Orbiter vehicle and crew. Flight and systems verification data collected included the following:

- Combined Shuttle vehicle performance;
- Shuttle vehicle aerodynamics, structural and systems characteristics;
- Orbiter and support system thermal response;
- Orbiter entry characteristics and performance;
- Guidance, navigation and control performance; and
- Inflight hardware and software checkout and performance.

As expected from any new space vehicle, several areas of concern were identified following the initial flight. Major concerns included the following:

- Solid rocket booster ignition overpressures of up to 3 psi (0.2 kg/sq.cm.) on Orbiter aft components;
- Debris from the external fuel tank during ascent;
- Variations in first stage performance resulting in vehicle lofting with pitch attitude error exceeding 5 degrees;

- External fuel tank rupture altitude lower than expected;
- Loss of Thermal Protection System (TPS) tiles on the Orbital Maneuvering System pods; and
- Orbiter landing touchdown approximately 3200 ft (975 m) beyond planned point.

Approximately 80 concerns were identified ranging from noise in the astronauts' communications headsets to difficulty in closing cabin stowage lockers. None of the concerns was considered hazardous to the crew or vehicle during the flight. Over 90% of the concerns were corrected as of July 1981, and all will be resolved prior to the second STS flight, scheduled for September 30, 1981.

A system providing fast and safe pre-launch egress for the Shuttle flight crew, passengers and ground close-out crew from the launch pad should hazardous conditions exist and danger be imminent is described in Reference 283, *Space Shuttle Slidewire Emergency Egress System*. The system, installed in the Orbiter launch site at the NASA Kennedy Space Center, consists of 5 multi-man baskets, six egress platforms, five slide wires, individual basket release devices, personnel safety net on the tower below the embarkation area, a deceleration system and landing zone, arresting safety nets (secondary backup stopping devices in case the deceleration system fails), and protective bunkers (Fig. 2). The slidewire system design is such that all personnel can be evacuated to a safe area within 2 minutes with winds as high as 34 knots from any direction. Additionally, the system design limits the forces imposed on the evacuees to less than ±2 g's.

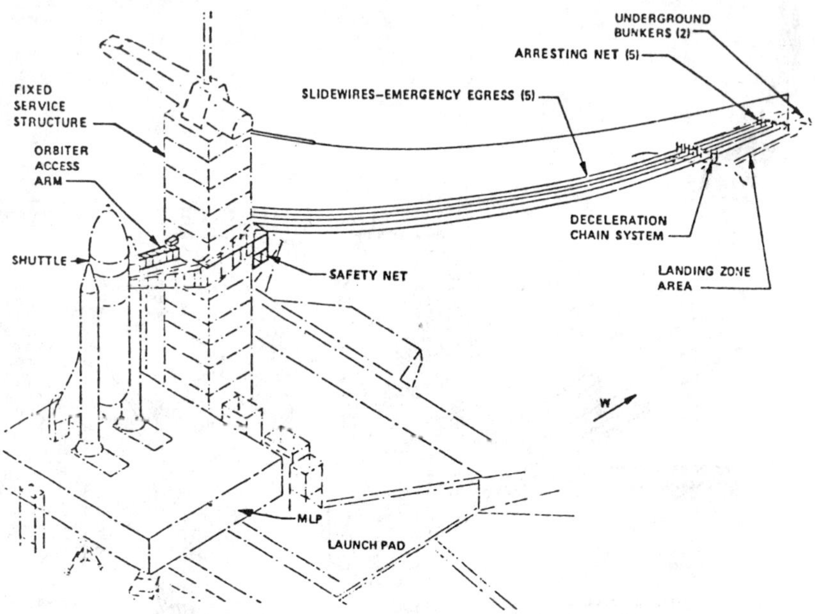

Fig. 2. Launch pad emergency egress system.

Each multi-man basket can accommodate up to 3 people. Should an emergency arise, the personnel would move from the Shuttle and White Room (immediately outside the Orbiter) across the Orbiter Access Arm (OAA) to the slidewire embarkation point on the service tower at 195 ft (59 m) elevation, where they would enter the baskets. After each basket is loaded, it is released to slide down the wire through the deceleration system to the landing zone (Fig. 3). The personnel exit the basket and walk or run to the bunker area. Evacuation and movement of incapacitated personnel are enhanced by a quick-release exit capability (i.e., a collapsible back panel).

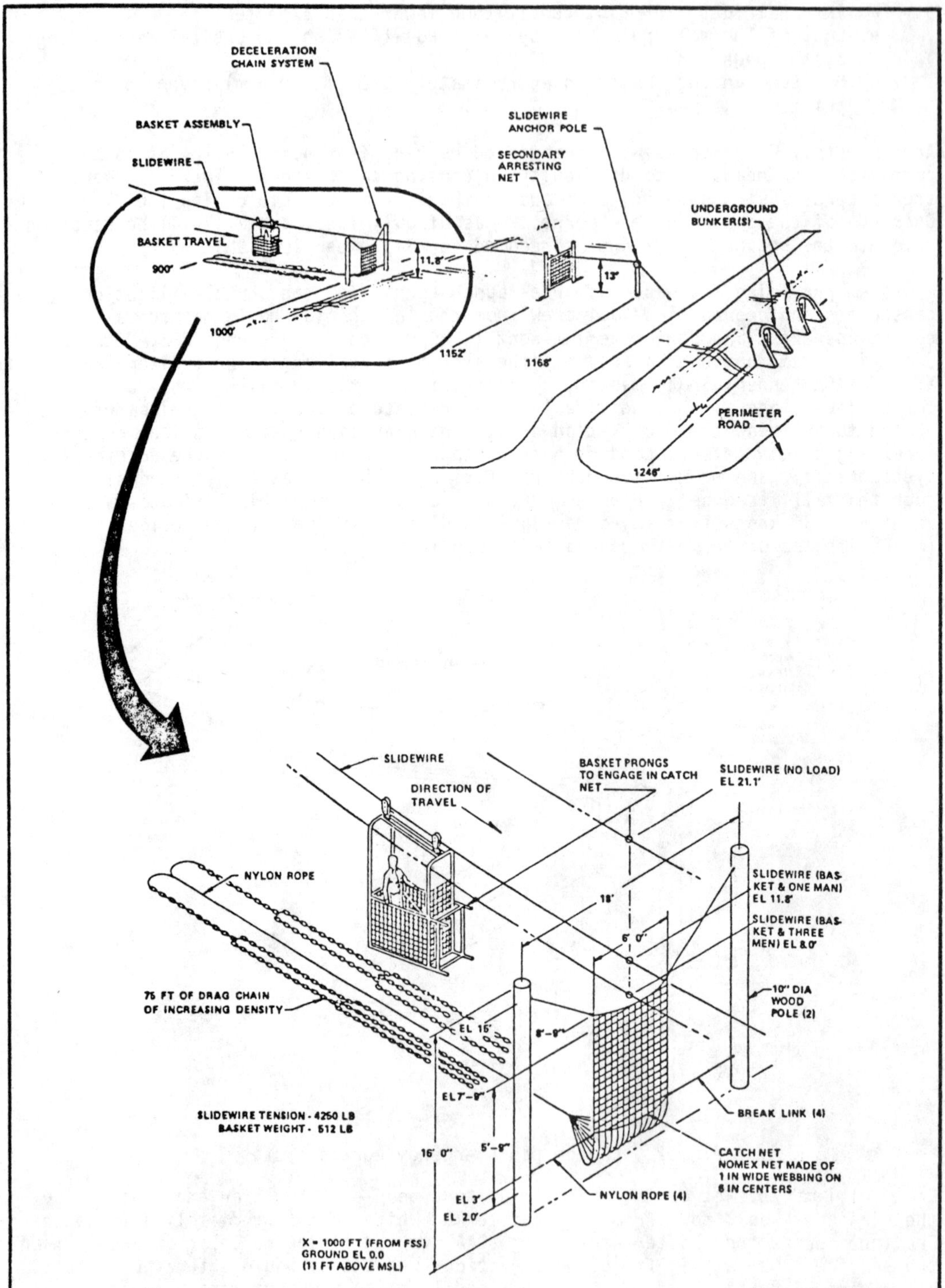

Fig. 3. Slidewire emergency egress landing zone and deceleration system.

Manned validation testing completed in 1980 indicate all design criteria were met. Safety and reliability were demonstrated by an absence of problems in the 30 test runs; the deceleration system performed as designed with maximum recorded g loads of ± 1.6. No parts or mechanisms failures were observed. Total egress times varied from 59 to 67 seconds, well below the 120-second program requirement. In addition to testing, the reference discusses math models describing the slidewire system which were used for analyses of the descent and deceleration phases of evacuation.

Testing activity at the Kennedy Space Center will continue with the arrival of the European Space Agency's Spacelab flight hardware in November 1981. Prior to its arrival, the Spacelab engineering model has undergone extensive testing, as described in *ESA's Spacelab Model To Begin Further Tests* (Reference 284). Engineering model hardware includes a Spacelab pressurized module and three Spacelab pallets. The hardware components are nearly identical to flight hardware and have already completed three months of mechanical tests at the Space Center. These tests involved checking the interfaces between the Spacelab and ground handling equipment, including the processing stand. Exercises checking installation and removal of experiments from both long and short Spacelab configurations (Figs. 4 and 5) are completed. The short module configuration allows more pallet-mounted instrumentation to be flown in the payload bay.

Fig. 4. Spacelab long module configuration.

Fig. 5. Spacelab short module configuration.

Electrical testing of the Spacelab is planned following checkout of the Spacelab automated test equipment (ATE). The engineering model will be tested with the ATE prior to electrical checkout of the flight hardware. After the Spacelab is tested, integrated Spacelab payloads will be placed in the Cargo Integrated Test Equipment (CITE) for hookup with the simulated Shuttle Orbiter--electrically and mechanically. Thus far, mechanical processing of the Spacelab engineering hardware has encountered no significant difficulties. These mechanical, electrical, and integrated tests are important in helping to ensure crew safety in orbit.

HUMAN CAPABILITIES AND CREWMEMBER ROLES

Reference 285, *Crew of Soyuz Repairs Salyut*, reports on repairs to the Soviet Salyut 6 orbiting space station by Leonid Kizim, Gennadiy Strekalov, and Olag Makarov. Extensive repairs were made to the Salyut 6 allowing extended use of the spacecraft. Repairing a hydraulic system was the most complex repair task and involved penetrating the fluid-based system, managing the hydraulic fluid, installing a new hydraulic unit, and incorporating four pumps in the Salyut's temperature control system. Several other repair tasks were conducted including:

- Installation of a new transducer in a power supply compressor that feeds Salyut's in-orbit refueling system;
- Replacing spacecraft control system and programming and timing devices; and
- Replacing electronic components in the air-to-ground telemetry system.

The three-man cosmonaut crew was in air-to-ground contact with Leonid Popov and Valeriy Ryumin during part of the repair activities for advice on specific characteristics of Salyut's performance as repairs were made. Special tools were developed for the repairs.

Prior to separating the Progress 11 tanker/transport vehicle from Salyut 6, the Progress 11 main propulsion system was used to boost the Salyut-Soyuz-Progress configuration to a 230 x 180 mi (370 x 200 km) orbit inclined 51.6 degrees with a 90.8 min. period.

Earlier on August 15, 1979, Cosmonauts Lyokhov and Ryumin donned their spacesuits and dislodged an entangled Cosmic Radio Telescope (DRT-10) antenna from the exterior of the Salyut 6. Demonstration of such maintenance and repair capability by the cosmonauts both inside and outside the spacecraft is an important step in the enhancement of crew safety in orbit.

Reference 286, *Assessment of the Effects of the Zero Gravity Environment on the Health and Safety of Space Workers*, published by the NASA Johnson Space Center, examines concerns for the well-being of crewmembers inflight:

> *The effects of null gravity in the various physiological systems of men are significant inasmuch as they impose distinct threats to work performance and safety. In planning and preparing for future flights, the potential impact of these effects must be assessed for each space venture, because mission success and safety in the presence of these adverse effects will depend upon what is required and who is required to do it. Thus, these physiological changes take on greater significance when work loads are greater, task demands are more versatile, flight durations are lengthened or intermittant, or increased g exposures are introduced, all of which are likely to characterize flight requirements of the future. Similarly, with the inclusion of more specialized personnel of both sexes and broader age groups as members of flight crews, greater susceptibility to these changes and less ability to sustain them may be anticipated.*

The report reviews currently available information relating to the adverse effects to health and safety that Space Power System (SPS) crewmembers may experience. The prime objective of the study was to assess the effects of all currently known deviations from normal of medical/physiological/biochemical parameters which appear to be due to zero gravity environment, including other stresses of working in the space environment. A second study objective was to prepare a research plan defining areas for further study regarding the health and safety effects of the SPS space workers.

Data available to date from manned space flight must be viewed in the light of differences of flight situations and conditions between past and SPS astronauts. Care during the study was taken to identify and define the nature and influence of these differences in formulating their prediction of adverse effects. Factors involved comparing past and future missions are shown in Table I. Some of the major differences in the program requirements relating to the SPS space crews vs. the astronaut crews are shown in Table II. Sources of information for the study included documented biomedical results of Skylab, Apollo, Apollo/Soyuz, and other manned space missions, documented results of extensive ground-based and flight research and testing, information from other programs involving isolated crew habitation and confined quarters, and direct contact with appropriate experts.

Table I. Factors compared for past and future crewmembers.

CREW (SPACE WORKERS)	HABITAT
• TYPE OF PERSONNEL • PREPARATION AND TRAINING • PRE, IN, AND POST FLIGHT ACTIVITIES • REGIMENTATION AND DISCIPLINE • ETC.	• ENVIRONMENTS • SHIELDING • RECREATION AND REST • FOOD AND NUTRITION • PRIVACY • ETC.
MISSION FLIGHT PARAMETERS	**CAREER***
• ORBITS • ACCELERATIONS • SOLAR ACTIVITY PERIODS • ETC.	• TOTAL TIME IN SPACE • CUMULATIVE PSYCHOLOGICAL/PHYSIOLOGICAL EFFECTS • JOB FATIGUE • SPACE TIME/GROUND TIME

*NOTES: During a 5-year career with a 90 day up/90 day down cycle, a person may suffer from space environment effects for 4 to 5 months out of each 6 months, resulting in a career situation of being in a deviate physical condition for 3 1/2 to 4 years of the 5 year total.

After the 84-day Skylab mission, two crewmen had not regained heel bone calcium by day 95 postflight.

Accomplishments of the study are described. A key finding is that this review of available information on previous manned spaceflights has revealed no physiological responses to the space flight enviornments considered in this study that would jeopardize the health and safety of SPS crewmembers. Other findings included the following: (1) Antigravity muscles lose mass, probably comprised of fluid surrounding the muscle fibers and protein from the muscle fibers themselves. Other skeletal muscles appear not to exhibit these losses, or at least not to the same degree. There is a small, reversible loss of strength and ability to perform work at maximal levels; (2) Cardiovascular adaptability or competence (orthostatic tolerance) is comprised as reflected by increased pulse rate and decreased blood pressure during stress tests (including LBNP--lower body negative pressure) and erect standing in normogravity following space exposure.

Organ systems and functions recognized as sensitive to weightlessness changes are the musculoskeletal, hemopoietic, cardiovascular, immune and endocrine systems and bioenergetic control. Organ systems minimally or not affected during space exposure include reproductive, digestive, respiratory, lymphatic, nervous (especially

Table II. Major differences in program requirements (space workers vs. astronaut crews).

1. TYPE OF PERSONNEL SELECTED	
SPACE WORKERS	SKYLAB ASTRONAUTS
• MALE-FEMALE • BROAD AGE RANGE • PHYSICALLY BASICALLY UNSCREENED • LARGE CREWS	• ALL MALE • LIMITED RANGE • PHYSICALLY SCREENED AND DEVELOPED • 3-MAN CREW
2. EXTENT AND TYPE OF CREW PREPARATION AND SPACE DUTIES	
SPACE WORKERS	SKYLAB ASTRONAUTS
• SHORTER PREPARATION TIME AND TRAINING, LIMITED PRIMARILY TO JOB RELATED ACTIVITY, WITH MINIMUM SPACECRAFT PHYSICS AND SYSTEMS, HABITABILITY, ETC.	• SEVERAL YEARS OF BROADBASED EDUCATION AND TRAINING IN ALL ASPECTS OF MISSION ACTIVITIES WITH EXTENSIVE EDUCATION IN FUNDAMENTALS OF ALL SCIENCES INVOLVED IN PROGRAM
3. NATURE OF MISSION ACTIVITY ASSIGNMENTS AND FREQUENCY AND DURATION OF FLIGHT TIME/GROUND TIME	
SPACE WORKERS	SKYLAB ASTRONAUTS
• BROAD VARIETY OF SPECIALIZED MANUAL, CLERICAL, STAFF SKILLS (WITH MINIMUM ENGINEERING AND SCIENTIFIC SKILLS)	• EACH CREW MEMBER CAPABLE OF ALL SCIENTIFIC, TECHNICAL, AND MANAGEMENT REQUIREMENTS
• WORK AT PEAK EFFICIENCY FOR MAXIMUM SAFE PERIOD DURING MISSION, RETURN TO SPACE IN SHORTEST SAFE AND PRACTICAL TIME	• WORK AT HIGHLY MOTIVATING JOBS AT CAREFULLY SCHEDULED TIME LINES BASED ON METABOLIC AND EXPERIMENT REQUIREMENTS, MISSION DURATION BASED ON CREWS' CONDITION (CAREFULLY MONITORED), RETURN TO SPACE NOT A PRESSING ITEM

psychomotor, behavior, judgement, problem-solving ability), sensory (except vestibular), and excretory. Crew performance may be adversely affected at work stations, during extravehicular activities (EVA), or upon return to earth. Countermeasures to these effects may include improved design of suitable work stations, restraint/mobility aids, and tools and equipment; continued development of life support systems with increased efficiency; and suitable diet and exercise regimens which suitably load both cardiovascular and musculoskeletal systems.

New areas of research suggested in the report include new design and operational approaches for life support and protective systems, habitability systems, and health care systems. Studies must be performed concerning large number and variety of types of people, task variety, length of stay time vs. ground time, span of career, and exposure to low-earth and geosynchronous-earth orbits. A reference/bibliography list is included.

EQUIPMENT REQUIREMENTS AND DESIGN FOR SPACE

Organizations using the Space Transportation System (Space Shuttle) to launch their payloads are responsible for assuring payload safety during all flight and ground operations. A December 1980 release of the NASA document NHB 1700.7A, *Safety Policy and Requirements for Payloads Using the Space Transportation System* (STS) (Reference 287), minimizes STS involvement in the payload design process while maintaining the assurance of safe operations. The document intent is to provide the overall safety policies and requirements while allowing for negotiations between the payload organizations and the STS safety officials in the method of implementing payload safety. The December 1980 issue of NHB 1700.7A permits the payload developer to pursue new design options to accomplish payload safety, relaxes monitoring requirements for payload parameters, and adds several new requirements.

The requirements in NHB 1700.7A are intended to protect flight and ground personnel, the STS, other payloads, ground support equipment, the general public, public and private property, and the environment from payload-related hazards. The safety policies and requirements to be implemented by the payload organizations are classified as (1) technical requirements and (2) systems safety requirements and are applicable to all payloads. Reporting the details of each safety requirement is beyond the scope of this survey; however, the major safety subject listing of NHB 1700.7A is provided:

(1) TECHNICAL REQUIREMENTS

a. Failure Tolerance
b. Control of Hazardous Functions
c. Retrieval of Payloads
d. Hazard Detection and Safing
e. Contingency Return of Payloads
f. Failure Propagation
g. Redundancy Separation
h. Structural
j. Hazardous Materials
k. Pyrotechnics
l. Destruct Systems
m. Radiation
n. Electrical Systems
o. Verification Requirements
p. Hazardous Procedures
q. Reflown Hardware
r. Extravehicular Activity
s. Series Payloads
t. Flammable Atmospheres

(2) SYSTEMS SAFETY REQUIREMENTS

a. Safety Analysis
b. Hazard Levels
c. Hazard Reduction
d. Safety Assessment Reviews
e. Safety Compliance Data
f. Accident/Incident/Mission Failures Investigation and Reporting.

The safety requirements apply under worst-case natural and induced environments, including STS abort and emergency landing conditions. The detailed interpretation and implementation of safety requirements will be determined on a case-by-case basis but must be consistent with hazard potential. Supplementary NASA payload safety related documents have been developed to assist payload organizations in complying with the technical and system safety requirements.

Safety requirements for high-pressure pneumatic pressure vessels and low-pressure propellant tanks for Space Shuttle payloads are described in Reference 288, *Parametric Study of Pressure Vessels for Space Shuttle Payloads*. In previous U. S. unmanned launch vehicles, the designers of pressure vessels for spacecraft were required to consider personnel safety only during the ground-operations phase of a mission. For the manned Space Shuttle, personnel safety relative to payload hazards apply throughout the flight (attached payloads) or until the payload is safely separated from the Shuttle Orbiter.

The NASA document NHB 1700.7A, "Safety Policy Requirements for Payloads Using the Space Transportation System (STS)," permits the payload pressure vessels to be designed to either the ASME (American Society of Mechanical Engineers) Boiler Code, MIL-STD-1522 (U. S. Air Force Military Standard), or NSS/HP-1740.1 (NASA Aerospace Pressure Vessel Safety Standard). The ASME Boiler Code requires a safety factor of 4.0 on the ultimate strength or 1.6 on the yield strength of the vessel. MIL-STD-1522 requires a safety factor of 2.0 on the ultimate strength of the vessel and two vessels tested to burst. Neither of the design criteria requires the use of fracture mechanics methods. NSS/HP-1740.1 requires a minimum safety factor of 1.5 on the ultimate strength of the vessel and requires that fracture mechanics techniques be used to preclude vessel failure resulting from pre-existing flaw propagation.

The additional parameters of materials, shape, weight, fabrication methods, cost and delivery schedule further complicate the choices available to the payload system designer. The effects of these parameters on low-pressure bipropellant tanks and high-pressure gas-pressurant tanks form the basis of this study. Materials studied were titanium alloys, aluminum alloys, stainless steel, and filament-overwrapped composite materials. Fabrication processes included forging, spin forming, hydroforming, cryoforming, and machining.

Pressure vessels for spaceflight are analyzed for all mission phases beginning with vessel fabrication and ending with flight completion. A variety of design conditions must be met simultaneously for overall mission reliability. The conditions do not always apply during phases that are hazardous to flight or ground personnel but must be met to ensure overall mission success. The conditions as described by NASA Handbook NHB 1700.7A and its referenced documents include:

1. Ultimate factor of safety;
2. Buckling due to external pressure;
3. Crack propagation due to cyclic stress;
4. Crack propagation due to sustained stress;
5. Buckling due to shear stresses;
6. Burst or leakage due to material creep;
7. Minimum fabrication thickness; and
8. Buckling due to handling.

The governing design criteria are the conditions that result in the greatest wall thickness.

This study used a hypothetical bipropellant propulsion system with nitrogen tetroxide (NTO) and monomethyl hydrazine (MMH). The oxidizer and fuel tanks were identical; therefore, the NTO properties determined the tank wall thickness. The evaluation parameters were weight, cost and delivery schedule. Cost and schedules were estimates obtained from various vessel manufacturers. The study results for propellant tanks and pressurant tanks are shown in Tables III and IV, respectively. The results could not specify preferred materials for a specific mission because of the many relative values for weight, cost and schedule. Aluminum off-the-shelf SCUBA tanks have been baselined for one payload with sufficient weight margin, while another payload is using all titanium tanks. The study shows a variety of design conditions for consideration and several material/fabrication combinations for satisfying the safety conditions.

The windows of a manned spacecraft which form a part of the habitable pressure vessel can present significant safety hazards to the crewmembers and vehicle should failure occur. Reference 289, *Structural Analysis and Test Space Shuttle Orbiter Window System*, describes the certification process for the Space Shuttle Orbiter three-pane, uncovered window system. The window system was analyzed for aerodynamic pressure and thermal loads on the external surfaces and for cabin pressure on the internal surfaces. Performance requirements of the window system are to withstand spaceflight conditions for 100 missions without rupture or collapse.

The Orbiter window system consists of six forward windows (two each forward, center and side), two overhead windows, a side egress hatch window, and two rear view windows for payload bay and aft viewing. All windows, except the rear view windows, consist of an outer thermal pane to carry aerodynamic pressure loads and thermal loads, an inner pressure pane to resist internal crew module pressure, and a redundant middle pane should the inner pressure pane fail. The rear view windows consist of only two panes since they are protected inside the Orbiter payload bay and are not subjected to launch and entry conditions.

Table III. Parametric study results for propellant tanks.

MATERIAL, HEAT TREAT CONDITION AND FABRICATION METHOD	REQUIRED WALL THICKNESS (IN.)					SELECTED DESIGN CONDITION	ESTIMATED WEIGHT PER TANK (LB.)	ESTIMATED COST PER TANK (1980 $)	ESTIMATED DELIVERY SCHEDULE (MONTHS)
	FACTOR OF SAFETY OF 1.5 ON ULTIMATE	EXTERNAL PRESSURE OF ONE ATMOSPHERE	CRACK PROPAGATION (FRACTURE MECHANICS)	MINIMUM FABRICATION THICKNESS	MINIMUM THICKNESS FOR HANDLING				
HEAT TREATED Ti-6Al-4V FORGINGS	0.018	0.060	0.016	0.010	0.010	FACTOR OF SAFETY ON ULTIMATE*	29	88K	59 34 MO. FOR FIRST TANK AND 4 TANKS PER MO. THEREAFTER
ANNEALED Ti-6Al-4V FROM SPUN PLATE	0.024	0.060	0.021	0.010	0.010	FACTOR OF SAFETY ON ULTIMATE*	39	63K	63 28 MO. FOR FIRST TANK; 4 TANKS/MO. THEREAFTER
HEAT TREATED 6061-T62 ALUMINUM FROM SPUN PLATE	0.073	0.070	0.066	0.010	0.010	FACTOR OF SAFETY ON ULTIMATE	72	18K	31 6 MO. FOR FIRST TANK; 4 TANKS/MO. THEREAFTER
HEAT TREATED 2219-T62 ALUMINUM FROM SPUN PLATE	0.067	0.068	0.079	0.010	0.010	FRACTURE MECHANICS	82	18K	34 9 MO. FOR FIRST TANK; 4 TANKS/MO. THEREAFTER
HEAT TREATED 2021-T62 ALUMINUM SPUN FROM PLATE	0.045	0.067	0.047	0.010	0.010	FRACTURE MECHANICS*	49	18K	34 9 MO. FOR FIRST TANK; 4 TANKS/MO. THEREAFTER
CRYOGENICALLY FORMED 301 STAINLESS STEEL	0.010	0.041	0.010	0.010	0.010	FRACTURE MECHANICS AND FACTOR OF SAFETY ON ULTIMATE	21	31K	43 18 MO. FOR FIRST TANK; 4 TANKS/MO. THEREAFTER
KEVLAR/EPOXY FILAMENT-WRAPPED COMPOSITE, ALUMINUM LINER	0.018 COMPOSITE 0.010 ALUMINUM	0.079 COMPOSITE 0.010 ALUMINUM	0.018 COMPOSITE 0.010 ALUMINUM	0.018 COMPOSITE 0.010 ALUMINUM	0.010 ALUMINUM	FRACTURE MECHANICS AND FACTOR OF SAFETY ON ULTIMATE	19	20K	40 15 MO. FOR FIRST TANK; 4 TANKS/MO. THEREAFTER

*REQUIREMENT FOR EXTERNAL PRESSURE RESISTANCE IS AN ARBITRARY REQUIREMENT.

Table IV. Parametric study results for pressurant tanks.

MATERIAL, HEAT TREAT CONDITION AND FABRICATION METHOD	REQUIRED WALL THICKNESS (IN.)		GOVERNING DESIGN CONDITION	ESTIMATED WEIGHT PER TANK (LB.)	ESTIMATED COST PER TANK (1980 $)	ESTIMATED DELIVERY SCHEDULE (MONTHS)
	FACTOR OF SAFETY ON ULTIMATE	CRACK PROPAGATION (FRACTURE MECHANICS)				
HEAT TREATED Ti-6Al-4V FORGINGS	0.147	0.128	FACTOR OF SAFETY ON ULTIMATE	29	10.5K	48
ANNEALED Ti-6Al-4V FROM SPUN PLATE	0.193	0.170	FACTOR OF SAFETY ON ULTIMATE	38	14.0K	48
CRYOGENICALLY FORMED .301 STAINLESS STEEL	0.084	0.083	FACTOR OF SAFETY ON ULTIMATE	30	9.5K	46
KEVLAR/EPOXY FILAMENT-WRAPPED COMPOSITE, THIN ALUMINUM LINER	0.144	0.144	BOTH EQUALLY	10	2.5K	15
HEAT TREATED 6351-T6 ALUMINUM SEAMLESS FORGINGS (COMMERCIAL SCUBA TANKS)	0.356 (2000 PSIA OPERATING PRESSURE MEETS MIL-STD-1522)	0.258	FACTOR OF SAFETY ON ULTIMATE	120 (3 TANKS REQUIRED FOR VOLUME)	0.5K	1

The outer thermal pane is an annealed glass (fused silica; Corning Glass Works 7940) with a low coefficient of linear expansion to minimize thermal stresses from entry heating. The redundant middle window uses the same material as the outer pane, but the inner pressure panes are fabricated of high-strength tempered glass--alumina-silicate. The redundant middle panes are subjected to stresses due to window substructure edge deflections. Since annealed glass exhibits static fatigue, the design criteria specified that each glass pane be evaluated for initial and final strength and have minimum safety factors (S.F.) as follows:

	Initial S.F.	Final S.F.
Annealed (outer, middle)	2.0	1.0
Tempered (inner)	2.0	2.0

For Orbiter missions, the total window pane stress is a combination of four sources --direct pressure load on the pane, substructure bending, substructure post out-of-plane deflection, and thermal stress. The stresses from mechanical loads are directly proportional to the pressure load and were computed using a finite element model of the window and its support structure. Thermal stresses were calculated using a finite difference thermostructural model. The forward windows were modeled using a 3-dimensional NASTRAN model shown in Figure 6. The total number of node points and elements used were 735 and 906, respectively. The panes are simply supported at the edges. The model was loaded with differential pressure, initial out-of-plane structural deflection at the corners, and adjacent panel loads as appropriate. The overhead window was modeled as a 2-dimensional model. The side hatch window was analyzed using classical methods due to its circular geometry.

The maximum load durations and ultimate strength requirements for the window system are summarized in Table V for a typical 7-day orbital flight. A stress history curve resulting from window system testing is shown in Figure 7 for a forward fuselage outer thermal window. The stress peaks resulted from wind gusts of short duration. Table VI defines the magnitude of the stresses contributed by the four sources on the forward outer thermal window.

Analyses were also made relative to natural environments (rain, hail, snow, fungus, ozone, humidity, salt spray, wind, lighting, sand and dust, solar radiation, and meteoroids) and induced environments such as vibration, shock, and acoustics. These environments were not considered critical because of testing, material selection, protective coatings, low probability of occurrence and using loads significantly lower than design loads. Only the meteoroid damage possibility showed a significant impact which will be handled by post-flight inspection and window replacement as required. All analyses and testing were completed and the window system certified for 100 mission cycles prior to launch of the first Space Shuttle Orbiter.

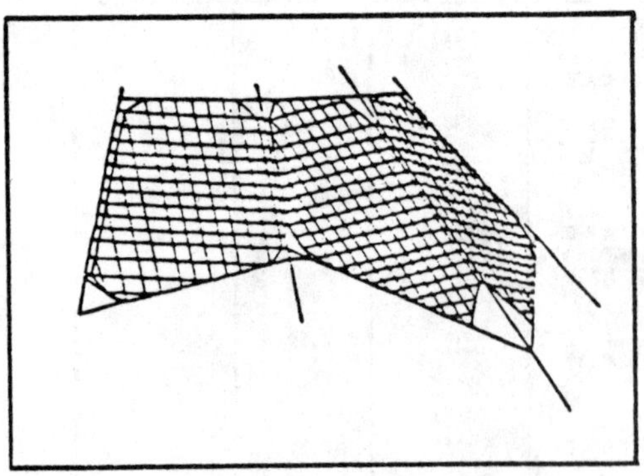

Fig. 6. Forward fuselage windshield, 3-D NASTRAN model.

An important phase of spaceflight preparedness underlying the overall safety of crewmembers is pre-flight testing conducted under simulated mission operational conditions, as summarized in Reference 290, *Effectiveness of Spacecraft Testing Programs*. This paper discusses the need for such testing and reviews the test results from a user viewpoint. Because of spaceflight profile changes in terms of sophistication and longer duration, methods for verification of design and workmanship prior to launch have become increasingly important. Comprehensive testing must begin with development,

Table V. Load duration and ultimate strength.

WINDOW	PANE	DURATION OF MAX. STRESS (SEC)	ULTIMATE STRENGTH (PSI)
FORWARD	THERMAL	10	5010
	REDUNDANT	500	4400
	PRESSURE	605,000	15,700
MIDDLE	THERMAL	10	5010
	REDUNDANT	500	4250
	PRESSURE	605,000	15,700
SIDE	THERMAL	10	5010
	REDUNDANT	500	4010
	PRESSURE	605,000	15,700
OVERHEAD	THERMAL	10	5010
	REDUNDANT	500	4460
	PRESSURE	605,000	15,700
REAR VIEW	REDUNDANT	605,000	15,700
	PRESSURE	605,000	15,700
SIDE HATCH	THERMAL	10	5010
	REDUNDANT	500	4460
	PRESSURE	605,000	15,700

Table VI. Stress breakdown--forward thermal window--point E (psi).

SOURCE	REGION			
	CENTER		EDGE	
	OUTBOARD	INBOARD	OUTBOARD	INBOARD
THERMAL	242	235	640	620
PERSSURE	409	1211	183	541
SUBSTRUCTURE BENDING	292	865	475	1403
POST OUT-OF-PLANE DEFLECTION	29	86	20	60
TOTAL	972	2397	1318	2624

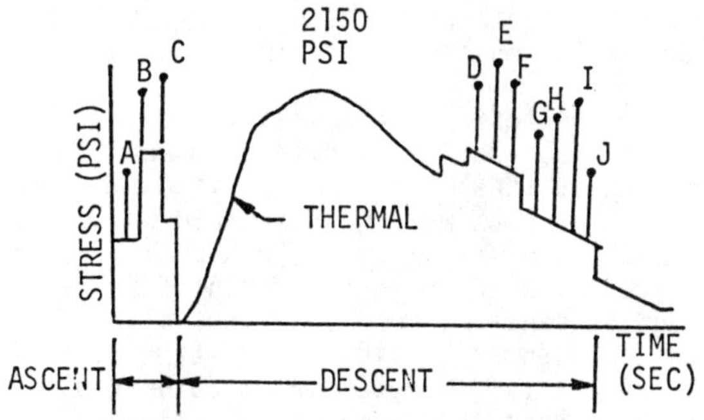

MAXIMUM PEAK STRESSES (psi)

	A	B	C	D	E	F	G	H	I	J
DURATION (SEC.)	13	3	2	30	10	10	20	10	10	20
CYCLE										
1	1002	1737	2004	2463	2624	2460	1804	1903	2006	1175
7	948	1717	1957	2330	2470	2320	1722	1802	1893	1141
24	842	1677	1864	2063	2163	2029	1544	1602	1666	1075
36	735	1636	1770	1795	1856	2093	1375	1402	1439	1003
24	628	1596	1677	1528	1548	1478	1201	1201	1211	941
7	521	1556	1583	1261	1242	1198	1023	1001	984	874

Fig. 7. Stress-history curve, forward thermal window.

through qualification and acceptance, and include flight readiness checkout at the launch site.

Testing is examined in terms of parts, components, and spacecraft. Piece-part testing procedures are based on voluminous engineering data and analyses of failure physics for the parts. The government and electronic parts industry have focused on standardization of test procedures. Procedures in military and industrial specifications and standards have resulted in a significant drop in part failures during subsequent unit level tests. Procedures for unit or component-level test are designed to verify all functional performance requirements over a range of environmental exposures without disassembly or change in the configuration of the component as installed in the spacecraft. Various test protocol is outlined for various type components, including electronic, digital, mechanical and electromechanical. Environmental testing for component-level qualification and acceptance are specified in MIL-STD-1540A, "Test Requirements for Space Vehicles." This source addresses functional testing before, during and after exposure to pyroshock, random vibration or acoustics, thermal cycling, vacuum, and EMI. Test levels and duration must be selected to accelerate detection of inherent defects without damaging or degrading good equipment.

Spacecraft tests include three separate phases: integration, environmental qualification or acceptance, and prelaunch verification. The sequencing of test events proceeds as follows: (1) components, subassemblies and harnesses (which have been tested at parts or component levels) are installed to form subsystems or groups; (2) physical inspection and functional tests are performed under ambient and environmental conditions to demonstrate that the full range of performance requirements is met; and (3) tests are performed at the launch site to verify functions and interfaces prior to launch. Discrepancies discovered during acceptance testing are corrected prior to launch. The author presents data correlating orbital performance to performance during acceptance testing. Lower defect rates during acceptance testing indicate better on-orbit performance.

In attempting to determine if the test program scenario described is really effective in locating all potential defects, additional data from actual projects are presented resulting in the following conclusions: (1) A majority of on-orbit discrepancies are due to subtle design defects which were not or could not be discovered during the qualification and acceptance test program; and (2) Repetitive room ambient testing over a variety of mission scenarios and using different test methods will disclose more defects than simple environmental exposure without electrical stress. The use of automated test equipment is seen as justified in bettering orbital performance. Driven by some spacecraft complexity, computer-based automated testing is identified as significant. Additionally, the author emphasizes designing spacecraft for testability and provides several appropriate design features: selection of test points and test connector location; accessibility for x-ray and visual inspection of critical connectors; equipment layout and clearances to enable replacement of faulty units; and selection of telemetry measurements.

Reference 291, *Extravehicular Mobility Unit Television System Program Requirements Document*, and NASA news releases describe a modified U. S. Air Force TV camera system that will aid in monitoring astronaut activities outside the spacecraft, evaluating Orbiter external systems/hardware status from both the cabin and ground, and assessing extravehicular crewmembers safety and well-being. A solid-state television system mounted on the Shuttle spacesuit helmet will transmit real-time scenes of what the astronauts are seeing during extravehicular activities to on-board monitors inside the Orbiter cabin. The transmission can also be taped or relayed to ground stations. The TV system is an RF (radio frequency) link from the Extravehicular Mobility Unit (spacesuit and life support system) to the Shuttle Orbiter using the Orbiter S-band hemi antennas, a TV receiver and the Orbiter closed circuit television system. The equipment mounted on the spacesuit helmet includes a mini-TV camera, electronics package, TV transmitter, antenna and a battery pack. Hardware located on the Orbiter consists of a TV receiver, video processing unit and associated cabling (see Fig. 8).

The camera field of view (FOV) is 32° horizontally (25° vertical FOV and 40° diagonal FOV) with a focal length of 19.7 mm. The lens range is preset to focus from approximately 15 in (38 cm) to any required distance. The transmitter has sufficient power to cover a minimum range of 250 ft (76 m). The battery pack consists of eight 3 1/2 volt batteries designed to last up to 8 hours; the batteries can be replaced during flight. The weight of the TV system mounted on the spacesuit helmet is approximately 10 lbs (5 kg). The TV camera is rigidly mounted on the helmet at the correct angle to view scenes as seen by the extravehicular astronaut. Pointing is accomplished by movement of the spacesuit since the helmet is not rotatable. During night operations, illumination will be provided by lights mounted on each side of the helmet. TV controls operated by the astronaut consist of an ON-OFF switch and an auxiliary flip-down lens for close-up viewing. Should conditions occur, the EMU TV system capabilities would be invaluable in relaying extravehicular on-orbit repairs and rescue information to the Orbiter cabin and ground operations for joint decision making.

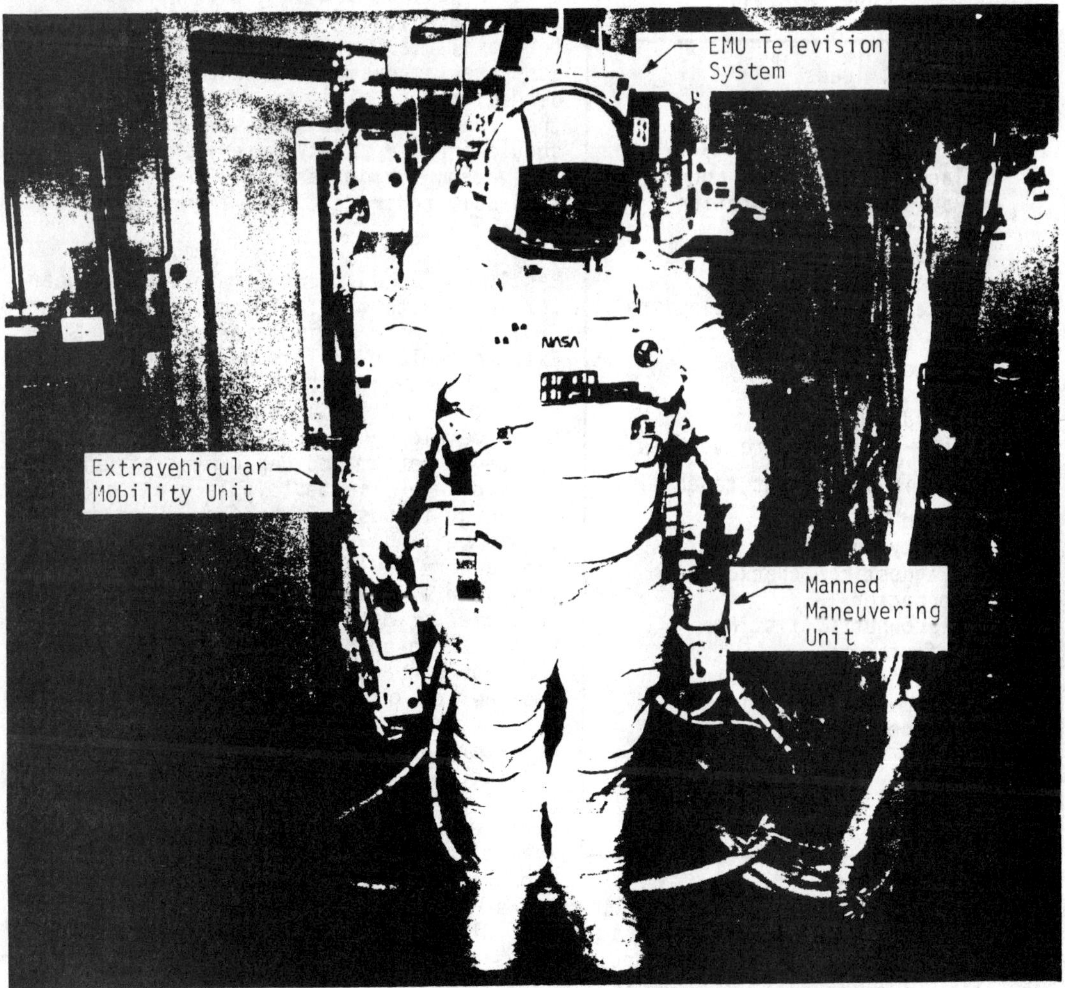

Fig. 8. Shuttle Extravehicular Mobility Unit (EMU) with television system and Manned Maneuvering Unit in test environment.

EVA *Equipment for Satellite Service* is the title of Reference 292, the second of a three-part article on engineering in space. As the Space Transportation System (STS) matures, a "new spectrum of manned space research and work activity" develops. Astronauts performing extravehicular activities (EVA) for maintenance and repair of satellites will become fundamental in achieving mission objectives via practical means. However, because the satellite servicing philosophy to date has been to design the spacecraft to stringent requirements for high reliability to satisfy long mission requirements, no inflight satellite servicing capability has existed. With the STS, these satellites will be "visitable" in low-earth orbits, and designers will have the opportunity to modify requirements permitting astronaut serviceability in orbit. To exploit this capability, service provisions must be designed into the satellite. Projected design requirements are summarized in Table VII. Safety during servicing is paramount not only for the EVA crewmember but also for the Orbiter and payloads. The design requirements will enhance crew safety, reducing hazards in terms of both payload and other hardware design and extravehicular operations. For example, the satellite configuration should permit easy crew access to maintenance areas and be designed to minimize risk of damage to the Orbiter and the spacesuit. Subsystems should include safety interlocks for deactivation during service and should be fail-safe to protect crewmembers against injury.

TABLE VII. Design requirements for satellite serviceability.

- Mechanical loads
- Safe surfaces & edges
- Accessible maintenance areas
- Replaceable subsystem modules
  - Payload instrumentation
  - Attitude control & propulsion
  - Power
  - Data processing & telemetry
- Fluid subsystems
  - Refueling
  - Safety venting
  - Fluid isolation
  - Fail-safe pressure vessels
- Diagnosis & checkout capability
- Standard interfaces
  - Safety interlocks
  - Diagnostic & checkout connector
  - Disconnects, fittings & fasteners
  - RMS and FSS* adapters
  - Crewmember restraints & handholds

*RMS--Remote Manipulator System
FSS--Flight Support System

Baseline STS equipment for performing near-term satellite servicing includes the Shuttle Orbiter, Remote Manipulator System (RMS), Extravehicular Mobility Unit (EMU), and the Manned Maneuvering Unit (MMU). A Flight Support System (FSS) platform, currently under development, provides a tilt-table device for automated deployment of payloads. Payloads are also deployable via the RMS. Servicing is possible by the RMS or an EVA crewmember. New EVA features and capabilities are called for: a rugged, high-tactility glove, an automatic visor wide-angle helmet suitable for use in varying lighting conditions, and hand-held power tool with interchangeable heads (e.g., drills, punches, fastener driver-bits).

Near-term STS mission planning calls for limited satellite retrieval. The RMS would grapple the free-flyer for berthing to the Orbiter. EVA could be useful in retrieval, assisting the RMS by nulling satellite motion, positioning the satellite for grapple, and manual guidance during berthing operations. New service equipment concepts are required to expand the capabilities such that all service tasks outlined in Table VIII are possible. These changes include several for the EMU: (1) expanded computer capability for surface temperature sensing, voice control of the MMU, range-rate-spin determination, transfer trajectory orbital mechanics, satellite service procedures, and life support system control; (2) regenerable CO_2 removal and no-prebreathe;

Table VIII. Mature satellite service capability tasks.

Service Tasks	PROPULSION	Tankage	Thrusters	POWER	Solar Cells	Batteries	Conditioning	DATA PROCESSING	Command & Control	Signal Conditioning	Telemetry	THERMAL CONTROL	Insulation	Passive Surface	Active Cooling	PAYLOADS	Active Sensors	Passive Sensors	Biomedical	STRUCTURE
• Manual & Hand-Tool Operations																				
Remove & install panels, covers & shields		●	●			●	●	●					●		●		●	●	●	
Inspect & photograph	●			●				●				●				●				●
Clean & refurbish surfaces					●												●	●	●	
Remove & install samples or modular components	●			●				●							●	●				●
Disconnect & connect electrical connectors			●	●				●							●	●				
Disconnect & connect fluid interfaces	●							●							●	●				
Repair electrical harness & connectors			●	●				●							●	●				
Repair leaking fluid lines & fittings	●														●				●	
Repair structural damage																				●
Assemble & install appendages					●						●				●					●
• Specialized Service Equipment Operations																				
Diagnose, checkout & calibrate	●			●				●							●	●				
Measure fluid quantities		●													●					
Detect fluid leakage	●														●					
Purge & refill fluid subsystems		●													●					

(Column headings grouped under "Satellite Subsystems".)

8-hour EVA capability; no-vent regenerable heat sink; long-life modular soft goods; wrist input terminal; and variable, over-garment hazards protection. Concepts for deployment and servicing are presented, including discussion of a fluid system repair facility, which would be located in the payload bay of the Orbiter and would provide satellites with electrical power for systems checkout, and support satellite propellant defueling, refueling, and pressurizing. The fluid service facility would be capable of metering fluid quantities transferred to the satellite. Computerized satellite subsystem diagnosis and checkout is also envisioned for assessing subsystem performance to the level of replaceable modules and for identifying problems in wiring. A computer for diagnostics could also provide for calibration of satellite equipment via software routines. Scenarios for EVA servicing of Long Duration Exposure Facility and Space Telescope payloads, as well as MMU retrieval, stabilizing of a payload, and collecting small space debris are presented.

The Shuttle has opened the door to large structures constructed in orbit via manipulators and crewmembers performing extravehicular activities to assemble modular components or elements. In *Comparative Evaluation of Operability of Large Space Structure Connectors* (Reference 293), eight different concepts for connectors for coupling structural members and utilities to form completed space platforms are evaluated. A test sequence is described in which operability by an EVA crewmember is evaluated during structural assembly. Man-in-the-loop tests were conducted in the NASA Marshall Space Flight Center Neutral Buoyancy Simulator with suited test subjects, who rated each connector with regard to five parameters: (1) connector simplicity, (2) ease of assembly, (3) ease of demate, (4) fragility, and (5) crew safety. All connector ratings were biased by any structural damage during the mating operation.

Each of the eight connectors is described and a sketch and picture included for details. For example, the Rockewll ball and socket connector, shown in Fig. 9, utilizes a ball at the end of a beam to couple with a socket in a cluster to produce a tetrahedral beam module. The ball in initially "soft-captured" in the socket, allowing for as much as 30° beam-socket misalignment. The EVA crewman rigidizes the interface by rotating manually a large retaining nut against the socket. The following connectors were evaluated: (1) Marshall Space Flight Center spike connector; (2) Rockwell International electrical connector; (3) Langley Research Center snap-joint union; (4) Rockwell International ball/socket connector; (5) Massachusetts Institute of Technology cluster slip-joint; (6) Essex lap-joint; (7) Vought module-to-module coupler; and (8) Vought automatic coupler.

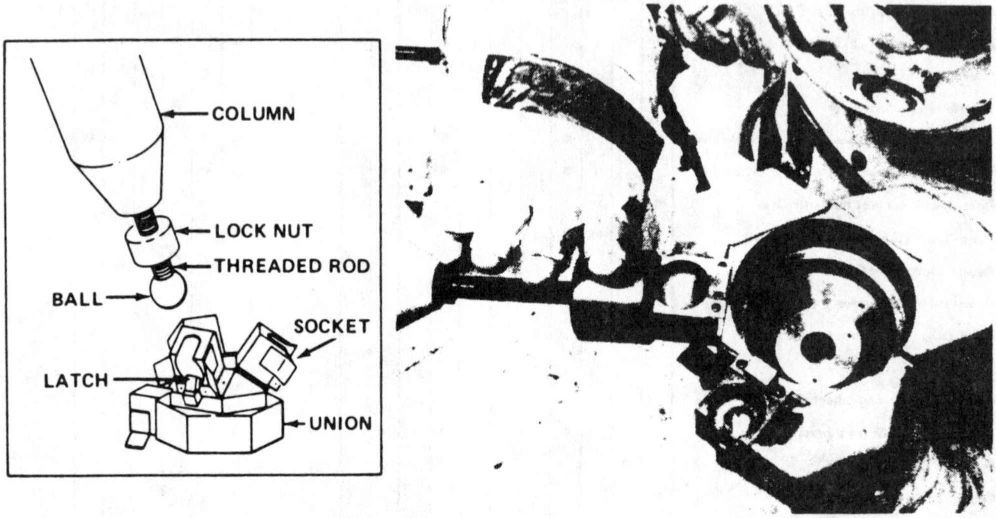

Fig. 9. Rockwell International ball/socket connector.

In terms of connector simplicity, subjects were asked to assess how obvious the connecting methods were and the amount of training required to mate the connector. Ease of assembly addressed the connector alignment prior to mating, ease of mating, crewman feedback for verification of completed assembly, and requirements for crewman restraints during the procedure. Ease of demating was evaluated even when not specified in the intended use of the connector for contingency reasons. Fragility testing determined the connector ruggedness during use and was related to the degree of flexibility between connector halves during mating. Crew safety, although discussed last, was felt to be one of the most important parameters.

Only one of the eight concepts was recommended to be eliminated from the list of those under consideration for large structure assembly: the spike connector, which was shown to be hazardous for pressure-suit operations. The spike throw during mating had sufficient force to penetrate a pressure suit. The paper concludes with general connector design recommendations including the following: (1) The connector shall be completely safe for crew operation. A design goal should be to minimize or eliminate stored energy in any of the components prior to, during, and after component mating; (2) The connector should require very low effort by the crewman to attach the components; (3) The crewman should have positive indication that the connection has been made; and (4) Mating should be a one-hand operation, without need for crew restraints or assembly aids.

LONG-RANGE SPACE PLANNING

Conceived in the late 1970's, the Space Operations Center (SOC) is proposed as the next major step in the development of the United States' manned space program. The SOC is foreseen as a permanently manned facility operating in low earth orbit and used for operational support of space activities such as space construction, unmanned and manned orbital transfer vehicle (OTV) operations, and on-orbit assembly, launch, recovery, and servicing of space vehicles. The SOC buildup, assembly, and resupply is planned via the Space Shuttle (Fig. 10).

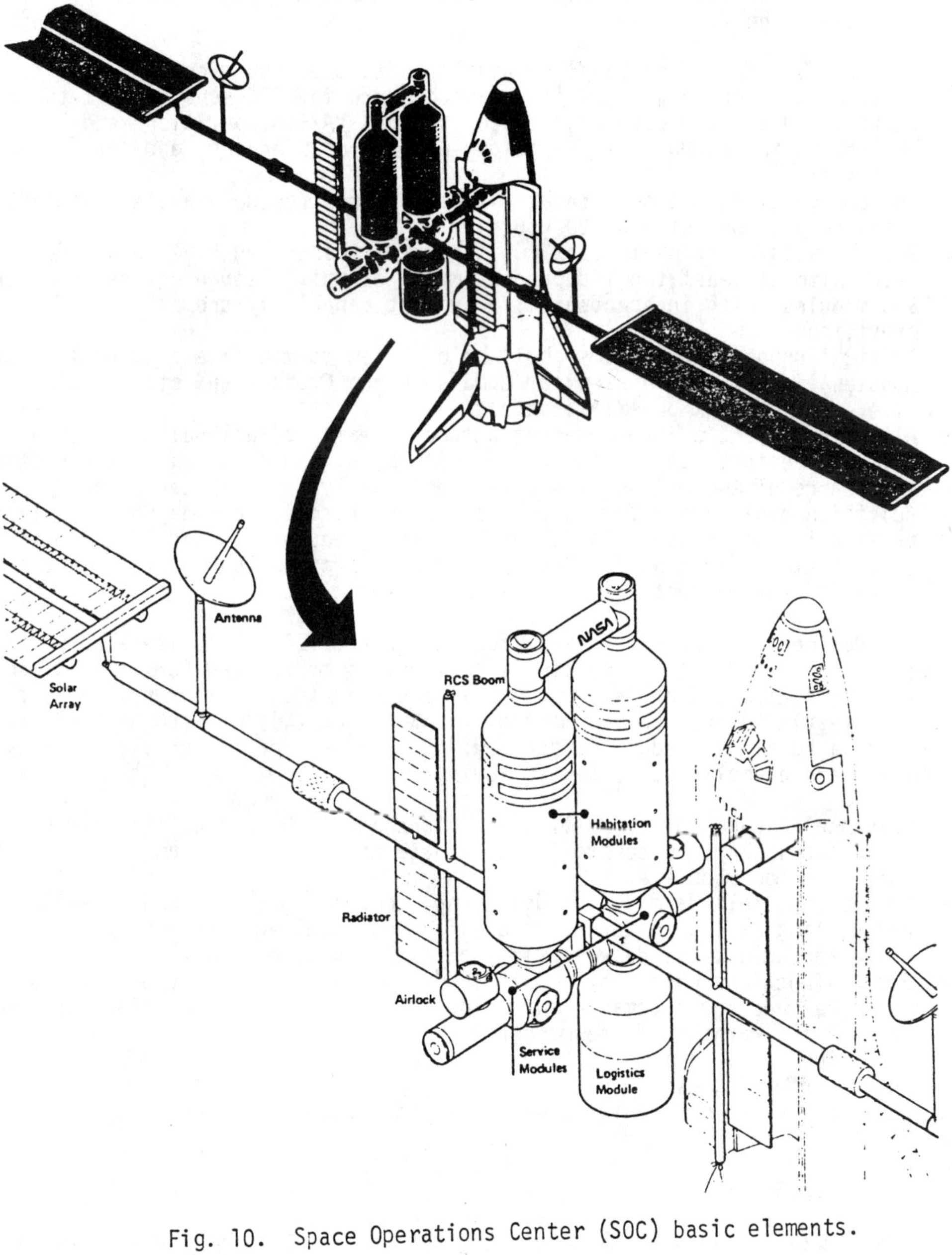

Fig. 10. Space Operations Center (SOC) basic elements.

Space Operations Center--Shuttle Interaction Study (Reference 294) summarizes an analysis of STS support to the SOC, including orbital altitude, berthing and/or docking, SOC assembly, resupply and fuel transfer, and flight support facility. Especially because of the requirement specifying that the SOC shall have the capacity for independent on-orbit operations with a full crew for at least 90 days, attention during SOC development to crew safety is paramount. Safety-related program, system, and subsystem requirements address availability of status information on-board, emergency conditions capabilities, basic system and subsystem design for safety, life support and radiation factors, maintenance and repair, and crew operations (intravehicular and extravehicular activities--IVA and EVA, respectively). Examples include the following:

- SOC status information shall be available to the ground continuously during critical or emergency operations, and onboard for SOC subsystem status and caution and warning display, for control of IVA/EVA operations and local logistics, for support of day-to-day operations planning, and for malfunction analysis;
- For emergency conditions, capabilities for crew rescue via the Shuttle Orbiter shall be provided within 180 hours;
- The SOC will contain at least two separate pressurized habitable modules (each with at least two independent means of shirtsleeved egress to other SOC modules) with independent life support capability and habitability provisions;
- Critical onboard subsystems shall be designed to minimize risk of loss of modules, injury to the crew, or damage to the Orbiter and other vehicles (i.e., fail-operational/fail-safe);
- All of the systems incorporating automated fail-operational capability shall be designed to provide crew notification and data management system cognizance of the malfunction until the anomaly has been corrected;
- Collision avoidance software and/or maximum torque override shall be incorporated in manipulators and other supporting equipment; and
- Contaminates resulting from manufacturing operations shall not affect the local SOC environment.

Goals, guidelines and design criteria for safety and reliablity are also specified. SOC design goals are that (1) no single malfunction or failure (or combination of malfunctions, accidents, or failures) shall endanger the lives or safety of crewmembers or result in crew abandonment of the SOC, and (2) catastrophic and critical hazards shall be eliminated or controlled. Guidelines and criteria for safety and reliability are exemplified by the following:

- Personnel escape routes shall be provided in all hazardous situations. A design goal shall be to provide alternate escape routes that do not terminate into a common module area.
- Provisions shall be made for detecting, containing (i.e., confining), and controlling (i.e., restoring to a safe condition) emergencies such as fire, toxic contamination, depressurization, structural damage, etc.
- For malfunctions which may result in time-critical emergencies, provision shall be made for automatic switching to a safe mode or operation and for caution and warning of crewmembers.
- Provision shall be made to return to the SOC crewmembers who are incapacitated while performing EVA.
- Subsystem or component failures shall not propagate sequentially.

REFERENCES

282. *STS-1 Mission Report*, NASA Johnson Space Center, Houston, Texas, July 1981.
283. Jeffcoat, Glenn B. and Emile S. Stephan. *Space Shuttle Slidewire Emergency*

Egress System, NASA Conference Publication 2181, 15th Aerospace Mechanisms Symposium, Marshall Space Flight Center, Alabama, May 14-15, 1981, pp. 47-62.

284. Covault, Craig. *ESA's Spacelab Model To Begin Further Tests*, Aviation Week & Space Technology, June 8, 1981, pp. 250-254.

285. *Crew of Soyuz Repairs Salyut*, Aviation Week & Space Technology, December 15, 1980, pp. 23-24.

286. *Assessment of the Effects of the Zero Gravity Environment on the Health and Safety of Space Workers*, NASA TM-81122, Lyndon B. Johnson Space Center, 18 February 1980.

287. *Safety Policy and Requirements For Payloads Using the Space Transportation System (STS)*, NHB 1700.7A, National Aeronautics and Space Administration, December 9, 1980.

288. Lewis, J. C. and R. K. Pugmire, *Parametric Study of Pressure Vessels for Space Shuttle Payloads*, 16th AIAA/SAE/ASME Joint Propulsion Conference, Hartford, Connecticut, June 30-July 2, 1980, AIAA Paper 80-1099.

289. Dote, H. and K. T. Theodore, *Structural Analysis and Test Space Shuttle Orbiter Window System*, 21st Structures, Structural Dynamics, and Materials Conference, Seattle, Washington, May 12-14, 1980, pp. 888-895.

290. Krausz, A. *Effectiveness of Spacecraft Testing Programs*, NASA Conference Publication 2150, 11th Space Simulation Conference, NASA Johnson Space Center, Houston, Texas, September 23-25, 1980, pp. 13-23.

291. *Extravehicular Mobility Unit Television System Program Requirements Document*, JSC-16808, Lyndon B. Johnson Space Center, August 1980.

292. Griswold, H. R. and R. C. Wilde. *Engineering in Space: Part 2--EVA Equipment for Satellite Service*, Mechanical Engineering, June 1981, pp. 40-49.

293. Stokes, Jack. *Comparative Evaluation of Operability of Large Space Structure Connectors*, NASA Conference Publication 2181, 15th Aerospace Mechanisms Symposium, Marshall Space Flight Center, Alabama, May 14-15, 1981, pp. 357-374.

294. *Space Operations Center--Shuttle Interaction Study*, Final Report, Executive Summary, SSD81-0076, NAS9-16153, Rockwell International, April 17, 1981.

WORLDWIDE DISASTER RESPONSE, RESCUE AND SAFETY EMPLOYING SPACE-BORNE SYSTEMS 1981

The following papers also appear in full in Volume 41, AAS Microfiche Series, supplement to this volume:

IAA 81-262, -263, -266, -269 and -271

The following papers, abstracted in this section, are available in full in Volume 41, AAS Microfiche Series:

IAA 81-265 and -267/268

IAA 81-266

SATELLITE ALERT WARNING - CATALYST FOR AN INTERNATIONAL DISASTER RESPONSE LEGAL REGIME?*

John T. Stewart, Jr.[+]

The emergence of satellite technology has made its application to disaster alert systems increasingly attractive. Legal regimes - that is systems of applicable legal principles - exist in a number of areas of disaster response. Most developed of these systems concerns search and rescue operations. Others are developing particularly in the satellite sensing field. There is however no single international legal regime for disaster response. The development of a satellite alert warning system could catalyze the coming into being of such a regime. The author explores some of the existing legal avenues available for the development of such a regime.

Keywords: Disaster response; search and rescue; disaster response legal regime; satellite warning system; space disaster legal regime.

INTRODUCTION

The past two decades have seen the emergence of space satellites as an integral component of the world's communication network. What was once a novelty has now become commonplace. It has been demonstrated that satellite systems are capable of alerting to such natural disasters as forest fires, floods, crop failures, cyclones, hurricanes, tornadoes and iceberg dangers.[1] Legal systems of a national, multinational and bilateral nature are developing to establish standards for the relationship, within, between and among governments concerning the use of this new medium of communication and guidance.[2] With the emergence of satellite sensor systems a more expanded technological use of satellite systems is inevitable even though the attendant legal issues associated with remote sensing from outer

---

* *The views expressed herein are those of the author and do not necessarily reflect those of the U.S. Government or any agency thereof.*

+ *AB Yale '49; LL.B. Columbia '55; Assistant Chief Counsel, Federal Aviation Administration, U.S. Dept. of Transportation, Washington, D.C. 20591*

space are causing some controversy.[3] One area now appears ripe for exploration. The development of sophisticated satellite systems capable of providing warnings of impending natural disasters may well catalyze a single international legal regime on disaster response acceptable to the nations of the world. There has already emerged a bilateral network of agreements between the launchers of information satellites and governments who act as receiving and disseminating nations. International frameworks are available in the aviation on shipping legal disciplines to serve as role models. It is the purpose of this article to explore some of the existing patterns of disaster relief response with a view to suggesting that it is possible there will emerge an international disaster response legal regime more responsive to the needs of the future.

DISASTER RESPONSE GENERALLY

The present international legal regimes encompassing disaster response are directed primarily at the relief of victims of natural and man-made disasters. The entire fabric of the International Red Cross system is private in nature, elaborate in its mechanism and successful in dealing with the situations which it is designed to alleviate - the suffering victims of natural disasters and the ravages of the man-made disasters of war. Multinational governmental structures have been somewhat less successful. While espousing high ideals and impelled by proper motives, the early endeavors of the League of Nations through the National Relief Union did not achieve any notable success. In a similar vein, the efforts of the United Nations while representing some forward movement in the area has had only limited success. The emergence of "International Human Rights Law" as reflected through the Universal Declaration of Human Rights may point in a direction of establishing the obligations of States with respect to the victims of natural disasters.[4]

Problems of availability of resources, State sovereignty and generally the lack of an "international will" loom large as obstacles.[5] More concrete and useful perhaps have been the various bilateral arrangements between States designed to establish the methods for assistance and cooperation during periods of natural disaster. These agreements at least point the way for immediate acceptable norms for State behavior in the disaster response environment. One such agreement exists between the United States and Mexico[6] which establishes a consultant committee on natural disasters as a vehicle of cooperation in planning for natural disasters such as hurricanes, floods, fires, earthquakes, tornados, landslides, snowfalls, volcanic eruptions and other catastrophies of a similar nature. It is designed to foster the protection of life, property, health and safety of inhabitants of the affected border areas between the two States. The committee also serves as a forum for the exchange of information and may establish joint working groups to carry out specific tests. Other arrangements dealing with specific forms of aid or the manner in which aid is given or dispersed have been entered into between governments.[7]

In the area of space exploration a form of disaster response has been achieved through the Agreement on the Rescue of Astronauts, the Return of Astronauts and the Return of Objects Launched into Outerspace.[8] Notable among its provisions are the following articles:

Article 2

If, owing to accident, distress, emergency or unintended landing, the personnel of a spacecraft land in territory under the jurisdiction of a Contracting Party, it shall immediately take all possible steps to rescue them and render them all necessary assistance. It shall inform the launching authority and also the Secretary-General of the United Nations of the steps it is taking and of their progress. If assistance by the launching authority would help to effect a prompt rescue or would contribute substantially to the effectiveness of search and rescue operations, the launching authority shall co-operate with the Contracting Party with a view to the effective conduct of search and rescue operations. Such operations shall be subject to the direction and control of the Contracting Party, which shall act in close and continuing consultation with the launching authority.

Article 3

If information is received or it is discovered that the personnel of a spacecraft have alighted on the high seas or in any other place not under the jurisdiction of any State, those Contracting Parties which are in a position to do so shall, if necessary, extend assistance in search and rescue operations for such personnel to assure their speedy rescue. They shall inform the launching authority and the Secretary-General of the United Nations of the steps they are taking and of their progress.

Article 4

If, owing to accident, distress, emergency or unintended landing, the personnel of a spacecraft land in territory under the jurisdiction of a Contracting Party or have been found on the high seas or in any other place not under the jurisdiction of any State, they shall be safely and promptly returned to representatives of the launching authority.

The launching of the space shuttle by the United States brought to the forefront a recognition of the beginnings of the merging of the space and aviation environments. Recognition of the necessity for arrangements in the event of space-oriented activity requiring assistance in the earth environment is demonstrated by the following pertinent provisions of an agreement between the United States and Japan:

The Embassy of the United States of America presents its compliments to the Ministry of Foreign Affairs of Japan and has the honor to refer to the treaty on principles governing the activities of States in the exploration and use of outer space, including the moon and other celestial bodies (Outer Space Treaty) ratified by the Governments of the

United States and Japan on October 10, 1967. The Government of the United States wishes to inform the Government of Japan that in furtherance of its goals of exploring outer space for peaceful purposes the United States is planing to launch its space shuttle in the 1980's to provide a low-cost reusable space transportation system to replace expendable launch vehicles for the Government of the United States, commercial and foreign users and to provide routine reliable and simplified operations in space. The space shuttle will make four to six orbital test flights. As these flights are experimental in purpose, the possibility exists that an emergency landing of the space shuttle may be required to ensure the safety of the astronauts. Because of trajectory constraints, it will not be possible to land the space shuttle on United States territory in all emergency situations. The Government of the United States therefore seeks to make such arrangements as may be necessary outside the United States. If such a landing should be necessary, United States personnel (both the National Aeronautics and Space Administration and contractor) would be dispatched to the landing site to prepare the orbiter for return flight to the United States.

The Government of the United States requests that, pursuant to Article V of the Outer Space Treaty, the Government of Japan render all possible assistance in the event of an emergency landing of the space shuttle on an airfield in Japanese territory to safeguard the lives of the astronauts.

To deal with such a contingency, the Government of the United States will provide planning information and consult with the Government of Japan to ensure the safest possible landing of the space shuttle. The Government of the United States is prepared to assume liability for damage caused by the space shuttle on Japanese territory under Article VII of the Outer Space Treaty and in accordance with the provisions of the Convention on International Liability for damage caused by space objects. [9]

These bilateral and multilateral efforts (both earth-and space-oriented) affecting cooperation in responding to disaster at least indicate an awareness of a need for governmental cooperation and the ability to establish some legal parameters within which to respond.

There exist provisions in multinational agreements which are special in nature. A form of disaster relief is characterized as "search and rescue". This is particularly applicable in the aviation and maritime arenas where international agreements and annexes thereto address these subjects. It would be useful to explore this area for it may provide a vehicle for thought for the development of a future albeit limited single multinational agreement centering around satellite alert warning systems.

SEARCH AND RESCUE REGIMES

The aviation and maritime environments provide examples in which a structure has been developed for international cooperation in disaster situations which relate not so much to the providing of assistance to victims of a disaster as providing a mechanism for extricating a disaster victim from his or her dire circumstances. Two international organizations have been particularly active in the area of search and rescue, one being the International Civil Aviation Organization (ICAO) and the other the Intergovernmental Maritime Consultive Organization (IMCO). The latter organization, IMCO, has been particularly active in sponsoring and developing an international conference on the establishment of an International Maritime Satellite System (INMARSAT).[10] The Convention On International Civil Aviation (Chicago Convention) provides:

> Each contracting State undertakes to provide such measures of assistance to aircraft in distress in its territory as it may find practicable, and to permit, subject to control by its own authorities, the owners of the aircraft or authorities of the State in which the aircraft is registered to provide such measures of assistance as may be necessitated by the circumstances. Each contracting State, when undertaking search for missing aircraft, will collaborate in coordinated measures which may be recommended from time to time pursuant to this Convention.[11]

ICAO has, for a number of years, been active in the aviation area in problems concerning search and rescue. As early as December 1946, recommendations were being made for standards and recommended practices for search and rescue. On the 25th of May 1950, the Council of ICAO adopted an annex to the Chicago Convention specifically dealing with the issue of search and rescue.[12] In December of 1965, an amendment was approved by the Council introducing material into that annex to accord with the International Convention for the Safety of Life at Sea which came into force on May 26, 1965.[13] Like its sister international organization in the maritime area, ICAO has created a group of experts to deal with the application of space techniques to aviation, which is known as the Application of Space Techniques to Aviation (ASTRA).

The regime adopted by ICAO in dealing with search and rescue generally provides that Contracting States to the Chicago Convention shall arrange for the establishment and provision of search and rescue services within their territories. Such services shall be provided on a 24-hour basis.[14] These ICAO standards also provide:

> Those portions of the high seas or areas of undetermined sovereignty for which search and rescue service will be established shall be determined on the basis of regional air navigation agreements. A Contracting State having accepted the responsibility to provide search and rescue service in such areas shall thereafter arrange for the service to be established and provided in accordance with the provisions of this Annex.[15]

> This phrase 'regional air navigation agreements' refers to the agreements approved by the Council of ICAO normally on the advice of Regional Air Navigation Meetings.[16]
>
> In providing assistance to aircraft in distress and to survivors of aircraft accidents, Contracting States shall do so regardless of the nationality of such aircraft or survivors.[17]
>
> Contracting States shall delineate the search and rescue regions within which they will provide search and rescue service. Such regions shall not overlap.[18]

Additionally, ICAO standards call for the establishment of preparatory standards and operating procedures in the area of air search and rescue.[19] Procedures are established for various phases of the developing potential disaster - the uncertainty phase, the alert phase and, finally, the distress phase.[20] ICAO also calls for States to establish a rescue coordination center in each search and rescue region which would have immediate communication with air traffic, services associated with rescue subcenters, appropriate direction-finding and position-fixing stations, and where appropriate coastal radio stations capable of alerting and communicating with surface vessels.[21] The sensitive problem of State sovereignty is recognized by ICAO which calls for States to co-ordinate their search and rescue organizations with those of neighboring States.[22] It also obligates States, subject to such conditions as may be appropriate, to permit immediate entry into its territory of rescue units of other States for the purpose of searching for the site of aircraft accidents and survivor rescue.[23] Further, it is recommended that the States enter into agreements with neighboring States setting forth conditions for entry of each other's rescue units into their respective territories[24] and in this regard it should be noted that the United States and Canada have entered into an agreement relating to search and rescue along the common border. The following text from that agreement illustrates the nature of the governmental arrangement contemplated in part by this ICAO recommendation:

> (1) ...public aircraft of Canada or the United States which are engaged in emergency Air Search and Rescue operations, be permitted to enter or leave either country without being subject to the immigration or customs formalities normally required by the Government of either country, provided that the Rescue Coordination Centre involved in the search or rescue, either directly or through some person delegated by it, assumes the responsibility of informing by telephone or telegraph:
>
> (a) The immigration office at the port of entry nearest to the territory over which any search or rescue is to be instituted, of the intended operation, furnishing it with details concerning the purpose of the flight; the identification markings of each aircraft; and the number of persons comprising the crew of each aircraft.

(b) The customs office nearest to the territory over which any search or rescue is to be instituted, of the intended operation, giving details concerning the territory to be searched; the possible duration of the stay of the aircraft; the identification markings of each aircraft; and the number of persons comprising the crew of each aircraft.

(2) That, should a landing be made by public aircraft of one country in the territory of the other in the course of such emergency search or rescue, an oral or telephonic report shall be made to the nearest Collector of Customs so that he may assist, in any way possible, in connection with any special importation required in the search or rescue operations. This report may be made by the Rescue Coordination Centre organizing the operation, or by the pilots concerned, whichever would best serve the interest of the rescue operations involved.

(3) That, should any merchandise carried, in the aircraft in question, from one country to the other in the course of such search or rescue, remain in the latter country on conclusion of an operation, such merchandise will be subject to the customs treatment normally accorded in that country to imported merchandise.

..."public aircraft", as used in this Note, refers to aircraft of the Canadian and United States Governments and such other aircraft of United States and Canadian registry as may be brought under the control of a Rescue Coordination Centre in either country for the purposes of an emergency search or rescue operation.[25]

While there may be bilateral arrangements such as that entered into between the United States and Canada, there is also a multilateral agreement on search and rescue which has been signed without reservation by Peru, Argentina, the Dominican Republic, Bolivia, Guatemala, Chile, Uruguay, Colombia, Costa Rica, Nicaragua and Paraguay.[26] The specific applicability of satellite technology to the air search and rescue regime is illustrated by the following commentary by way of a summary in a paper entitled "The Search and Rescue Satellite (SARSAT) System Project" by H.L. Werstiuk and A.E. Winter, Communications Research Centre, Department of Communications, Ottawa, Canada:

For the past several years, Canada has been studying the application of satellite technology as an aid to search and rescue (SAR). The work has included examination of SAR requirements and the investigation of various system concepts culminating in proof-of-concept experiments using the AMSAT OSCAR satellites during 1975 and 1976. In the experiments, signals from simulated Emergency Locator Transmitters (ELTs) were received by the polar-orbiting OSCAR 6 satellite and relayed to a ground station at the Communications Research Centre (CRC). By the processing of Doppler information contained in this relayed signal, the position of the ELT could be determined with an accuracy of 5-15 km.

In the past SARSAT-related studies and experiments have been conducted separately by both Canada and the United States. In the fall of 1976 a joint program was proposed to demonstrate a satellite-aided search and rescue system. This joint program has been approved by both countries and calls for a pre-operational demonstration of the system commencing in 1981 using suitably instrumented weather satellites operated by the National Oceanic and Atmospheric Administration (NOAA). France has recently taken a decision to participate in this project. The satellite system under consideration would be capable of detecting and locating ELTs and EPIRBs (Emergency Position Indicating Radio Beacons) operating at 121.5, 243 and 406 MHz. [27]

The foregoing illustrates that at the bilateral level agreements of a research and developmental nature are implementing agreements concerned with jurisdictional and institutionalization issues.

As aviation lawyers have addressed search and rescue cooperation there is a comparable body of legal arrangements in maritime law in the shipping environment. They include the Convention with respect to assistance and salvage at sea, Brussels, 1913;[28] Convention on the High Seas, Geneva, April 1958;[29] and the Convention on the International Maritime Search and Rescue which was recently ratified by the United States but is not yet in force.[30]

The search and rescue legal arena with its multilateral and bilateral regimes already in existence might serve as a model for comprehensive multilateral accommodation of disaster response. For instance, concepts contained in the aviation environment which obligate cooperation among neighboring States, exchange of information, establishment of disaster response units and procedures, as well as the willingness of States to accede to intrusion in their territory by other governmental entities in the course of disaster response might usefully be applied in a broader context. Can and should some of these principles be applied to disaster response for floods, earthquakes, tidal waves and similar natural catastrophies? Should air, sea and space[31] search and rescue legal regimes be combined into a single multilateral system? Should such a system be internationally institutionalized? These are questions which must eventually be answered in the context of satellite technology as it creates an atmosphere of rapid and total communication.

DISASTER DISCOVERY AND WARNING REGIMES

Although there are some well established multilateral and bilateral legal regimes designed to aid and assist in extricating individuals from specific disasters through search and rescue mechanisms, with the development of a sophisticated satellite technology, the world is venturing into an era in which it may be possible to detect and forecast the course of potential and actual natural disasters. The most dramatic indication of a technology presently available for the benefit of forecasting potential disasters emerged with the aid of satellite technology in the meteorological environment. The ability to detect and predict meteorological phenomena as demonstrated by cyclone and hurricane forecasting has established both the viability and practical benefit of a disaster warning system in which satellite technology is the prime ingredient. The impact of satellite technology in this area has been particularly noticeable through the activities of the World Meteorological Organization (WMO). Such programs as the World Weather Watch (WWW) and the Global Atmospheric Research Program (GARP) have demonstrated the feasibility of satellite disaster warning systems.[32] Beyond even this horizon, however, lies the potential capability of remote sensing satellite technology and its contribution to a world-wide system for disaster alert.

The utilization of remote space sensing technology, even in its present infancy has opened up the possibility of developing a global disaster warning system in such areas as crop failure forecasting, snow cover management, forest infestation and fire hazards, water resource assessments, problems associated with ice and glaciology. The practical uses of this technology in part has been demonstrated by the United States National Aeronautics & Space Administration through its launch in 1972 of an experimental Earth Resources Technological Satellite (ERTS-1). These series of experiments, now called LANDSAT, have involved the cooperative participation of a number of countries and international organizations.[33] Since this system employs ground stations for processing the satellite data, a number of bilateral agreements have been entered into between the United States through NASA and other governments.[34] There is, therefore, already established on a bilateral basis a preliminary legal regime for utilizing this media of space technology data. Recognition of the usefulness of centralizing such a data distribution system was noted by a United Nations Scientific and Technical Subcommittee. It commented in connection with the need for further study in this area:

> The Subcommittee noted that while a network of ground stations optimally distributed could provide a satisfactory coverage of the world's land surface and thereby enable all countries to receive data in respect of their territories, the possible development of a global centre, complementary to regional centres, should be studied in the light of such advantages as, for example:
>
> > Provision of maximum international cooperation, thus ensuring benefits to all countries and to the international community as a whole;

Coordination, interaction and data exchange between regional centres;

Dissemination of all data and information to all countries on an equal and non-discriminatory basis;

Provision of the basis for an approach to world-wide problems and of those which go beyond regional interest and capabilities. [35]

Although remote sensing capabilities have excited visionaries in terms of its potential benefit for mankind, including disaster control, it is nonetheless fraught with a number of sensitive legal issues.[36] Not the least among the issues are the cost and the control of such a system. The issue of control raises questions of the invasion of sovereign privacy. A number of commentators have discussed these issues in detail.[37] Suffice it to say here that while these remain real problems it may well be that the existence of the satellite sensing capability for long-range disaster assessment may yet stimulate an international accord somewhat along the lines that have been developed in the areas of search and rescue. There would have to be agreement by States as to funding, access and distribution of information, as well as commitment to utilize resources to alleviate or combat natural disasters either taking place or predictable for the future. If the concept of disaster response and relief can be segregated from other issues which entwine remote sensing such as resource exploration and military uses there is a possibility that some form of accommodation can be achieved among nations on a multilateral basis, thus affording some hope of utilizing this new technology for the benefit of mankind. For instance, sight should not be lost of satellite sensing potential for such uses as monitoring possible depletion of the ozone. This kind of activity might lend itself to a multinational agreement. One such venture was initiated as a result of the commencement of Concorde flights to the United States in the form of a tripartite agreement between the United States, the United Kingdom and France to monitor the stratosphere with a view to determining any changes which might be detrimental.[38]

While most of the efforts for disaster discovery and warning have been focussed on the earth's atmospheric and surface environment, there is the possibility of disaster affecting the earth emanating from outer space in the forms of meteor showers and phenomena which would affect the earth's atmosphere. A multilateral agreement to monitor this potential disaster is one which needs further exploration. With the possibility of space travel and, indeed the creation of space platform inhabited by humans, a system designed to detect potential space environment disasters is one which is rapidly approaching the need for consideration. As pointed out earlier, the recognition of space disaster has already been evidenced with respect to astronaut activity. Finally, it would be worth at least mentioning in passing that any disaster warning system also possesses a malfunction possibility, thus resulting in misinformation or a lack of information at a critical point in time. This in the world of legal regimes raises the spectra of potential liability with damages resulting therefrom.[39] Any multilateral system for disaster warning would have to take this into consideration. This could take the form of damage agreement analogous to that achieved in connection with space objects or a system which either permits the damages to be resolved by municipal law or because of the enormity of the potential monetary damage possible from such disasters as floods, crop failures, and similar catastrophies, a system which envisions a form of internationally agreed limits on the amount of recoverable damages as has been done in aviation,[40] and domestically in the United States in connection with nuclear damage.[41]

Satellite systems designed to predict on a long-range basis the onset of disasters could catalyze on a bilateral and multilateral level the beginnings of a legal regime designed to accommodate disaster warnings and response. Some elements for such a regime already exist in a primitive bilateral network formed for collecting and assimilating information. The more critical elements of such a multilateral regime - those directed at requiring or stimulating collective governmental response to impending disasters - have yet to be developed. Patterns for resource response may well lie in the so-called country "aid programs" established by individual governments and the United Nations.[42]

PRIVATE SECTOR ALERT WARNING REGIMES

While governments have been active in satellite development technology the development of a private sector system which could be utilized, on a compensated basis, for disaster warnings and response - "space alarms" owned and operated by private industry - is not unimaginable.

Most of the current space and satellite development technology has been by governments or public international organizations. There has been, however, a growing increase by the private sector in both interest and capability in space activity and satellite technology. Nearly sixteen years ago COMSAT launched the world's first private enterprise communications satellite. Since that time telecommunications satellites have become viable and potentially profitable for private companies. The successful launching of the United States space shuttle with its private payload capability has opened new vistas for entrepreneurial activity related to space. It is not beyond imagination to envision the possibility of privately owned and operated satellite sensing systems.[43] The combination of telecommunications and sensing satellites could offer lucrative activity for those willing to invest the capital necessary to implement such a system. For a fee, such systems could be used for example by lumber companies, private shipping organizations and agricultural concerns to monitor their particular holdings and to make available on an individual basis analyses which could permit these companies to take appropriate steps to avoid potential disastrous situations such as forest fires, forest infestations, crop failures, blocked shipping lanes, and so forth. However, a private system, should it be developed, would have to exist in the context of both public and private law regimes. Liabilities, while they could perhaps be limited by contractual arrangements, would be subject to municipal laws while the use of communication "wave lengths", safety and security matters would continue to be regulated by governmental regulatory agencies or in accordance with agreed multinational standards.

The situation is somewhat different should the system not be oriented earthward but oriented to a pure space environment. These systems would envision the day when space activity is no longer the sole domain of governments but has become a new frontier for man's social and economic activity. In this case the municipal law which would govern earthward-looking activity may be nonexistent. In its place, at the present time, there is a network of international conventions concerning space and moon activity.[44] While these conventions contemplate the possibility of private activity in space, they are not specifically designed to create a legal environment to accommodate purely private activity. There are no standards for such activity, no parameters of enforcement of standards developed and no means of protecting purely private investment rights existing in an exclusive space environment. As man's activity in space increases, initial concern as it has always been for ventures into the unknown is for the safety of the individuals involved, the vehicles employed and ventures utilized. Thus, a disaster response legal regime would seem ripe for development and satellite early warning systems would be the catalyst encouraging development of such a regime. The potential for space platforms with space colonies, colonies on the moon or on other terrestrial bodies, admittedly in the realm of fantasizing at the moment, nevertheless are possible. The development of systems for disaster response involving such colonies seems inevitable. Here private enterprise may, in fact, demand some form of international governmental disaster response before it is willing to invest enormous amounts of resource necessary.

CONCLUSION

Satellite technology has emerged from its embryo state and is rapidly moving into the realm of full maturity. In so doing, its use is being employed with a greater frequency and in a wider range of environments. While its cost continues to be a limiting factor, as it demonstrates its capability for utilization, it becomes a more viable technical force for surveillance and communication. Through the history of mankind it has been demonstrated that as man invents, society responds by creating the social structures in which inventions may be utilized in accommodation with human needs and desires.

Simply put, inventions, when they demonstrate a practical use, catalyze the formation of legal regimes. One need only look at the history of aircraft to illustrate this truism. The first penetration by man-made devices into outer space has catalyzed a series of international agreements creating a legal regime designed to govern the use of a yet unexplored dimension. The existence of satellite technology has already served to initiate agreements between and among States either on a multilateral or bilateral level concerning the transmitting of information and its potential utilization. In the area of search and rescue, which is the most dramatic in the disaster response environment, satellite technology has been added to and caused adjustment of existing multinational legal regimes in the aviation and maritime areas. The emergence of satellite technology in the remote sensing arena has stimulated bilateral arrangements designed to facilitate receipt and dissemination of certain kinds of information. It has stimulated thought in the multilateral agreement atmosphere.

There are a number of international organizations, some of which have been described or alluded to herein which are presently in existence designed to deal with problems of natural disaster response. The problems of balancing a global need with the sovereignty and resources of individual States is one with which the world community is presently struggling, not only in the terrestrial environment but also in that of outer space. Indeed, at least one commentator has posed the question, "Is there any real value in clearly establishing general legal obligations concerning a global response to natural disasters?"[45] As the availability of space technology proliferates, as it becomes an incentive for private entrepreneur activity and as governments continue their exploration and probing of outer space that question may, by virtue of circumstance, have to be answered in the affirmative. As in aviation and shipping the more sophisticated the vehicles became the more clearly was the perception for the need to develop legal regimes within which they could be utilized and developed further. There are a number of initiatives within a number of institutional structures, both national, international and private sector, which are confronting the problems inherent in responding to or searching out in advance natural disasters. Satellite technology which is rapidly developing the capability of alerting institutions to the potential for natural disasters will serve as the catalytic vehicle for molding the various legal regimes presently in existence into a single multilateral regime to deal if not with the totality of natural disaster relief response, certainly some elements thereof.

REFERENCES

1. See generally "Remote Sensing Satellites - What Do They Actually Measure and How Sensitive is the Information", Morley, Lawrence W; *Legal Implications of Remote Sensing from Outer Space*, edited by Matte and De Saussure, A.W. Sijthoff-Leyden, 1976, p. 13 *et seq*.

2. Not specifically addressed in this article but of some relevance are national laws designed to enhance disaster response within and by States. Examples of United States law in this regard can be found in the United States Federal Aviation Act of 1958 as amended requiring installation on certain U.S. aircraft of Emergency Locator Transmitters (49 U.S.C. 1421(d)).

3. See generally *Legal Implications of Remote Sensing from Outer Space*, edited by Matte and De Saussure, A.W. Sijthoff-Leyden, 1976.

4. See generally *Disaster Assistance, Appraisal, Reform and New Approaches*, New York, New York University Press, 1979.

5. *Id*.

6. See 19 U.S.T. 4810, as amended 23 U.S.T. 290. (Terminated upon provisional entry into force of new agreement on cooperation in cases of natural disasters signed at Mexico, Jan. 15, 1980.)

7. See generally the "Funding of International Relief Assistance in Disaster Situations", Pordes, Richard, *Disaster Assistance, Appraisal, Reform and New Applications*, New York, New York University Press, 1979, p. 49 *et seq*.

8. 19 U.S.T. 7570; TIAS 6599; 672 UNTS 119.

9. TIAS 9915.

10. See "INMARSAT: The International Maritime Satellite Organization - Origins And Structure", Doyle, Stephen E., *Journal of Space Law*, Vol. 5 Nos. 1 & 2, 1977, p. 45 *et seq*.

11. 61 Stat. 1180, Art. 25.

12. 61 Stat. 1180, Annex 12.

13. *Id*.

14. *Supra* n. 12, Ch. 2; Sec. 2.1.1.

15. *Id*. Sec. 2.1.1.1.

16. *Id*. NOTE

17. *Id*. Sec. 2.1.2.

18. *Id*. Sec. 2.2.1.

19. *Supra* n. 12, Chs. 4 & 5.

20. *Supra* n. 12, Ch. 5, Sec. 5.2.1., 5.2.2 & 5.2.3.

21. *Supra* n. 12, Ch. 2, Sec. 2.3 & 2.4.

22. *Supra* n. 12, Ch. 3, Sec. 3.1.1.

23. *Id*. Sec. 3.1.3.

24. *Id*. Sec. 3.1.5.

25. 63 Stat. 2328; TIAS 1882; 43 UNTS 119.

26. Aeronautical Agreements and Arrangements, ICAO Doc. 9307-LGB/347, p. 5.

27. Scientific and Technical Aerospace Reports, NASA Accession Number N79-18115.

28. 37 Stat. 1658.

29. 13 UST 2312; TIAS 5200; 450 UNTS 82.

30. On June 25, 1980 the U.S. Senate Foreign Relations Committee recommended Senate advise and consent to ratification of this convention, adopted in Hamburg on April 27, 1979, under the auspices of IMCO.

31. For a general discussion of search and rescue relating to space activities see "Search and Rescue in Space Law," Dr. I.H. Ph. Diederiks-Verchoor, 27th International Astronautical Congress, Oct. 10-17, 1976.

32. For a general discussion of activity in the meteorological environment see *Aerospace Law, from Scientific to Commercial Utilization,* Matte, Nicolas M., 1977, pp. 47-51.

33. See generally *Id.* pp. 117-121.

34. *Id.* p. 129.

35. Report of the Scientific and Technical Subcommittee, United Nations Committee on Peaceful Uses of Outer Space, A/AC. 105/150 May 7, 1975, pp. 9-10.

36. *Supra* n. 3.

37. *Id.*

38. TIAS 8255.

39. See generally Catastrophic Events Leading to De Facto Limits and Liability, Solomon, K.A. and Okrent, D., Prepared for the National Science Foundation under Grant OEP 75-20318, UCLA-ENG 7732, May 1977.

40. The Warsaw Convention (49 Stat. 3000).

41. Pub. L. 85-256, Sept. 2, 1957, 71 Stat. 576.

42. *Supra* n. 7.

43. See International Space Cooperative: Participation of Private Enterprises, Cassidy, Daniel E., Counsel, Committee on Science and Technology, U.S. House of Representatives, *Proceedings of the Twenty-third Colloquium on The Law of Outer Space,* IISL, Sept. 21-28, 1980, Tokyo, Japan, p. 133 *et seq.*

44. See generally Space Law, Selected Basic Documents, 2d Ed., Committee on Commerce, Science and Transportation, United States Senate, Committee Print, 95th Congress, 2d Sess., 1978.

45. See "The Relevance of International Law in The Prevention and Mitigation of Natural Disasters, Samuels, J.W., *Disaster Assistance, Appraisal, Reform and New Approaches,* New York, New York University Press, 1979, p. 249.

IAA 81-269

ANALYSIS OF THE NOVEMBER 23, 1980 EARTHQUAKE AS A DESIGN BASIS FOR SATELLITE EMERGENCY COMMUNICATION

G. Berretta*, S. Rossignoli**

P. Daly, H. Fromm, D. Tits***

An analysis of the most recent earthquake in Italy, together with a survey of possible satellite means to allow earthquake prediction, localization, and delivery of aid are addressed by this paper. A proposal is made for utilization of television broadcasting satellites to ensure localization and emergency communication.

Keywords: Earthquake, satellite emergency communication, earthquake prediction, earthquake localization, television broadcasting satellite, satellite TV terminal, remote sensing, distress signal, data collection.

HISTORICAL BACKGROUND

Italy was probably born from an ancient ocean between 50 and 100 million years ago as a consequence of the approaching and collision of the European and African continental masses. Looking at its bended shape we have immediately the feeling of the strains to which the peninsula is subjected in its anti-clockwise movement, with the Adriatic Sea being closed at the bottom and the Tirrenian being stretched (Fig. 1). This tectonic instability is clearly visible looking at the long mountain chains (Alps and Apennines) and to some of the most striking phenomena like earthquakes, active volcanoes, coastal erosion and slow movements to which some areas are subject. The earthquake striking the area of Irpinia in southern Italy on November 23, 1980 was only the last of a long series of catastrophic events in the southern regions of Puglia, Campania and Basilicata. First recorded event, dated back to 63 A.C., as we are told by Tacitus and Svetonius, happened only a few years before the eruption of Vesuvius in 79 A.C. burying Pompei and Ercolano. Since then, analyzing many different sources, such as chronicles of monasteries, municipal, episcopal and notary records, private letters of people happening to see the seismic events, inscriptions and printings, public advices and gazettes, as well as seismic catalogs (Fig. II) and maps with damage indications (Fig. III), 46 major catastrophic events with intensity of over VIII MCS are recorded in this area (Table I).

* *European Space Agency, Paris, France*

** *European Space Research and Technology Centre, Noordwijk, The Netherlands*

*** *FIAR - Fabbrica Italiana Apparecchiature Radioelettriche, Milan, Italy*

Reprod. E. Guerrera

THE ITALIAN PENINSULA

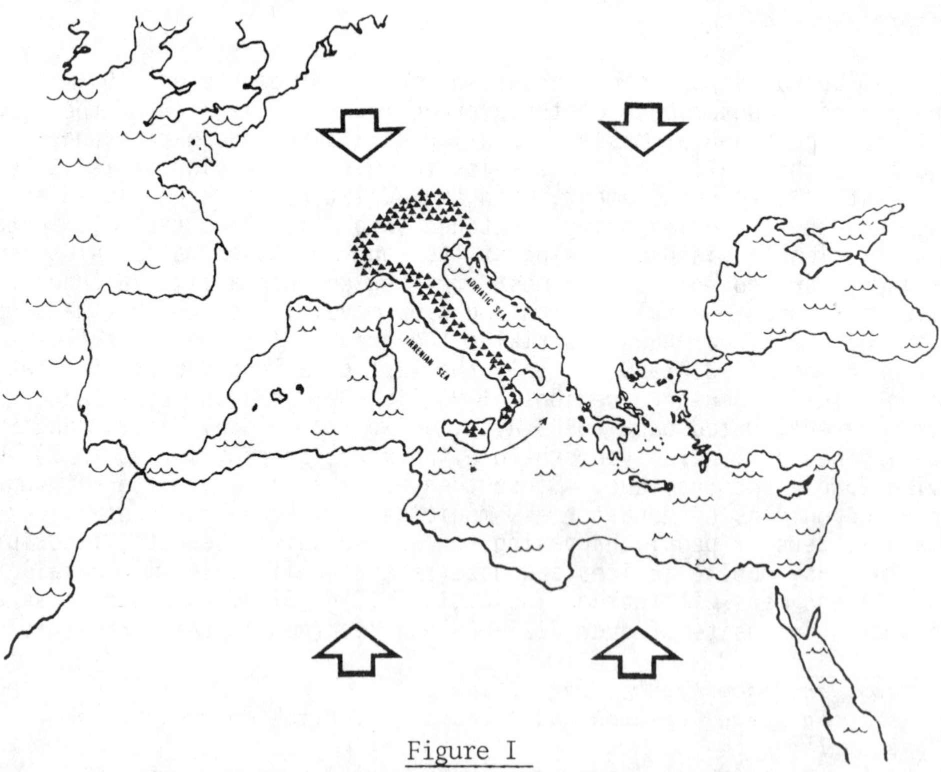

Figure I

FRONT PAGE OF ITALIAN FIRST SEISMIC CATALOG BY G. MANETTI (1457)

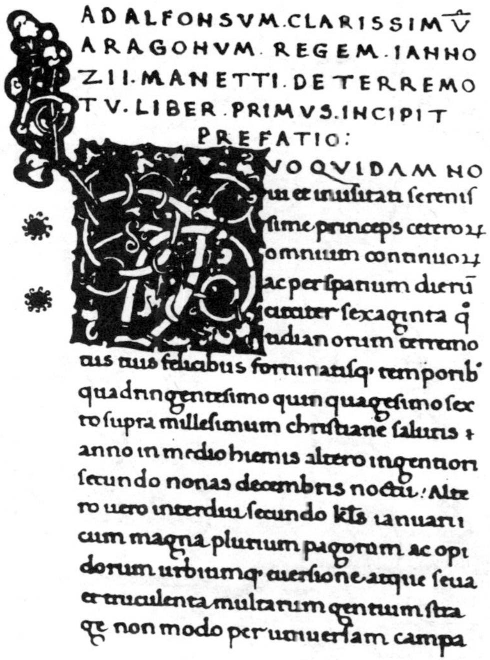
ADALFONSVM.CLARISSIMV
ARAGONVM. REGEM.IANNO
ZII.MANETTI.DE TERREMO
TV.LIBER.PRIMVS.INCIPIT
PREFATIO:
VO QVIDAM NO
ui et inuisitati serenis
sime princeps cetero
omnium continuo
ac per spatium dierū
circiter sexaginta q
udianorum terreno
tus tuis felicibus fortunatisq; temporib'
quadringentesimo quinquagesimo sex
to supra millesimum christiane salutis
anno in medio hiemis altero ingentiori
secundo nonas decembris noctu. Alte
ro uero interdiu secundo kls ianuarii
cum magna plurium pagorum ac opi
dorum urbiumq; euersione atque seua
et truculenta multarum gentium stra
ge non modo per uniuersam campa

Figure II

RECORDED EARTHQUAKES WITH $I_o \rangle$ VIII MCS(°) IN SOUTHERN ITALY (EXCLUDING CALABRIA AND SICILIA).

	DATE		COORDINATES			
YEAR	MONTH	DAY	LAT.	LONG	I_o	LOCATION
63	2	5	40.8	14.5	X	Napoletano
324	8	6	41	14	X	Campania
344	8	6	41	14	X	Campania
369	-	-	41.1	14.8	X	Benevento
847	6	-	41.5	14.2	X	Benevento-Isernia
853	-	-	41.5	14.5	X	Campobasso
930	-	-	41	14	IX	Napoli (?)
990	-	-	41.2	14.7	X	Benevento-Ariano
1087	-	-	41	17	IX	Puglia
1095	1	1	41.2	14.7	IX	Benevento
1125	10	11	41.2	14.7	X	Benevento
1139	1	22	41.2	14.7	IX	Benevento
1180	-	-	41.2	15.0	X	Ariano-Napoli
1223	-	-	41.5	15.7	X	Gargano
1273	-	-	40.5	15.7	X	Basilicata
1282	-	-	40.7	14.5	IX	Napoli (?)
1309	7	11	41.5	14.5	IX	Molise
1361	7	17	41.2	15.5	IX	Foggia
1414	-	-	41.5	16.2	IX	Gargano
1448	-	-	40.7	14.2	IX	Napoli
1456	12	5	41.2	14.7	XI	Campania
1550	8	25	40.7	15.0	IX-X	Vallo di Diano
1560	5	11	41.2	16.2	IX	Puglia
1561	7	31	40.5	15.5	IX	Vallo di Diano
1561	8	19	40.5	15.5	X	Vallo di Diano
1627	7	30	41.7	15.2	X	Gargano
1627	8	9	41.7	15.2	IX	Gargano
1646	5	31	42.0	15.7	X	Gargano
1654	9	8	41.0	15.7	IX	Potenza
1688	7	5	41.2	14.7	X	Benevento
1694	9	8	40.7	15.5	X	Lucania-Irpinia
1702	3	14	41.2	15.0	X	Benevento
1731	3	20	41.5	15.5	X	Foggia
1732	11	29	41.1	15.1	X	Avellino
1805	7	26	41.5	15.5	XI	Molise
1826	2	1	40.6	15.7	IX	Tito (Lucania)
1836	11	20	40.1	15.7	IX	Lagonegro (Lucania)
1851	8	14	41.0	15.7	X	Volture
1853	4	9	40.8	15.2	IX	Avellino
1857	12	16	40.3	15.9	X	Basilicata
1881	4	3	40.7	13.9	IX	Ischia
1883	7	28	40.7	13.9	IX	Ischia
1893	8	10	41.7	16.1	IX	Gargano
1910	6	6	40.9	15.4	X	Irpinia
1930	7	23	41.1	15.4	X	Irpinia
1962	8	21	41.1	15.1	IX	Irpinia

(°) MCS: Mercalli-Cancani-Sieberg, 1917

Table I

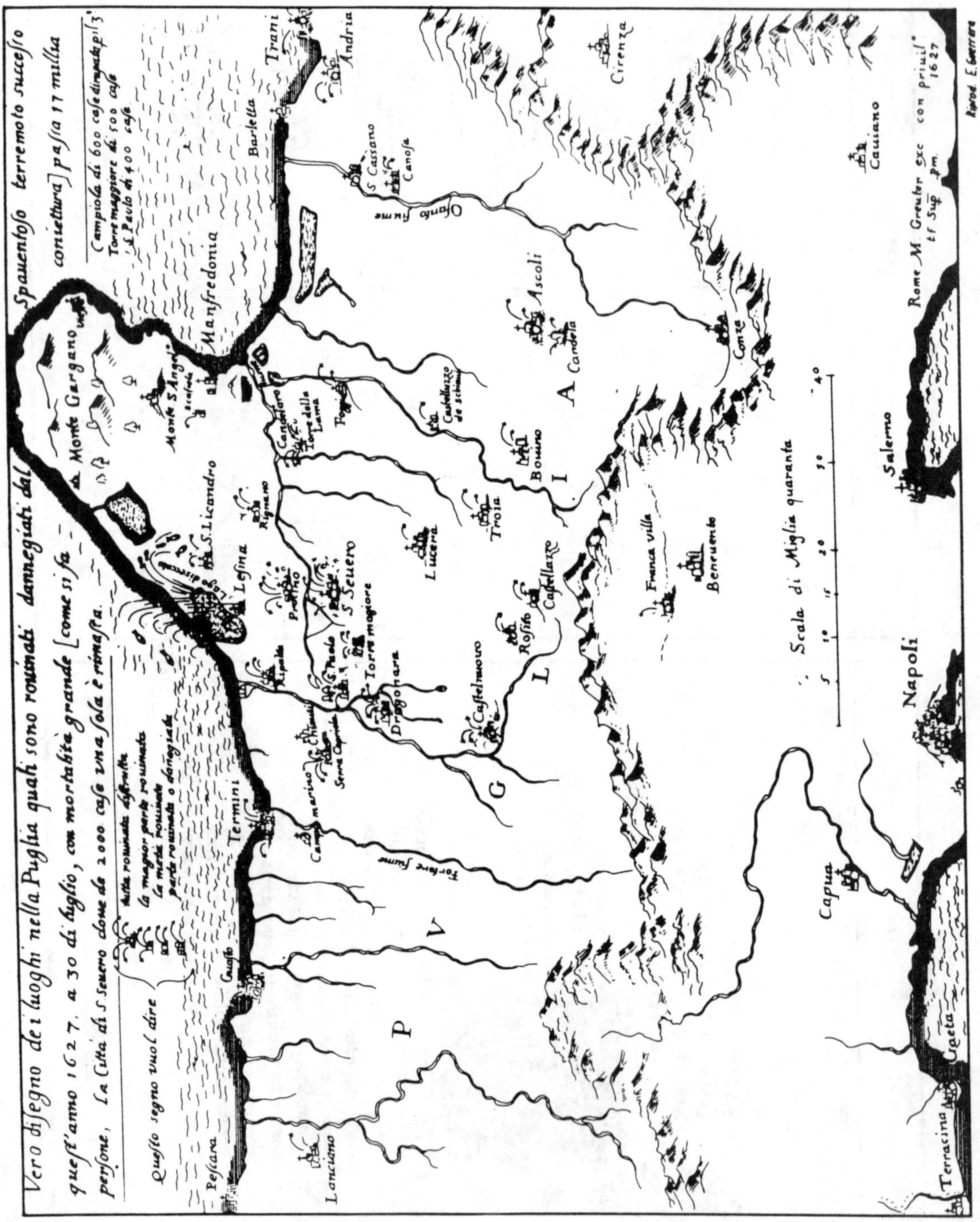

Figure III FIRST SEISMIC MAP WITH 4 GRADE INTENSITY BY M. GREUTER AND H. MARINARI (1627)

THE IRPINIA NOVEMBER 23, 1980, CASE

The earthquake under consideration hit severely an area of approximately 100 km long by 50 km wide, which included 280 communities and four major cities at its edges. According to the Trieste Wood-Anderson the epicenter was located at 40° 81' N latitude and 15° 38' E longitude and the magnitude (M_1) of the main event was computed at 6.5. In Fig. IV the isoseismic lines are depicted and recorded accelerations are given, the maximum of which was 0.33 g. According to official reports, 2,735 people died and 8,848 people were injured, but the most dramatic situation was with the survivors and with providing aid as we are going to describe in this paper. About 280,000 people had their houses destroyed or seriously damaged and had to be provisionally evacuated and put into tents, trailers, containers, ships, hospitals, public buildings, railway coaches, hotels and all kinds of dwellings available with the aid of about 43,000 military personnel. To date about 31,000 pre-fabricated houses are being installed in Irpinia, while another 20,000 are foreseen just for Naples which was subsequently shaken on February 14, 1981. The preliminary cost for reconstruction of private and public properties damaged by the earthquake is estimated today at $9.000 M.

Although these might seem very impressive data, an analysis of the world-wide situation shows that since 1960 there have been 26 major earthquakes with about 200,000 deaths (Table II). There are no firm indications about the cost for the reconstruction but we can guess that it must have been of the order of several billion US dollars.

To describe what happened to the population after the event is a very difficult task, because every village had its story and many dramatic pages have been written by reporters who sometimes reached the sites before any official aid arrived. Coordination was the most difficult job for several reasons. A broad localization of the area, for instance, was not defined until some hours after the event, because it was Sunday and because of lack of automatic communication links between the various seismic observatories. The various prefectures, under the Ministry of Interior, charged with providing aid had very limited understanding of the happenings because of lack of information from the area and poor global coordination, until the day after, when aerial surveys started giving an overall view. A very chaotic situation lasted for several days to the point that, as an example, a living 60-year old man was found under the ruins of his house six full days after November 23. Unfortunately Italy being a highly seismic area, responsibility for prediction, localization and, most important, house building standards, according to the seismic risk, lies with three different Ministries. At the same time university laboratories, the Electric Power Utility (ENEL) and the Nuclear Research Center (CNEN), are competent in these areas because of studies made for the location of nuclear power plants.

Great hopes lie in the future, due to the decision of the present Italian government to establish a ministery for coordination of efforts during natural calamities. And, as a result of studies made by the National Research Council (CNR) since 1976 responsibilities will include also

ISOSEISMIC LINES (MSK 64) AND ACCELERATIONS (g/10)
23.11.80

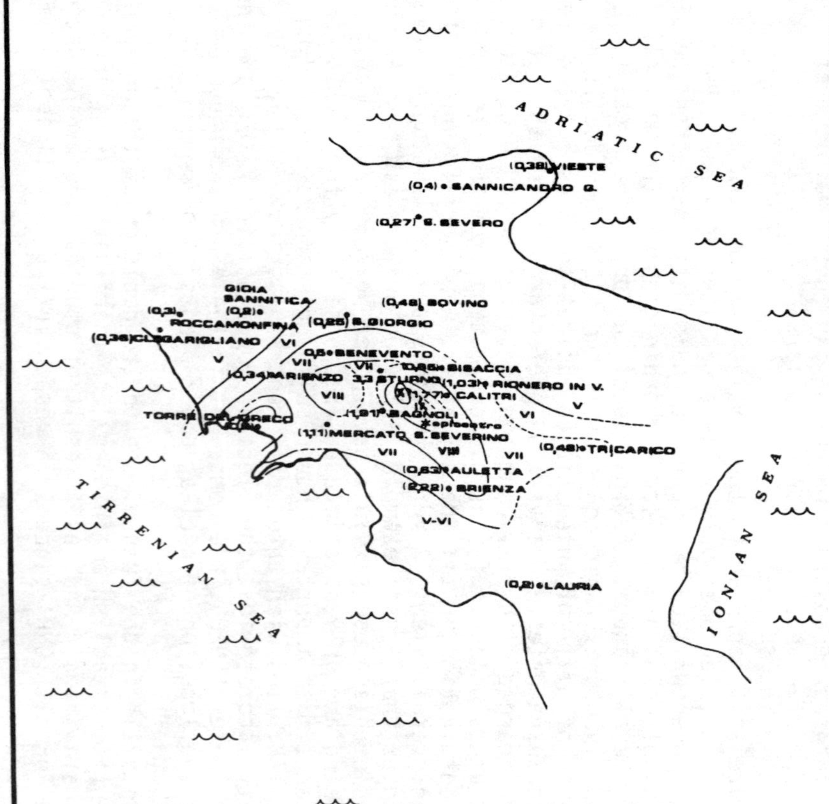

Figure IV

MAJOR WORLD EARTHQUAKES SINCE 1960

COUNTRY	DATE	DEATHS
Morocco	29.2.1960	12.000
Iran	1.9.1962	11.000
Yugoslavia	26.7.1963	1.078
Turkey	19.8.1966	2.529
Venezuela	29.7.1967	300
Italy (Sicily)	14.1.1968	337
Iran	31.8.1968	12.000
China	26.7.1969	3.000
Turkey	28.3.1970	1.100
Peru	31.5.1970	66.000
Iran	30.7.1970	200
Turkey	22.5.1971	800
Iran	10.4.1972	5.000
Nicaragua	23.12.1972	12.000
China	11.5.1974	20.000
Pakistan	28.12.1974	5.300
Turkey	6.9.1975	3.000
Guatemala	4.2.1976	22.836
Italy (Friuli)	6.5.1976	925
Rumania	4.3.1977	1.570
Iran	22.3.1977	167
Iran	6.4.1977	300
Indonesia	19.8.1977	185
Argentina	23.11.1977	100
Iran	20.12.1977	600
Iran	16.9.1978	11.000

Table II

prediction and localization of quakes in the Italian territory. In the case of Irpinia, neglecting the above problems, it is very interesting to analyze the performance in an industrialized country of the main networks supporting, in one way or the other, the delivery of relief to the population. The main networks considered in this analysis are: Transportation, Telecommunication and Electric Energy.

Transportation

Irpinia is a rural area with a large number of tiny villages on the top of the mountains with a very limited railway network and a very dispersed network of country roads. Five railway lines were interrupted with minor damage as a result of the quake and gradually repaired and utilized for locating hospital trains and coaches to give shelter to the population. Four national roads and many smaller roads were damaged but since no bridge completely fell, circulation was rapidly restored, by just repairing all the cracks. But most of the difficulties in supplying the most immediate aids were due to the weather conditions, which prevented an intensive use of air launches, and to traffic congestion because of lack of information given to the first-aid columns, which sometimes were held up on the road unable to proceed owing to inadequate planning of life lines.

Telecommunication

As to the land telecommunication network, six telephone districts were interested in the event, having something more than 100,000 telephone subscribers, 30% of which were out of operation. Out of 200 telephone exchanges connecting the users, about 100 were left without electric supply and operated provisionally on batteries. Forty-four urban exchanges suffered serious damage to equipment or to buildings. In the most severely hit districts, all traffic in and out was interrupted for a few days and provisionally re-activated between two and four days after the earthquake. Emergency generators were made available at those exchanges lacking electric power and emergency communication links were made available to the civilian authorities (200 lines on 25/11, 1125 lines on 30/11 and 2600 lines by mid-December) also to support mobile exchanges. Irpinia is a country of very large emigration. Notwithstanding, the traffic to and from the area was extremely heavy and congested until the end of December and required also use of military equipment to restore direct communications with the Center of Coordination in Naples.

Electric Energy

Concerning the electric power network, the earthquake caused an outage on 415 communes in five provinces. Most of the failures were in the medium/low voltage transportation network involving 840 transformation stations which had to be restored to operation or rebuilt. Still 58 communes after 12 hours and 13 communes after 48 hours were left without electric energy at low voltage, while all supply was restored four days after the event. Some 1650 km of medium voltage lines were to be inspected, 1300 km of low voltage lines had to be repaired or rebuilt, while supply for 50,000 subscribers was completely destroyed. Yet the most difficult problem was not to supply customers no more existing, but to ensure electric energy for the new provisional installations of the population and to establish new medium and low voltage lines and with 110 new transformer stations, which was accomplished by the Italian Electric Utility by the end of December 1980.

PHASES OF INTERVENTION

From earthquake analysis, it is possible to identify at least five phases where appropriate human intervention could reduce the tragic consequences of this phenomenon, i.e.:

- Prediction of earthquakes
- Location of earthquake focus
- Evaluation of damage and identification of life lines
- Localization of survivors in distress
- Communications during emergencies

The use of satellites can be foreseen in all the phases. In some cases they can play a minor complementary role to other technical means, but more often they could be an essential part of an Alert and Rescue System, taking advantage of their capability of earth observation and communication. In general, the peculiar advantages of satellite systems with respect to most other systems are the rapid installation and continuous availability of the service and the possibility to deliver the collected information, earth observations or human communications, to any point in a large coverage area wherever there are adequate receiving stations. These superior characteristics are very important in case of natural disasters, when it is essential to act rapidly and to have a continuous picture of the situation at a central point for the most efficient organization of rescue. We shall now go briefly through some examples of satellite use in the different phases to finally concentrate on a specific original use of direct TV broadcasting satellites for the phases of distress signaling (identification of survivor groups in need) and particularly of emergency communication.

Prediction of Earthquakes

To date very little is known about crustal dynamics to allow precise earthquake predictions, but several means are being studied and put into operation to gain better knowledge of these phenomena. There are at least ten types of precursor phenomena which are related to build-up stresses in the sub-surface. Some of these may occur some years before a major earthquake and their behavior versus time, as well as their spatial distribution, is related to the magnitude of the impending quake. Others increase in numbers several hours before a quake. Continuous monitoring of such phenomena, which is today a powerful tool for scientific research, will become the most important means of initiating timely warning of earthquake hazards when the accuracy of prediction will be improved. In the prediction phase, satellite systems can be used in two ways: directly and indirectly. In the first case, the satellite is used in its function of earth observation, with direct measurements from space;

in the second case, in its function of communication, transmitting information generated by instruments installed on the ground to a central point of data collection and interpretation.

Direct Measurements. Eventually as the facility for highly accurate position fixing develops, satellites will play a major role in the direct determination of crustal movements. Ultra-precise point-positioning with cm-accuracy appears to be the most promising feature of a future earth research satellite, particularly for regional distances. Several approaches are being studied: for example placing a laser in orbit around the earth and passive retroflectors on the ground, or using signals from GPS satellites as VLBI sources. One or both of these space-derived systems may have economic and operational advantages compared to ground-based lasers and quasar-based VLBI for determining relative position changes over the range of distances of about 20-100 km. Spaceborne laser ranging techniques have been studied since 1976 in Europe and, encouraged by the preliminary results from this work, the European Space Agency is currently performing a concept study accompanied by a technical plan for such a system. This work anticipates as its principal investigation area the Mediterranean orogenic belt, and comprises a particularly selected test area in the Friuli fault zone. Satellites can also play an important role in obtaining gravity field data which are relevant parameters for the understanding of crustal-dynamic phenomena. Because of the long correlation lengths for important gravity field variations, it would be very difficult to obtain the necessary information over the more remote areas of the world by ground measurements.

Indirect Measurements. The role of satellites is seen as a means of telemetering data from ground instrumentation placed along relatively remote areas, such as the Central Apennines, or the Calabrian arc. Ground-based instruments are more suitable than direct space methods to keep track of movements that occur over short periods in a limited area of high seismic risk. Other relevant data, such as radon gas concentration, electric resistivity, seismic velocity, water level and number of percursor shocks can be detected only by means of ground instruments. Meteosat, with its Data Collection System, could be used to relay these geophysical data for real-time interpretation with sophisticated equipment in a geodynamic center, far from the region of measurements. This system allows regular reporting on request from dedicated platforms and immediate transmission of short messages whenever measurement thresholds are exceeded (Alert).

Location of Earthquake Focus (Epicenter)

For the great majority of earthquakes, the location is determined from the time taken by seismic waves to travel from the focus to a seismograph. If arrival times at three observatories are available, then triangulation can be used to determine latitude and longitude of the earthquake focus and the time of occurrence of the earthquake. To increase the accuracy it is common practice to use the readings from many observatories in International Seismological Centers around the world. Satellite systems can play the role of relaying in real-time Measurements to the International Centers. Moreover, in this case, a precise clock only at the Central Observatory would be needed. In order to use the Meteosat DCS system it would be necessary to develop data compression algorithms for seismographs and the corresponding software for on-site data processing. With data relay communication satellites, these compressions would not be necessary but the cost of the link would be higher.

Evaluation of Damage and Identification of Life Lines

The evaluation of damage and the identification of accessible and most appropriate life lines could be accomplished through the interpretation of images taken by earth resource satellites. A very important parameter is the resolution of the pictures which is linked in a complex manner to the capability to detect, for instance, bridge destruction or road blockages. The capability to take land pictures in cloudy conditions is also an almost essential requirement. Other important characteristics are the angle of view and the cycle of repetition of images of the same area. Future earth resource satellites with Synthetic Aperture Radars will satisfy both the requirements of high resolution and capability of imaging in cloudy conditions and during the night. ERS-1, the European Satellite presently under study, should have a resolution of 20/25 meters with a Synthetic Aperture Radar System. Present satellites in orbit such as Landsat 2/3 have a resolution of 80 meters with Multi-Spectrum Scanners, and of 40 meters using a Return-Beam Vidicon.

Localization of Survivors in Need or Distress

The satellite approach is independent of the extension of the disaster area and permits the information to be delivered to a center for coordination of rescue operations. However, to perform a distress function, it is necessary to have all the high-risk areas equipped with appropriate transmitters, and this could prove to be rather costly in many cases. Vesting a TV Broadcasting Satellite with this task is in principle a rather inexpensive solution, as it will be explained in the next section. Another solution could be to use the dedicated distress channel of a Maritime Communication Satellite, such as MARECS. All the community at risk should be equipped with Emergency Position Indicating Radio-Beacons, but this solution seems to involve difficult implementation and acceptability by the maritime community. Moreover, the maritime distress channel is designed to support only 20 simultaneous transmissions of EPIRB'S in a 10-minute period.

Communication in an Emergency

This phase can last for a relatively long period of time, until normal communication links are re-established. Even after the telecommunication network has been restored, it could be convenient to maintain a separate emergency communication system considering that, during the emergency period, the traffic requirements from the disaster area might be much greater than normal. For emergency communications, the satellite approach is certainly the most flexible one, if transmitting-receiving stations can be made available in the disaster area.

There are several interesting developments in the field of small transportable stations. Among others, the Indian experiment with Emergency Communication Terminals which has been operational since 1977 with the Symphonie Satellite System in the frequency range 4/6 GHz, using SCPC-FM. The emergency terminals are air-liftable and jeep-transportable. The power amplifiers deliver 20 watts to a 3-meter antenna, assuring very good quality voice transmission. This system has been used during the 1977 cyclone in Andhra Pradesh and experiments will be continued with Indian satellite APPLE. In this latter case, a smaller terminal will also be available(SCOT-Small Communication Terminal), with an antenna of 1,38 meter, and lower quality requirements, but still sufficient for emergency communications. In Europe, systems of this type could be conceived at 11/14 GHz, or 12/14 GHz, using ECS.

Yet in the system briefly described and in others of the same type it is necessary to deliver the terminals to the disaster area with a small team of specialists for the installation, while in most cases this can hardly take place in the first emergency phase, when the roads are cut off and weather conditions are bad. Operationally, terminals of this type can be readily installed only in proximity to the disaster area to relay information to a relief center. Moreover, there are problems at the other extreme of the link, where it is necessary to have a relatively large coverage area of communication satellites. In most cases, such stations will not exist at the local relief coordination center, which could be the regional main city. For all these reasons, solutions of this type seem to be better suited to the second phase of emergency, in order to have additional and separate telephone links from the emergency area in a broad sense, for example, to a national coordination center. Yet other solutions could be envisaged, using national communication satellites at 20/30 GHz, where some of the above problems could disappear. The use of a forthcoming Direct TV Broadcasting Satellite System, as described in the next sections, is the solution that we see as the most promising.

USE OF A DIRECT BROADCAST SATELLITE SYSTEM FOR EMERGENCY COMMUNICATION

The Approach

Broadcast satellites, by their very nature, are conceived for operation with small ground terminals. This is well known for the receiving function with home receiver antennas below one meter in diameter, but it can also be realized for a transmitting function. The latter requires a certain operating mode of the satellite broadcast repeater which is normally provided anyway or which can be added at very little extra cost. The proposal is the following: A certain number of receiving terminals in any potential earthquake area will be equipped with distress transmitters and facilities for voice communication. In case of an earthquake (or actually any other emergency) short distress messages can immediately be released and relayed via a repeater of a national Direct Broadcast Satellite. Compatibility with the ongoing broadcast service is ensured by a proper choice of the modulation characteristics of the distress signals and by placing them at the band edges of the frequency-modulated TV signal. On the ground, a central emergency surveillance station permanently observes the broadcast downlink spectrum for the presence of distress signals. Upon reception of such signals the central station will have immediate information on the extent of the disaster and will also have identified survivors in need. It can then, if deemed necessary, establish voice communication links through a broadcast satellite repeater. Such links are provided on a time-sharing basis with broadcast transmissions, and will normally be limited to the early rescue phase as discussed in a previous section. The temporary interruption of the broadcast service in one channel is believed to be acceptable, in view of the difficult situation shortly after an earthquake.

The Signal Performance Requirements

Distress signals can be of the same type as proposed for the emergency Position Indicating Radio Beacon System of Maritime Satellites. An overall carrier-to-noise density ratio C/No of 26 dB (Hz) and a data rate of some 10 bits/sec, have been assumed for designing that system and shall also be taken for the purpose of this paper. The 2-way voice links must be usable by non-experienced people and the voice quality need be comparable with that obtained during normal maritime or land mobile communication links. For the purpose of this paper, narrow-band frequency modulation employing single channel per carrier techniques was found to be a reasonable design guideline; however advanced modulation techniques, with the aim of minimizing transmitting power, could be further investigated. An overall link quality, expressed as carrier-to-noise density of 52 dB (Hz), should produce a satisfactory voice quality.

The Link Design

Two specific phases need to be distinguished: the transmission of distress signals frequency multiplexed with the broadcast signal and the transmission of 2-way voice signals through a dedicated broadcast satellite repeater. The transmission of the distress signal is radio-frequency interference-dominated, rather than noise-dominated. To minimize the effect of this interference, due to the spectrum of the broadcast signal, the distress signal should be placed at the edge of the 27 MHz wide broadcast channel. Normally the broadcast spectrum is confined to within some $\pm$ 10 MHz around center frequency. The worst case spectrum is typically represented with a 100% color bar signal, an example of which is illustrated in Figure V. At band edge, adjacent channel interference may contribute to the spectral power density, but the worst case total power density should never exceed the satellite receive noise density by more than 20 dB. The broadcast carrier is dispersed with 600 KHz peak-to-peak and it appears permissible to treat the spectral power noise-like. On this assumption a typical link budget can be constructed as shown in Table III. Due allowances have been made for propagation margin and small signal compression in the satellite repeater, while the other assumptions are as typical for a national Italian Direct Broadcast Satellite System. The required transmitting power of the 18 GHz distress transmitter is well below what can be achieved with semiconductor technology, even considering the smaller antennas below one meter in diameter. The approach for 2-way voice communication is that one broadcast satellite repeater will be made

RF-SPECTRA OF 100% COLOUR BAR SIGNAL

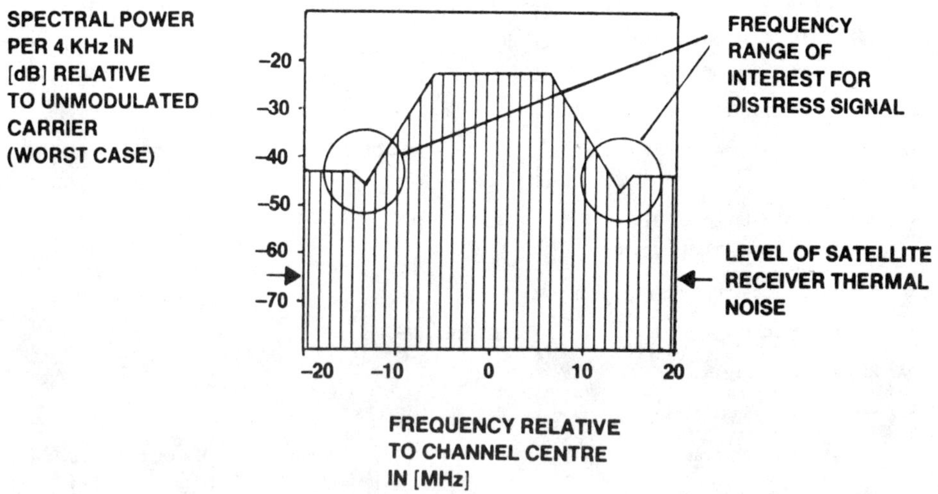

FIGURE V

BLOCK DIAGRAMME OF AN EMERGENCY TRANSMIT/RECEIVE STATION

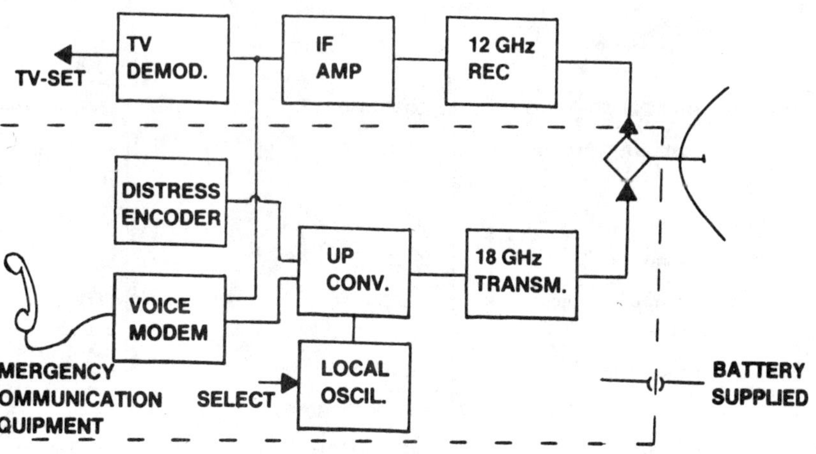

FIGURE VI

TABLE III-I LINK BUDGET "DISTRESS SIGNAL TRANSMISSION" (UPLINK, 18 GHz)

		VALUE		NOTES
1.	Required C/I_o at satellite receiver in distress channel	30 dB (Hz)		Incl. 4 dB margin for compression
2.	TV carrier level at sat. receiver	-98.3 dB (W)		For C/N = 26 dB
3.	Max. spectral density at band edge	-143.3 dB (W/4KHz)		45 dB relative to TV-carrier
4.	Req. carrier level of distress signal at satellite receiver	-149.3 dB (W)		
5.	Satellite receiver antenna gain	38 dBi		Typical for Italy
6.	Spreading loss	-162.7 dB (M^2)		
7.	Propagation margin	5 dB		
8.	Req. earth transmit EIRP	24.4 dB (W)		
9.	Effective earth transmit antenna gain	36 dBi	43 dBi	Estimate for 0.9 and 2.0M 12/18 GHz antenna
10.	Required transmit power	-9.6 dB (W) 0.1 Watt	-16.6 dB (W) 0.02 Watt	Includes 2 dB post-transmitter loss

TABLE III-II LINK BUDGET "DISTRESS SIGNAL TRANSMISSION (DOWNLINK, 12 6 Hz)

		VALUE	NOTES
1.	Sat. Transmit EIRP TV-signal	60 dB (W)	Edge of coverage
2.	Sat. Transmit EIRP distress signal	10 dB (W)	54 dB below TV-carrier incl. 3 dB suppression
3.	Spreading loss	-162.7 dB (M^2)	
4.	Propagation margin	3 dB	
5.	Rec. ant. gain of emergency control centre	47 dBi	Eff. gain for 3.0 M antenna
6.	Receive noise temperature	600K	
7.	Receive K.T	-200.8 dB (W/Hz)	
8.	Distress carrier-to-noise ratio	49.1 dB (Hz)	TV spectra not considered
9.	Minimum distress carrier-to-interference ratio	27 dB Hz	Due to TV-spectrum, thermal noise negligeable

TABLE IV-I LINK BUDGET "VOICE COMMUNICATION" FORWARD, 18 GHz
EMERGENCY STATION TO SATELLITE

		VALUE		NOTES
1.	Required transmit power of emergency station	0.7 Watt -1.9 dB (W)	0.13 Watt -8.9 dB (W)	Includes 2 dB post-transmitter loss
2.	Effective emergency transmit antenna gain	36 dBi	43 dBi	Estimate for 0.9 M and 2M 12/18 GHz antenna
3.	Required minimum transmit EIRP	32.1 dB (W)		
4.	Spreading loss	162.7 dB (M^2)		
5.	Propagation margin	5 dB		
6.	Satellite receive antenna gain	37 dBi		Italy, edge of coverage
7.	Sat. receive noise density	-198.6 dB (W/Hz)		Thermal noise only
8.	Required minimum uplink carrier-to-noise density ratio	55 dB (Hz)		
9.	Required minimum carrier level at satellite receiver input	-143.6 dB (W)		
10.	Permissible maximum carrier level at satellite receive input	-138.6 dB (W)		

TABLE IV-II LINK BUDGET "VOICE COMMUNICATION" FORWARD, 12 GHz
SATELLITE TO CONTROL CENTRE

		VALUE	NOTES
11.	Satellite repeater gain	122 dB	Fixed gain mode
12.	Satellite transmit antenna gain	37.5 dBi	Italy, edge of coverage
13.	Minimum satellite EIRP per voice carrier	15.9 dBW	
14.	Spreading loss	162.7 dB (M^2)	
15.	Propagation margin	3 dB	
16.	Effective receive antenna gain of control centre	47 dBi	3M 12/18 GHz antenna
17.	Minimum receive carrier level	-145.8 dB (W)	
18.	Receive noise density	-200.8 dB (W/Hz)	T = 600 K
19.	Receive minimum carrier-to-noise density ratio	55 dB (Hz)	
20.	Overall carrier-to-noise density ratio	52 dB (Hz)	Incl. uplink contribution

TABLE IV-III LINK BUDGET "VOICE COMMUNICATION" RETURN, 18 GHz
CONTROL CENTRE TO SATELLITE

		VALUE	NOTES
1.	Required transmit power of control centre per voice carrier	-2.3 dBW 0.6 Watt	300 W for 50 carriers and 8 dB output backoff
2.	Effective transmit antenna gain	47 dBi	3M 12/18 GHz antenna
3.	Required minimum transmission EIRP per voice carrier	44.7 dB (W)	
4.	Spreading loss	162.7 dB (M^2)	
5.	Propagation margin	5 dB	
6.	Satellite receive antenna gain	37 dBi	Italy, edge of coverage
7.	Satellite receive noise density	-198.6 dB (W/Hz)	Thermal noise only
8.	Required minimum uplink carrier-to-noise density ratio	67.6 dB (Hz)	
9.	Required minimum carrier level at satellite receiver	-131 dB (W)	
10.	Permissible maximum carrier level at satellite receiver	-125 dB (W)	Less propag. margin

TABLE IV-IV LINK BUDGET "VOICE COMMUNICATION" RETURN, 12 GHz
SATELLITE TO EMERGENCY STATION

		VALUE		NOTES
11.	Satellite repeater gain	122 dB		Fixed gain mode
12.	Satellite transmit antenna gain	37.5 dBi		Italy, edge of coverage
13.	Minimum satellite EIRP per voice carrier	28.5 dB (W)	21.5 dB (W)	
14.	Spreading loss	162.7 dB (M^2)		
15.	Propagation margin	3 dB		
16.	Effective receive antenna gain of emergency station	36 dBi	43 dBi	Estimate for 0.9M and 2M 12/18 GHz antenna
17.	Minimum receive carrier level	-144.2 dB (W)		
18.	Receive noise density	-199.2 dB (W/Hz)		NF = 6 dB
19.	Minimum receive carrier-to-noise density ratio	55 dB (Hz)		
20.	Overall carrier-to-noise density ratio	52 dB (Hz)		Incl. uplink noise contribution

available for these links only (in time-sharing with the broadcast service). Hence, the voice link design is normally thermal noise-dominated although adjacent channel interference may constrain the use of the band edges for voice communication. Voice signals are transmitted on pre-assigned frequencies. To optimize for the maximum number of channels, while respecting the permissible carrier-to-intermodulation product, due to repeater non-linearities, operation below saturation of the satellite power amplifier is essential. This can be readily implemented by commanding the satellite repeater into a fixed gain mode (automatic gain control is normally applied for broadcast). The optimization of forward and return link transmit power requirements implies different transmitting power levels for these links at satellite output, the maximum transmit level being required for the return link from satellite to emergency station. Typical link budgets, assuming again Italian specific data, are summarized in Table IV. The required transmitter power of the emergency station is higher than that needed for the distress signal but still below the one-watt level. This suggests the application of available Field Effect Transistor technology as further discussed below.

The Emergency Transmit/Receive Station and System Operations

Such a station needs to be of low-cost design, easy-to-operate and should be immediately available and usable just after an earthquake. These factors suggest the integration of emergency communication functions into some of the future broadcast receive terminals. The 18 GHz transmitter, distress encoder and voice modem could be added as illustrated in Figure VI. The stations would then be installed in one or two locations in any potentially endangered location. Protection against possible damage due to earthquakes would be ensured by choosing a proper support for the antenna in particular. This would then allow the dispatch of a distress signal, in a simple push-button operation, immediately after the distress. The transmitted message would carry a station identifier and brief information about the extent of the disaster. The latter is obtained by selection from a choice of a few pre-programmed messages. If judged necessary by the central emergency control center, 2-way voice communication would be initiated by transmitting a specific carrier through the satellite which would be automatically detected by an emergency station and would serve as an announcement for the voice communication mode. The low power consumption of such emergency stations (of the order of a few tens of watts at the most) permits battery operations, thus making the system independent of power supply interruptions.

A review of the basic elements of the emergency transmit/receive stations reveals that available components can be used throughout. The 18 GHz transmitter can be realized with a Field-Effect-Transistor amplifier. Current prices for such transistors are still high, but significant reductions are predicted for the next few years. The requirements on the local oscillator are dependent upon more detailed trade-off studies; too stable a frequency would entail too high a price, while too great a tolerance would complicate the overall system design. A stability of the order of 10^{-6} appears to be a reasonable compromise. Distress encoder and voice modem designs are available for several similar applications and even more advanced modulation and access techniques should be feasible, considering the current pace in the development of micro-electronics.

Finally, a few words about the central emergency control station. The link budgets in Tables III and IV demonstrate that a 3-meter antenna supplied by a 300-watt amplifier should be sufficient for a multi-carrier operation. Such stations could also be installed in regional centers in order to organize the necessary help in a more direct way.

System Capacity and Compatibility Aspects

Compatibility with international regulations could be a major requirement, while compatibility with national broadcast transmissions is believed to be of lesser importance. The carrier of a distress signal will be some 60 dB below the broadcast signal carrier. Even if 100 distress signals were transmitted simultaneously, the total power would still remain 30 dB below the broadcast carrier, a level which is considered to be permissible for co-channel interference. During 2-way voice communication the spectral power density of the satellite EIRP should not exceed the level which is obtained for broadcast transmissions. The requirement of multi-carrier operation implies an output back-off of some 5 dB; this would allow the simultaneous transmission of about 50 pairs of voice links using single carrier narrow-band frequency modulation. The spectral power density of such a multicarrier arrangement is well below the permissible level at channel center, while it might become marginal at the edges of the 27 MHz wide broadcast channel. The investigation of more advanced modulation and access techniques, such as spread spectrum, might prove useful for a further system optimization.

A Possible Field Experiment

As from 1985/1986 the European Large Satellite L-SAT will provide a pre-operational broadcast service in one channel over the Italian territory. The transmit characteristics of this channel are as recommended in the final texts of the 1977 Satellite Broadcasting Conference (WARC BS-77). The satellite receive function is provided through a common transmit/receive antenna, thus offering the high receive gain as required for the emergency communication links. Actually the link budgets in Tables III and IV are based on L-SAT performance data. The obtainable coverage is illustrated in Figure VII. For reference, the centers of major previous earthquakes are indicated by little dots. Such a field experiment would require a central earth transmit/receive station and a few emergency terminals distributed over an Italian region. For the control station, partial use could possibly be made of a broadcast earth transmitting station. Of course such a system concept requires further studies, particularly on modulation and access techniques, before an experiment can be designed.

A Feasibility Assessment by Hardware Simulation

In order to demonstrate the viability of operating a low-level emergency link within the bandwidth of a normal TV transmission, a series of tests

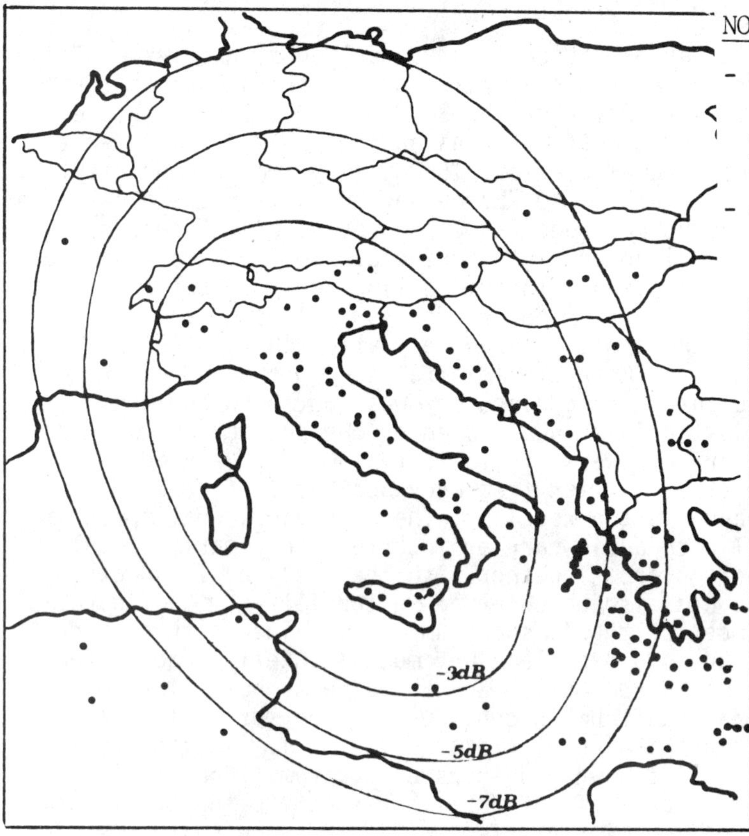

NOTES

- contours refer to -3dB, -5dB, -7dB, relative to peak gain.
- dots represent epicenters for major earthquakes in the last decade.

FIGURE VII - ITALIAN BROADCAST SATELLITE COVERAGE AREA

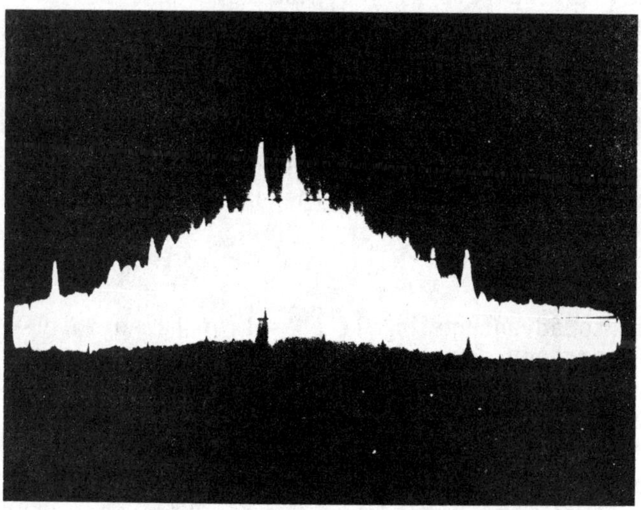

NOTES

- vertical: 10 dB/division
- horizontal: 5MHz/division
- audio/video carrier power: -30 dB

FIGURE VIII - SPECTRUM OF RF-MULTIPLEXED TV AND SOUND SIGNALS

was carried out at ESTEC earth station in station loop with driver TWT in saturation and normal receive levels as relevant for transmissions through the Orbital Test Satellite (OTS). For the 36 MHz bandwidth TV picture, without energy dispersal, both test patterns and normal program material were used. The emergency signal carrier, 15 MHz away from channel center, was FM-modulated to a bandwidth of 40 KHz by spoken English via a cassette recorder. For a range of voice/video carrier power levels, recordings were made of the received video and interfering carrier to establish the limits of operability of the link. The video link began to suffer severe distortion with audio/video carrier power of -5 dB. At the other extreme the audio was still comprehensible but severely affected by noise at a power ratio of -35 dB due to FM threshold effects when the audio carrier was around -45 dB. The interesting discovery was that both audio and video waveforms were substantially unaffected by one another over a wide range of audio/video carrier powers (from -10 to -30 dB). A gradual reduction of audio and video carrier frequency separation from 15 MHz made no significant change in the results. The received signal spectrum is shown in Figure VIII with audio carrier interference clearly visible 15 MHz above video band center. Together with the received TV spectrum, intermodulation effects can also be observed in the left of the figure. It is noticeable that most of the TV spectrum is contained with a bandwidth of around 20 MHz and could explain why mutual interference between video and audio waveforms is so small. Such a voice transmission in the presence of the broadcast spectrum is considered to be more critical than the actually intended transmission of a distress signal. In fact the distress signals will present a very dispersed spectrum. Considering 100 distress signals at a level of -50 dB with respect to the broadcasting signal, they will be confined within a few MHz if the same system is used as in the maritime satellite. The present test proves also that even in the presence of the TV signal it is possible to transmit a single voice carrier without appreciable interference with the TV signal. Considering the most critical link, which is the return link, the voice channel will be 31.5 dB below the TV signal which is within the considered range of the tests. For the forward link the power of the voice carrier at the satellite is 44 dB below TV carrier. If we consider a 2-way system, the third order intermodulation near the TV signal is the most dangerous product. However this product will be more than 50 dB below the carrier. Note that 50 dB is the value of spurious specification for TVBS. More tests will be performed with several carriers in the near future.

CONCLUSIONS

Satellites can be applicable to advantage in all phases of human intervention from prediction to emergency communication. The proposed system for emergency communication, using a Direct Television Broadcasting satellite, presents very interesting features:

- The communications link to the disaster area is immediately available.
- No installation with specialized personnel is required.
- The TV receiving antenna reused for emergency communication is already pointed to the satellite. Even in case that the particularly dedicated antenna is out of service, considering the large number of TV receiving antennas, there is very high probability that a working antenna will be found in the area.

- The system is easy to use and there is no need of specialized personnel to operate it.
- The overall system cost is expected to be low in comparison to other systems.
- Flexible communication links can be set up between regional centers and the emergency area.
- The coverage in all European cases is at least national.
- In the particular Italian case the TVBS beam covers also regions in other neighboring countries of very high seismic risk (e.g. Yugoslavia and Greece) and the satellite could be made available also to those countries.
- The proposed system requires negligible modification to TVBS satellites.
- It should be possible to set up a demonstration experiment with L-SAT.
- The tests carried out using OTS ground station in station loop increase confidence in the feasibility of the proposed approach.

ACKNOWLEDGEMENT

The authors would like to express their gratitude to Mrs. M. Dalla Volta of FIAR for the support given in collecting information on the Irpinia earthquake from various sources. Thanks are due also to Prof. G. LUONGO (Osservatorio Vesuviano), Mr. F. SPINOZZI (ENEL - Compartimento di Napoli), Mr. D. MOLIN (CNEN), SIP (Ufficio Stampa) and Mr. S. HIEBER (ESA) for their help.

REFERENCES

Bruce A. Bolt. *"Earthquakes" a primer*

R. Ciuni. January 1981. *Quei giorni delle macerie, della paura e della rabbia.* Napoli

COUNCIL OF EUROPE. January 1980. Proc. of 2nd Workshop. *European earthquake prediction programme.* ESA SP-156. Strasbourg

EUROPEAN SPACE AGENCY. June 1976. *Dumb-bell Mission for Gravity Field and Magnetospheric Studies.* DP/P6 (76)22. Paris

ISRO-SPACE APPLICATION CENTRE. Telecommunication Research Centre. P. & D. Delhy. March 1979. *Proc. of seminar of STEP experiences.* Ahmedabad

ISRO-SPACE APPLICATION CENTRE. May-June 1981. *SAC COURIER.* Vol. 6 M-3. Ahmedabad, C.H. Jones. June 1981. The Radio and Electronic Engineer. *A manpack Satellite Communication Earth Station.* Vol. 51, N.6, 259-271.

G. Magri and D. Molin. May 1981. *Sulla raccolta e catalogazione delle notizie di sismica storica.* Udine

NASA Technical Paper 1464. August 1979. *Application of Space Technology to crustal dynamics and earthquake research.* Washington D.C.

R. Robson and H. Houet. January 1980. Proc. 2nd Workshop by ESA and Council of Europe on European Earthquake prediction Programme. *The Meteosat data collection system and its potential as a data relay for geophysical measurements.* ESA SP-156. Strasbourg

R. Rogard. September 1980. Discussion of the Safety Capability of a Maritime Space Segment. A new proposal for a satellite EPIRB system. *Proc. of IAF Congress.* Tokyo

P.L. Ward. 1978. *Geophysical predictions.* National Academy of Sciences. Washington D.C.

IAA 81-263

POTENTIAL FOR DETECTION OF NATURAL DISASTERS VIA METEOSAT

A. Robson, J. Morgan*

R.W. Herschy**, J. Zschau***

The Meteosat geosynchronous satellite, positioned over the equator and prime meridian provides: a) an earth imaging capability; b) a broadcasting service for images; c) a data collection service.

The imaging and broadcasting capability make it possible for a large number of authorities to monitor weather phenomena which could give rise to disasters. The data collection provides a cheap and reliable means of relaying environmental measurements from 40% of the earth's surface, i.e. within about 80° great circle arc of the subsatellite point. As such, the service can be used in systems to detect natural disasters.

System characteristics are described as well as current data collection system applications in: a) hydrology; b) storm surge prediction.

Both applications provide for the detection of natural disasters and indicate how the system could be utilized on a wider scale for disaster detection.

Keywords: Data Collection, METEOSAT, environmental monitoring, satellites, storm surges.

1. INTRODUCTION

The Meteosat series of satellites of the European Space Agency are geosynchronous satellites with three missions to fulfill:

(1) The acquisition of images of the earth's disc every half hour in the visible and infrared parts of the spectrum. From these images a variety of meteorological parameters were to be determined e.g. wind vectors, sea surface temperatures, cloud top heights, etc.

* *European Space Agency, European Space Operations Centre, D-6100 Darmstadt, West Germany*

** *Water Data Unit, Department of Environment, GB-Reading RG1 8PS, England*

*** *Institut für Geophysik, University of Kiel, D-2300 Kiel, West Germany*

(2) The dissemination of corrected and annotated images to users' ground stations after retransmission by the satellite.

(3) The collection and distribution of environmental data gathered by terrestrial platforms.

The first of these satellites was launched in November 1977. It suffered an onboard failure in November 1979 preventing the missions 1 and 2 from continuation but leaving the data collection mission intact.

The second Meteosat was launched in June 1981. During its in orbit checkout, it was discovered that the transponder system used for data collection was defective but the other mission capabilities were intact. Hence together Meteosat 1 and Meteosat 2 is able to provide the full missions capability.

The operation of the two satellites is funded until the end of 1983 but, barring onboard failures and given further funding, they could supply services until 1985/1986.

The two satellites are considered as part of a pre-operational system for the meteorological agencies of the participating states. The groundwork to set up an operational programme up to the mid 1990s is presently under preparation.

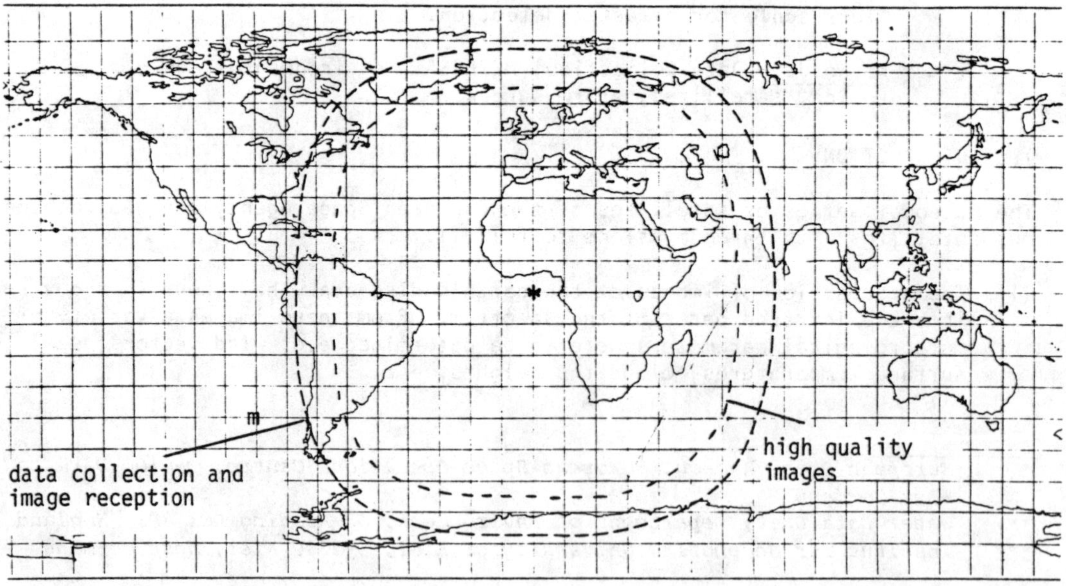

Fig. 1. Coverage area of Meteosat.

Regarding the use of Meteosat for the detection of natural disasters, all three missions can contribute in their different ways to this field of activity.

1. The imaging mission with its 65°* great circle arc coverage from the intersection of the equator and the prime meridian and its image frequency of 48 per day permits a near real time monitoring of the exceptional weather phenomena which can have disasterous consequences.

2. The dissemination mission where the received images are processed and sectorized and then retransmitted at high power by the satellite so as to permit reception by authorities and other users of images relevant to their areas. These images can be received by rather simple and therefore cheap stations thus permitting virtually all authorities in the Meteosat coverage zone to set up image stations for the monitoring weather phenomena.

3. The data collection mission which provides for the relay of conventional environmental measurements via Meteosat to the data user. This service is suitable for relatively low data quantities for locations within about 80° great circle arc of the subsatellite point. This covers about 40% of the earth's surface.

The operation of these three missions can provide a large capability for the detection of natural disasters given the appropriate organisational implementation throughout the coverage area.

2. IMAGING AND IMAGE BROADCASTING

2.1 General

The main payload of the Meteosat satellite is an imaging radiometer which has the capability of generating earth images in three spectral bands at intervals of half an hour for the full earth disc or more frequently in the case of smaller areas. Provision is also made for use of the satellite as a relay, enabling transmission of sectors of the full image to relatively inexpensive user stations.

The three spectral channels include the Visible channel, of which an example is shown in Fig. 2.1, and two channels in the Infra-Red part of the spectrum. The first of these has a response centred around 11 microns, and is generally known as the IR or "window" channel (Fig. 2.2), because radiation in this part of the spectrum is absorbed only slightly by the atmosphere and consequently provides a view of the Earth's surface on a 24 hour basis, unlike the Visible channel which of course is useful only during the daylight hours. The third channel, near 6 microns, is associated with radiation from water vapour in the atmosphere - this is of great interest to meteorologists and other atmospheric physicists but as this channel does not provide a view of the surface it is of limited usefulness in the context of disaster monitoring. However, the first two channels, providing earth views at intervals of half an hour, have a capability of observing events larger than the resolution of the individual picture elements. This pixel size is a square of size 2.4 km at the subsatellite point in the case of the visible channel and about 4.8 km for the IR window channel. With this resolution it is clear that the images cannot be used to detect very localised events, but there are disasters for which this information is of use, particularly when coupled with the satellite's capability for relaying this data to local user stations.

---

* The image actually covers the full earth disc its potential for detection of natural disasters is limited to about the 65° arc.

Fig. 2.1. Full disc image from Meteosat-2 in the Visible Channel. Weather systems are clearly delineated by the cloud patterns. Surface features such as rivers, mountains, lakes, variations in vegetation, are also visible, particular when computer enlargements are made.

2.2 Observable Events

The disaster associated phenomena which can be observed by the Meteosat radiometer are mainly, but not entirely, meteorological in nature. In many cases it is not possible to see the final impact of such phenomena by only that, for instance, severe weather is expected or in progress. For example the images can be used to track hurricanes and to estimate their intensity but cannot directly see the resultant damage (except in the case of widespread flooding). However the potential scale of damage can be inferred in many instances. The events which can be observed include the following:

Fig. 2.2. Full disc image from Meteosat-2 in the IR 11 micro "window" channel used to monitor surface conditions and weather systems 24 hours each day.

Hurricanes. These are defined as tropical storms with sustained wind speeds in excess of 64 kt. The wind speeds close to the central core can be much in excess of this value and their potential for devastation of coastal areas is well known.

Tropical storms can be observed readily in Meteosat Images, and can be easily tracked from image to image. This enables forecasts to be made of their progress, and their characteristic cloud patterns shown in Meteosat imagery also enables estimation of their violence and potential for destruction. Three areas within the Meteosat field of view suffer from tropical cyclones of this type - the Eastern seaboard of the U.S.A. and the Caribbean, the Arabian Sea, and the Southwest Indian Ocean.

Most awareness of such cyclones is related to those affecting the densely populated areas of the U.S.A. These have their origin near the West coast of Africa and hence Meteosat can be used (Fig. 2.3) to provide the earliest warning of potential danger which is then monitored by U.S.A. satellites and aircrafts. Less publicised are those affecting S. Arabia and Madagascar, where Meteosat images are possibly the only source of information.

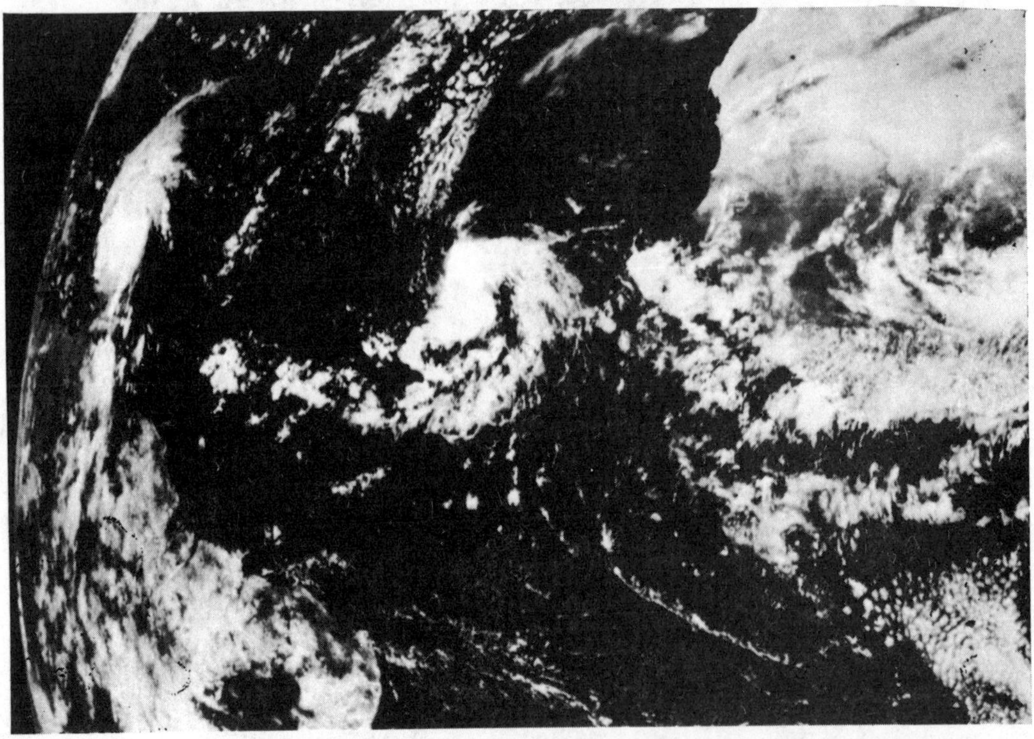

Fig. 2.3. Hurricanes Davis and Frederick over the Atlantic, 29 August 1979

Figs. 2.4 and 2.5 show, respectively, examples of tropical storms detected on Meteosat images in those two areas. Early availability of images provides the necessary warning to ensure that all possible precautions are taken before the arrival of the storm. Use of the images by Aid Centres in areas remote from the storm can estimate the areas of potential maximum damage, and in the case of widespread flooding can easily identify the affected areas after the storm has passed. This must help in the direction of aid relief to areas with shattered communications systems.

Extra-Tropical depressions. The latitude belts between 40 and 60 degrees are characterised by the mid latitude depressions which all too frequently produce poor weather near Western Sea boards in these latitudes. Most are not associated with natural disasters, just poor weather, but a very few can be classified in that way. Typically such depressions are very vigorous disturbances which grow very rapidly and move quickly. Their rapid rate of growth means that they can easily slip through the networks of conventional observing stations, especially in their early stages.

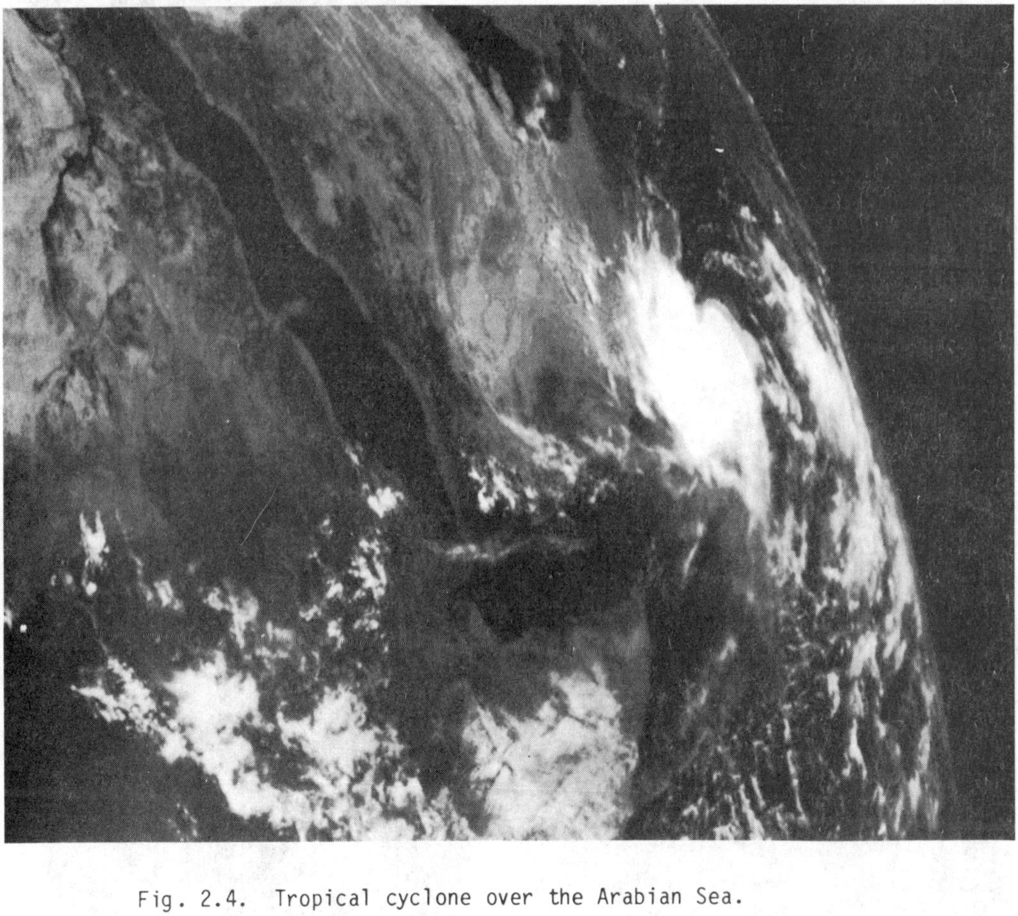

Fig. 2.4. Tropical cyclone over the Arabian Sea.

Alternatively the conventional data hints at their existence but cannot adequately monitor their growth and speed. The half-hourly images from Meteosat avoid this - disturbances can be monitored much more easily thus improving forecasts. Because of the rarity of such events the potential use is not quite so obvious as in the case of tropical cyclones, but one may nevertheless quote instances which must be classed as exceptional weather and with a potential for being classified as disasters. Examples include exceptional storms at sea (for instance the Fastnet storm of August 1979) and storms causing exceptionally heavy rain in North West Europe with resultant damage to bridges and associated wind damage to property.

Drought. The Meteosat images can also be used to infer drought situations, which on a large scale are classified as disasters. Areas of marginal agricultural activity

Fig. 2.5. Tropical cyclone in the Southwest Indian Ocean centred north-east of Mauritius.

can be monitored for the presence of rain bearing clouds and their absence over long periods is a clear precursor of possible drought and famine. Furthermore the appearance of the surface of the Visible image actually changes in these areas according to the vegetation cover. The infra-red image also plays a part, since this measures the temperature of the surface, and the diurnal rate of temperature change can be linked to soil moisture content - helping to delineate drought stricken regions. On a smaller scale the area of lakes can be monitored - large changes being a danger signal as regards irrigation projects, failure of which can also lead to at least local famine.

Flooding. Conversely, Meteosat images can also be used to detect extensive flooding, or even change from merely marshy conditions to full water cover, near coasts after cyclones or inland after long periods of rain. The potential for flooding after severe local storms in mountainous regions and following thaw of snow covered areas can also be assessed and monitored.

2.3 Distribution

The image data of course cannot be utilised unless made available to those concerned. In the case of Meteosat this is achieved by the provision of a comprehensive and rapid dissemination system. Raw images received at the central ground station (European Space Operations Centre - ESOC) in Darmstadt, Federal Republic of Germany are pre-processed in real time and then sent back via the satellite to user stations. Thus National Authorities and Aid Centres throughout the coverage zone of Meteosat can get rapid, cheap, access to these images. The broadcasting of these data is made according to a routine schedule whereby some 400 image sectors are transmitted daily. Schedules can be obtained, at present free of charge, from ESOC and the user can acquire the data in either digital or analogue formats. The former lends itself to computer processing of the data, for example in the present context to monitor the extent of arid conditions automatically or for the implementation of algorithms for rainfall prediction. The analogue formats are better suited to the reproduction of the image data, either as hard copy or on a television screen, for more qualitative applications.

At the time of writing there were around a dozen of the more expensive digital receiving stations in operation, but the cheap analogue devices were in widespread use throughout the Meteosat field of view at more than 200 locations (Fig. 2.6). Thus the capability for disaster monitoring by use of the image data exists and hopefully more use will be made of this existing facility to help ameliorate the effects of the described phenomena.

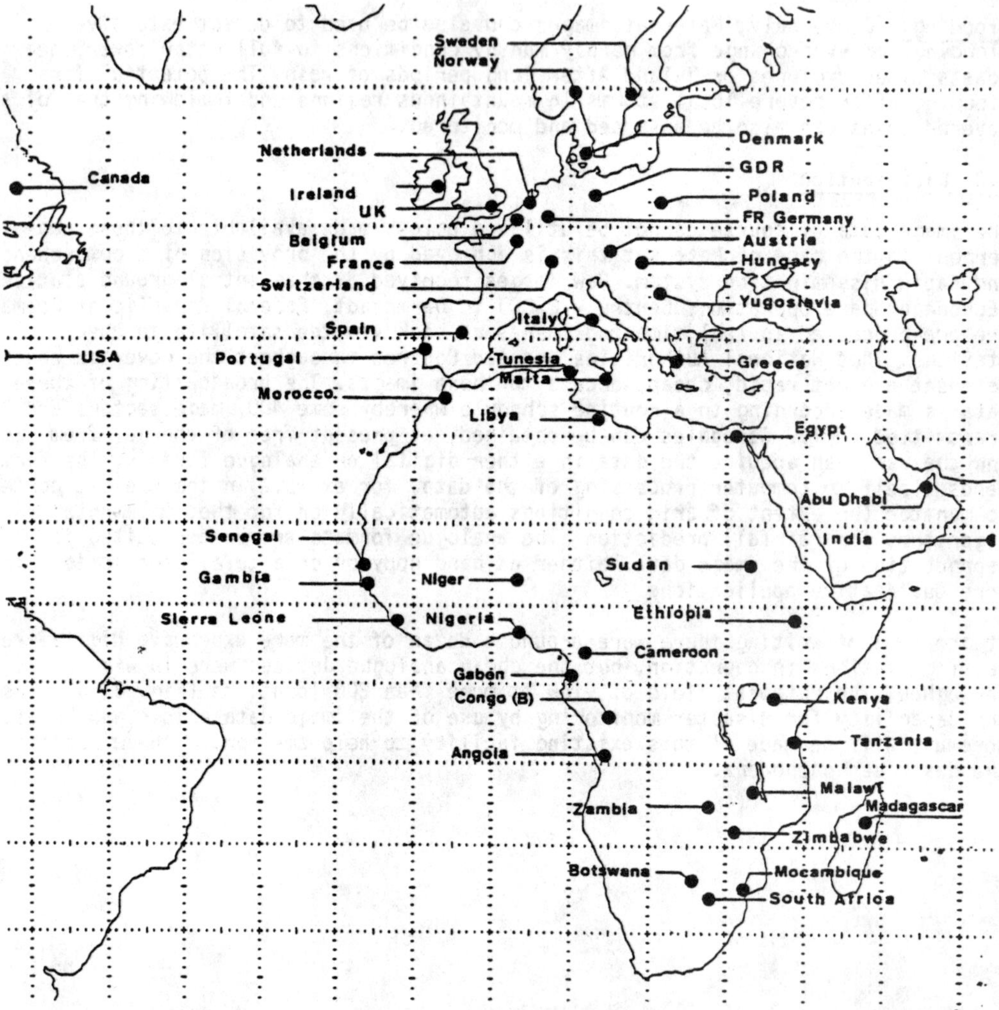

Fig. 2.6. Distribution of known user stations equipped to receive analogue transmissions (December 1980).

3. DATA COLLECTION

3.1 The Data Collection System (see fig. 3.1)

Because Meteosat is a geostationary spacecraft, it provides a link that is permanently available to all compatible data platforms within its field of view (about 80˚ of great circle from subsatellite point). The use of this link is restricted for data related to the earth and its natural phenomena (ITU regulations). The data link is generated by small UHF transmitters, which when connected to environmental sensors are known as data collection platforms (DCP).

There are three basic types of DCP:

- 'self-timed', which transmit their data at regular intervals based on an internal clock;
- 'interrogated', which transmit their data upon reception of a request signal from the satellite;
- 'alert', which transmit either a small amount of data or request for interrogation when a particular parameter has been exceeded.

The DCPs transmit to Meteosat on any one of sixty-six 3 kHz reporting channels in the 402 MHz band. After conversion to the 1675 MHz band, the satellite retransmits the messages to its central ground station, in the Odenwald in Germany. There, the signals received are checked for quality before being transmitted to ESOC in Darmstadt for processing and distribution.

At ESOC, messages from self-timed and interrogated DCPs are collected, processed and disseminated every 30 minutes. Any message received from an alert DCP is processed immediately and made available for transmission to the platform's owners.

To interrogate a DCP, an interrogation message is generated in Darmstadt and sent to the Odenwald ground station. From there it is transmitted in the 2122 MHz band to Meteosat, which then retransmits the message on one of two possible frequencies in the 468 MHz band. The message includes the unique address of the DCP to be interrogated and only that particular platform will respond by transmitting the data it has collected.

Currently, the platform data received in ESOC are provided to the platform owners and others who wish to receive the data, either via the World Meteorological Organisation's Global Telecommunications System (GTS), by telex, or on magnetic tapes for non-real-time applications.

In general, this data distribution service is slower than the needs of many users, especially where critical conditions can exist. Up to the moment, the only possibility for a user to receive data in real-time was to build a receiving station capable of acquiring the data directly from the satellite. The cost of such a station is however normally prohibitive for the majority of users. For this reason it has been decided to relay the data via the satellite's high power transmitters being used for the image broadcasting service. The technique will be to receive the data messages at the Odenwald station and store them until the next break in the analogue image transmission schedule whereupon the messages will be transmitted to the satellite for broadcasting. These breaks occur every 4 minutes and last for 23 seconds. The data transmission rate will be about 7 kbs which should allow 20 or more messages to be transmitted every 4 minutes.

For the users to receive their messages, they will have to procure a ground station

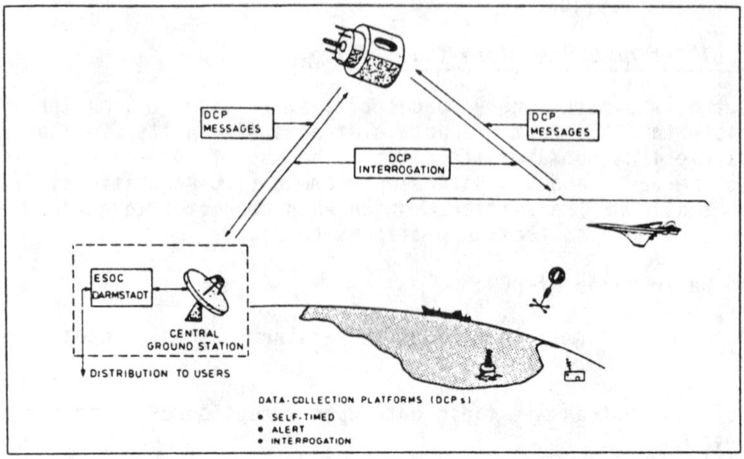

Fig. 3.1. The Data Collection System

based upon the cheap analogue image reception station. The station will require a 2 m diameter antenna with preamplifier and downconverter, a conventional receiver with F.M. demodulator and a microprocessor for extracting the user's own data messages for display or printing.
The cost of such a station could be under $ 20,000.

3.2 Data Collection Platforms

Although the system has been designed for the three types of platform, interrogation, self-timed and alert, there has been a reluctance by users to procure interrogation platforms since their complication makes them rather expensive hence the following refers to self-timed and alert platforms (they can be combined into a single platform).

The DCP consists of two main elements

1. sensors: a wide range is available for many aspects of environmental monitoring

2. radio sets: UHF transmitter and antenna, timing circuits, data storage.

The radio sets have to be certified by ESA before being allowed to use the system but many manufacturers now offer these certified designs in their equipment range. The radio sets come in various shapes and sizes, but generally they are no more than 50 x 40 x 20 cm^3 and weigh less than 20 kg, excluding sensors and antenna. The size of the antenna depends upon the transmitter output power needed from the DCP; with a 5 W transmitter, a 1.5 m long helical antenna suffices.

D.C. Power requirements are modest and can be satisfied by a variety of different energy sources, e.g. rechargeable lead-acid batteries, solar cell power generators etc. The cost of a radio set is around $ 5,000 and to this, one must add the price of sensors before the DCP is complete.

3.3 Real-Time Data Collection for Water Management

Floods are now causing more damage to life and property than ever before and as development continues within the flood plain, this trend will increase. Numerous fatalities are recorded regularly and damage amounts to many millions of dollars annually. Whilst most of this damage is to agricultural land and to domestic or industrial premises, dam failure alone caused by flooding can lead to many thousands of deaths as well as to costly replacement of water or hydroelectric services. Flood damage is not restricted to the large rivers of the world and serious damage can be caused by even the smallest of rivers. The April 1981 flood in a small area of the U.K. for example caused damage to agricultural land alone amounting to some 3,000,000 pounds sterling. Nevertheless the most catastrophic floods affecting both large areas of population and agricultural land occur in the basins of the world's largest rivers such as the Yangtze, the Yellow River, the Ganges and the Mississippi. In the July 1981 Monsoon floods in India at least 60 people were drowned and some 3 million people affected. Again in July 1981 the Yangtze suffered its highest flood this century when some 4,000 people were killed, some half a million made homeless and a total of at least 2 million people affected.

A reliable flood warning system may prevent or at least reduce loss of life and damage. A warning received too late however is useless and warnings given too often when no flooding occurs have a decreasing value. Flood forecasts must therefore be based on a reliable maximum rainfall value or maximum river level value or both and the transmission of these data to the flood forecasting centre is crucial. In small catchments the time between the rainstorm and the flood peak is short (1 to 2 hours) so forecasts are not usually practical and an alarm is necessary. In large catchments there is normally sufficient time to collect the necessary rainfall and river level or flow data to prepare a forecast and issue a warning based on this forecast. No matter what method is used in preparing and issuing this forecast, the initial 'alert' or 'emergency' warning is crucial to the system. It is in this area that satellite transmission offers a reliable cost effective alternative to terrestrial telemetry transmission.

The use of geostationary satellites for hydrological data collection is currently under investigation, particularly in the United States, Canada and the United Kingdom. In the United States and Canada the satellite being used is GOES and in the United Kingdom METEOSAT. These investigations are concerned with the possible future implementation of a satellite based data collection system to replace existing systems which rely on a semi-automatic computer system. The existing systems involve considerable manual and transportation costs in routine data collection on a monthly basis.

The Data Collection Platforms used are therefore self-timed, for the general collection of historic data, but with an alert facility for real-time data. Interrogable DCPs are also under consideration.

Over 300 self-timed hydrological DCPs are presently being operated in the U.S. and Canada and about 10 DCPs with alert facility are currently operated in the U.K.

All 3 investigations have shown that the Data Collection Platform is a low cost reliable device and that Satellite Data Collection System offers a decrease in present operational costs, reduced manpower, less travel to sites and improved data capture.

The system is particularly reliable in transmitting alarm or emergency conditions. In the U.K. study alarm levels are set at both flood danger levels and drought emergency levels for raingauges and river water level recorders.

The present methods of flood warning generally in use throughout the world rely on either manual or telemetry means. Clearly the former is unsatisfactory and will not be discussed here. In the case of telemetry the 2 systems in common use rely on land-line (telephone) or line of sight radio communication. Land-lines are prone to damage during floods and are thus unreliable just when the data are required and because rivers are located in valleys, line of sight radio requires expensive repeater towers or relay stations. Probably the most important disadvantages of land-line or line of sight radio systems however are the high installation and maintenance costs and hardware redundancy all of which increase with time. Furthermore such systems usually are designed as local independent systems having distinct disadvantage in national coverage leading to additional hardware and maintenance costs.

Satellite systems, providing launching and commissioning has been successful, do not suffer from any of these disadvantages and because of the large potential international market are expected to become more cost effective with time. Furthermore such systems are virtually free of interference and in the U.K. investigation the only minor problem encountered was the vandalising of an antenna at one station which is located in the heart of a city. This problem was solved by placing the antenna inside the hut and no further trouble has been encountered.

On the other hand one DCP is situated in the open on a remote mountain slope in the Scottish Highlands and this DCP has been in trouble free operation since installation 2 years ago.

DCPs are able to monitor certain station faults and detect malfunctions which are then dealt with on a call-out basis rather than on a routine visit as with the existing data collection system.

The alert facility on the United Kingdom DCPs has proved reliable in use and although as yet not an operational system on one occasion actually transmitted the emergency signal missed by a line of sight radio system.

There is little doubt that based on the United States, Canadian and United Kingdom studies that governments considering installing flood warning or flood forecasting systems should give serious consideration to a satellite system. It offers reliability even under extreme climatic conditions and, provided a geostationary satellite is available, appears to offer the most cost effective system.

3.4 Real-Time Data Collection for Application in Earth Movement

Data Collection Platforms for transmitting surface data to an orbiting satellite for relaying to a central data center are of high potential use in studying Earth crustal movements of endogene as well as of exogene origin. The first category covers all kinds of movements connected to the dynamics of plate tectonics, and to the distribution of strain and its propagation within lithospheric plates, including earthquakes as well as volcanic eruptions. The second category comprises landslides, Earth deformations due to tidal forces as well as crustal loading and unloading caused by marine tides, by storm surges, by seasonal sea level changes, by regional and local air pressure variations, and by seasonal changes in the global air pressure distribution.

On the one hand such phenomena may be studied for purely scientific reasons, on the other hand the investigation of these phenomena has applications in hazard prediction for reducing losses due to major natural catastrophies such as earthquakes, volcanic eruptions, storm surges and landslides. In both cases the crustal movements are monitored continuously or at least quasi-continuously. This in general implies that the measuring equipments have to be visited once or twice a week to replace the chart paper or just to assure that the instruments are performing as planned. This is extremely time consuming, especially if data are to be collected from a dense network of geophysical instruments such as necessary for instance for an operational earthquake prediction system. Data transmission via phonelines is very often not possible because the geodynamic measurements are usually carried out far away from any industrial noise, and sometimes in extremely unaccessible, remote and underdeveloped regions where phonelines are practically non-existent. In these cases data transmission via satellite has its greatest value. It provides the means

1. for real-time acquisition of geodynamic data from geophysical networks in isolated regions where in general no other data retrieval method is feasible, and

2. for real-time but still complex interpretation of such data in a modernly equipped Geodynamics Center far away from the region of measurements.

This in turn is

3. the basis for immediate decisions necessary before a threatening disaster such as an earthquake, a volcano eruption or a storm surge. Warning times in these cases may be not much more than several hours.

Satellite relaying of Earth movement data has scarcely been practised so far, mainly because it is only recently that the cost for a satellite system has become competitive with ground systems such as phonelines and radio links. However Ward et. al. already in 1974 report about the USGS prototype volcano surveillance system on 15 volcanoes in Alaska, Hawaii, the contiguous United States, Central America and Iceland. Information on the number of earthquakes per day as well as ground tilt in the neighbourhood of the monitored volcanoes was transmitted in these cases via Landsat. Allenby et. al. (1977) report about successful transmissions of tiltmeter data from the Lesser Antillian Arc to Goddard/U.S.A.

In the following we will briefly describe two Kiel projects where satellite relaying of ground motion data is already performed or will be carried out in the near future, respectively. One project is concerned with storm surge prediction in the German Bay, and the other with earthquake prediction research in Turkey.

1. Storm Surge Prediction in the German Bay

It is well-known that the North Sea coast of Northern Germany, especially the German Bay at the mouth of the river Elbe, suffers from heavy storm surges which sometimes are catastrophical. The most recent storm surge disaster occurred in January 1976, and hence is only five years old. The Geophysical Institute at the University of Kiel/Germany has therefore started a project to investigate whether these storm surges can be forecasted from the crustal tilt induced by the change in the water load distribution during and before the surges. It was found that even 70 km inland from the sea, the Earth's surface is tilting by some tens of milliseconds of arc, when a heavy North Sea surge approaches the coast. This is a several hundred times the accuracy of the borehole-tiltmeters used for these investigations. Furthermore it was observed that the measured crustal tilts preceeded the waterlevel changes in the German Bay by up to 12 hours in advance (Zschau, 1977, 1979, Zschau et. al. 1979, Zschau and Kümpel, 1979, Kümpel and Zschau, 1979a, 1979b), suggesting that the development of storm surges in the German Bay can be predicted from tiltmeter readings. The Geophysical Institute in Kiel has therefore installed three such borehole-tiltmeters in this region (see figure 3.2). Two of these tiltmeters are equipped with a data collection platform and regularly send data via Meteosat. The third tiltmeter will be equipped with the same kind of satellite system soon. The DCP's collect tiltmeter data every two minutes and relay the data during a short time slot once every hour. The data are then received at ESOC in Darmstadt and passed on to the Geophysical Institute in Kiel via Telex. Satellite relaying of these tiltmeter data is working successfully since several months. Within the next three months a minicomputer will be installed which will automatically record the data coming from Darmstadt, and to carry out necessary real-time analysis. The satellite system has so far proved to be extremely worthwhile for the regular control of the tiltmeter stations each of which is more than 50 km apart from the others. It is expected that with this kind of satellite relaying the tiltmeter data will effectively contribute to accurate real-time storm surge predictions in Northern Germany.

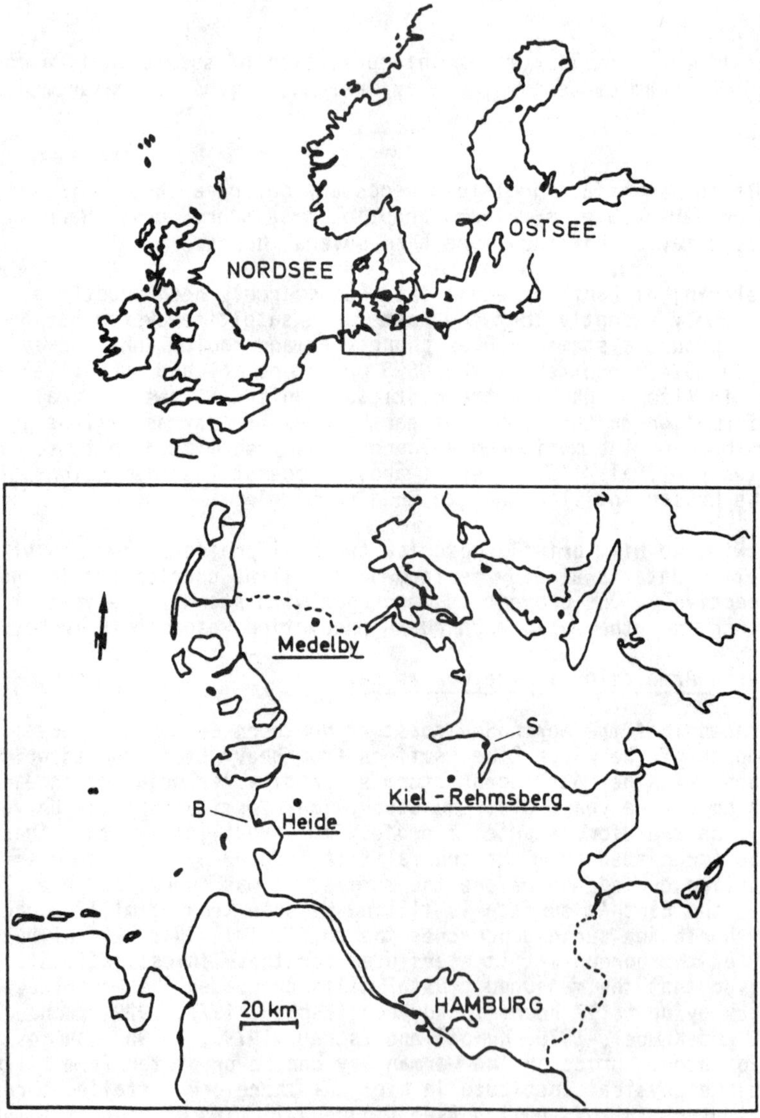

Fig. 3.2. Tiltmeter sites for storm surge prediction: Kiel-Rehmsberg, Heide and Medelby. B and S are reference tide gauges.

2. Earthquake Prediction Research in Turkey

The Geophysical Institute at the University of Kiel together with the Geophysical Institute at the University of Frankfurt and the Institute of Theoretical Geodesy at Bonn University in close collaboration with the University of Istanbul have prepared to start a long-term project on earthquake prediction research at the western part of the North Anatolian Fault Zone in Turkey. In a first step five experiments are planned, including the monitoring of seismicity in that area, the regular control of the seismic propagation velocity, its anisotropy and absorption along a fixed line crossing the fault zone, crustal tilt measurements at both sides at the fault, high precision distance and -angle measurements, and waterlevel monitoring in wells. The geodynamic network will be set up in remote area. Therefore satellite relaying of measured data via Meteosat to a center in Germany is envisaged. Most of these experiments will provide a data density well suitable for the present capability of Meteosat, while the amount of data from the seismic experiments will have to be reduced and selected in order to stay below the capacity threshold of Meteosat. Considerations concerning the data reduction and -compression are presently carried out in the above-mentioned institutes. The project is supposed to begin in spring 1982.

4. CONCLUSIONS

From the text it can be seen that the potential of Meteosat to aid the detection of natural disasters over a large area of the earth's surface is very high. The basic information can come from images or conventional instrumentation making measurements and relaying them via Meteosat. The value of disaster detection, or better still, prediction depends upon the information being presented within a time span to allow alleviatory measures to be implemented and this is a telecommunications function. Meteosat has been designed with this function in mind, and particularly to permit operation with cheap user stations. Up to the present this has been confined to image transmissions but within the next year, it should be possible to receive data collection data in near real-time from the satellite and therefore give users an almost immediate access to information originating in remote areas. With this possibility, the responsible authorities and agencies will be given another tool to help ameliorate the effects of natural disasters.

REFERENCES

To 3.2:

Halliday, R. A., 1978, The use of satellites in hydrometry: Principles and Practices. John Wiley, 427-447.

Herschy, R. W., 1980, Hydrological data collection by satellite. Proc. ICE. Part 1, Nov.

Shope, W. G., and Paulson, R. W., 1980, Real-time data collection via satellite for water management. Proc. ASCE, Oct. 80-561.

To 3.3:

Allenby, R. J., Webster, W. J., and Painter, J. Satellite Relaying of Geophysical Data. Goddard Space Flight Center, Document X-922-77-273, Greenbelt, 1977.

Ward, P. L., Endo, E. T., Harlow, D. H., Allen, R., Marquez, D., and Eaton, J. P. Development and Evaluation of a Prototype Global Volcano Surveillance System Utilizing the ERTS-1 Satellite Data Collection System. Final U.S.G.S., Rpt. for NASA Contract S-70243-AG-Z, Feb. 1974.

Kümpel, H. J., and Zschau, J. Das Gezeiten-Vertikalpendel in der Sturmflutvorhersage. Wasser und Boden 10, 284-287, 1979a.

Kümpel, H. J., and Zschau, J. Crustal Tilt Induced by Storm Surges. DGK, Reihe B, Nr. 231, 95-106, 1979b.

Zschau, J. Sturmflutvorhersage mit Vertikalpendeln. Christiana Albertina, University Kiel, No. 7, 5-17, 1977.

Zschau, J. Prediction of Storm Surges from Marine Loading Tilt Mesured Inland from the Sea. Proceed. 8th Int. Symp. on Earth Tides held at Bonn 1977, 787-804, 1979.

Zschau, J., and Kümpel, H. J. Prediction of Storm Surges Using Vertical Penduluus. Geophys. Astrophys. Fluid Dynamics, Vol. 13, 245-252, 1979.

Zschau, J., Kümpel, H. J., Meißner, R., and Carow, U. Eine neue geophysikalische Methode zur Vorhersage von Sturmfluten. Die Küste, No. 34, 71-78, 1979.

IAA 81-271

A NEW SHORE-TO-SHIP-ALERT CHANNEL VIA MARITIME SATELLITES (SHOSAC)

W. Goebel*

In order to complete search and rescue communications by a shore-to-ship alert channel for contacting ships in the vicinity of a vessel in distress, a very low G/T ship earth station (G/T = -29 dBK) to be operated in a receive-only mode is described.

Both coherent and non-coherent modulation schemes and the respective implication on acquisition and tracking are discussed. Typical bit rates and transmission times have been estimated. Multipath effects can be largely compensated by use of channel encoding. The addressing to selected ocean areas can easily be achieved by an area code of 14 bits calling one or more out of 3000 square areas in total.

Keywords: Distress radio communication; alerting; area code; selective broadcast; forward error correcting code; navigational warnings; INMARSAT.

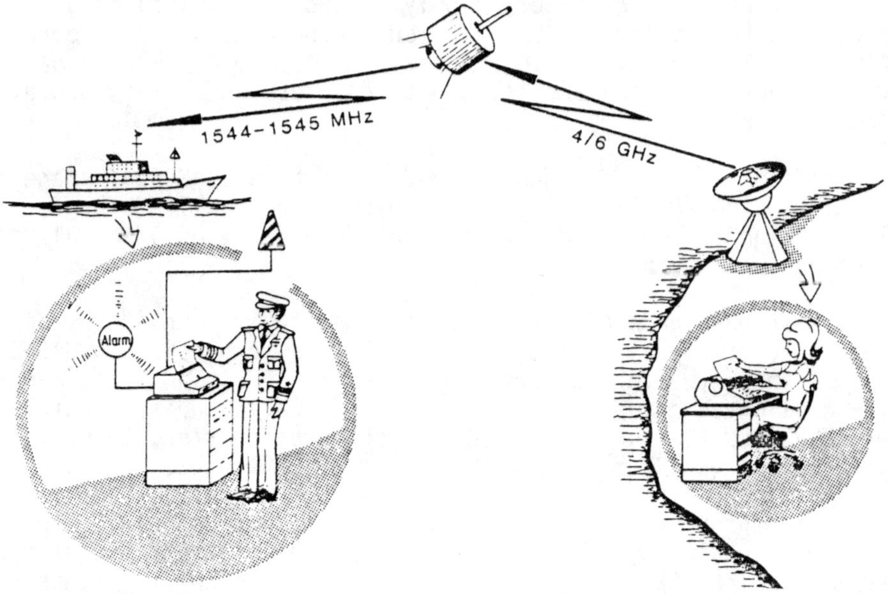

The Shore - to - Ship Alert Channel (SHOSAC)

* *DFVLR-German Aerospace Research Establishment, Institute for Communications Technology, D- 8031 Oberpfaffenhofen, West Germany*

1. INTRODUCTION

Today there exist a number of proposals (1) for maritime emergency and distress alerting systems using float free Emergency Position-Indicating Radio Beacons which transmit via satellite (Satellite EPIRB). As a necessary complement to these ship-to-shore alerting systems, IMCO requested for its "Future Global Maritime Distress and Safety Systems" that the shore based Rescue Co-ordination Centre (RCC) should be enabled to quickly and effectively contact ships in the immediate vicinity of the vessel in distress. This ship-to-shore communication, however, is the weak link in the maritime SAR communications, especially when the ships are not equipped with INMARSAT A standard ship earth station.

A simple and reliable radio link, a Shore-to-Ship Alert Channel (SHOSAC) is lacking to date.

Therefore, it is being proposed to connect even those vessels which do not posses a ship earth station by employing an inexpensive receiver and operating at low bit rates via satellite. Ships could be addressed individually or selectively in certain areas. The link could be used for distress alerting as well as for navigational and meteorological warnings.

2. SYSTEM DEFINITION

The system must be inexpensive to own and operate; therefore, it can only have an omnidirectional antenna. In order to overcome sea surface multipath effects, the polarization should be circular. Since this service is in the shore-to-ship direction only, the ship terminal is a Receive-Only Ship Earth Station (ROSES). It should be used in the frequency band 1544 to 1545 MHz which is allocated exclusively to the Mobile Satellite Service (space-to-Earth) and limited to distress and safety operations by footnote 728.

The operational satellites MARISAT and first phase INMARSAT space segments INTELSAT/MCS and MARECS do not have implemented this 1-MHz-band. The implementation of this band into the second generation INMARSAT space segment has been proposed.

The small antenna gain of ROSES, in conjunction with an inexpensive receiver and a limited satellite power, allow a very narrowband SHOSAC-transmission only, i.e., a very low bit rate. Assuming that a SHOSAC would be allotted the same effective isotropically radiated satellite power (EIRP) as the public correspondence voice carrier, then this value would be about 18 dBW for both MARECS and INTELSAT/MCS.

One predominant problem is the signal acquisition, especially for phase shift keyed modulation (PSK), as opposed to non-coherent frequency shift keyed modulation (FSK), which is 4,0 dB less efficient. In this context, the accuracy and stability of the carrier frequency radiated from the spacecraft is of importance. As an example, the following frequency inaccuracy values of the MARISAT system may be considered (2):

	Frequency Error	Compensation by Automatic Frequency Control
Satellite Translation	25 565 Hz	Yes
Shore Station up-converters	630 Hz	Yes
Shore-Satellite Doppler (inclination 5°)	420 Hz	Yes
Ship Motion Doppler (30 knots)	82 Hz	No } 551 Hz
Ship-Satellite Doppler (inclination 5°)	219 Hz	No } 551 Hz
Ship down-converters	250 Hz	No } 551 Hz

After automatic compensation of the first three contributions in the coast earth station, there remains $\pm$ 551 Hz or 1102 Hz peak-to-peak inaccuracy. In the following, a frequency inaccuracy of 2 kHz has been assumed, to allow for residual compensation errors.

3. DISCUSSION OF SUITABLE MODULATION AND CODING TECHNIQUES

Unlike the link between a buoy and the satellite, with its substantial path blockage and multipath effects, the link between a ROSES antenna, mounted on a high point of the vessel, and the satellite, does not experience any path blockage and only reduces multipath effects. This could allow the use of coherent modulation.

Signal fades can be largely compensated by using forward error correcting codes (FEC). For a data frame of 1023 bits, e.g., approximately half of which are parity bits (520), 58 errors per frame can be detected and corrected (which amounts to a bit error rate of 11.5 percent that can be entirely corrected).

FSK, on the other hand, does not require continuous carrier acquisition; signal fades can again be largely compensated by using FEC.

Methods such as time diversity, i.e., superpositioning of several frames of identical content for purposes of improved carrier-to-noise ratio, do not seem to be necessary for SHOSAC, because they are applicable for more severe propagation environments and which would need longer transmission times.

4. LINK BUDGET

The following link budget is based on INTELSAT/MCS and MARECS figures. It is assumed that SHOSAC EIRP allocation is that of a voice carrier. The multipath fade assumed is 6 dB; the FEC coding gain employed here also to 6 dB.

		MARECS/INTELSAT MCS
S/C EIRP (5° elevation)	dBW	18,0
Space Loss (1544 MHz; 5° elev.)	dB	- 188,5
Antenna Gain; Ship Terminal	dBi	1
System Temperature (1000 K)	dBK	- 30
Boltzmann's Constant	dBW/HzK	228,6
Multipath Fade	dB	- 6
C/N_o-down path	dB	23,1

Calculation of the bitrate:

		PSK	FSK
C/N_o-downpath	dB-Hz	23,1	23,1
Required E_b/N_o (BER 10^{-4})	dB	- 8,3	- 12,3
FEC improvement (approx.) (Code Rate = 0,5)	dB	6	6
Modem Losses	dB	- 2	- 2
	dB-Hz	18,8	14,8
Informations Bitrate (32 & 16 B/s)		- 15,1	- 12,1
Margin	dB	3,7	2,7
Transmitted Bitrate	B/s	64	32
Transmission Time for 500 Informations Bits	s	15,6	31,2

The FEC coding gain assumed in the budget is a coarse approximation and needs further refinement.

Problems arising from acquisition and tracking, have not been considered in this link budget and shall be discussed in the following section.

5. ACQUISITION AND TRACKING

There are four alternative proposals being discussed, beginning with a completely coherent system and leading to a totally incoherent system. To avoid mutual intra-system interference between SHOSAC's in the overlap regions between two adjacent satellites, a 5 kHz carrier frequency offset is being proposed for adjacent satellites. A secondary intermediate frequency (IF) of 200 kHz is being proposed, preliminarily as a standard for all ocean regions, for which cost effective filters and voltage controlled oscillators (VCO's) with bandwidths of a few kilohertz are readily available commercially. Furthermore, it is being assumed that at the beginning of each transmission the carrier remains unmodulated for a period of time, thus allowing the ROSES to acquire lock.

5.1 BPSK with PLL for Acquisition and Demodulation (Fig. 1a)

Binary PSK (BPSK) using phase lock loop (PLL) has been based on a carrier-to-noise density ratio (C/N_o) of 23.1 dB-Hz, corresponding to the MARECS and INTELSAT/MCS parameters and taking multipath effects into consideration. On that basis a PLL of second order has a loop bandwidth of 25 Hz and a maximum sweep rate of about 200 Hz/s. Sweeping in the worst case, over the entire range of 2 kHz as derived above, would take some 10 seconds. Tracking can be accomplished using the same PLL even for carrier power reduction (phase deviations of 45° reduce the carrier power by 3dB) because fades caused by multipath effects can be compensated for using the fly-wheel technique so that these types of fades should not be subtraced from the C/N_o in total.

5.2 BPSK with Aided PLL for Acquisition, Using a Beacon Receiver (Fig. 1b)

If a satellite beacon is available, ROSES could retain lock on it permanently. It would be irrelevant how long the initial acquisition would take; an addition sweep phase of the SHOSAC signal would be avoided completely. Present planning, however, does not foresee satellite beacons in future INMARSAT space segments; the possible use of INMARSAT's AFC pilot frequencies needs further study. If so, these beacons could be, for example, 25 kHz above the corresponding SHOSAC-signal. All other system parameters in this case are identical to case 5.1. Another approach would be to continue carrier transmission after message stop until the next message begins. This would reduce the complexity to one PLL identical to 5.1 but without losing lock.

5.3 BPSK with Noncoherent Acquisition and Coherent Demodulation (Fig. 1c)

Using a bank of narrow bandpass filters with the constant bandwidths of, say, 1.2 times the bit rate, an envelope detection could be performed with the outputs of the filters, so that, after integration with integration time of about 1 s for rise and 10 s for decay, DC levels could be achieved per filter channel. These DC levels could be scanned and fed into a comparator. In case of levels being higher than the noise floor, which means an incidental signal identification, the PLL could be preset, thus obtaining lock. All other system parameters in this case are identical to case 5.1. The tertiary down conversion to about 3 kHz prior to the filter bank is required to allow for ready implementation of the filters in the filter bank (approx. 80 Hz bandwidth).

5.4 BPSK with Non-coherent Acquisition and Non-coherent Demodulation (Fig. 1d)

Again a bank of narrow bandpass filters is proposed with a filter center frequency of some 3 kHz. Each filter has two different outputs: as described in 5.3 with the integration time quoted, and one output without the integrators, for data detection purpose. For binary FSK, always pairs of filters have to be combined according to the frequency separation chosen. The filter passband equals approximately the bit rate (approx. 32 Hz).

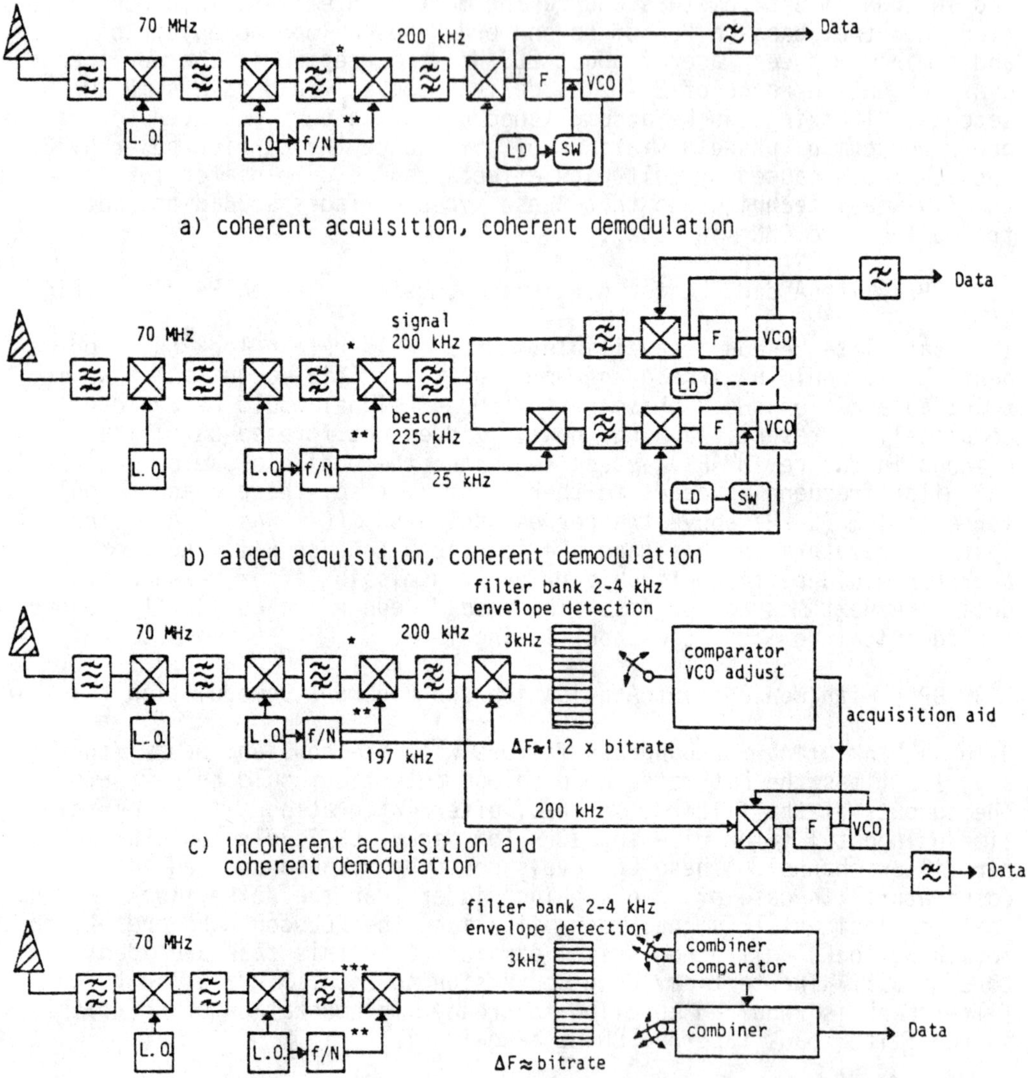

LD = Lock detector; SW = sweep; F = filter
L.O. = local oscillator;
f/N = frequency divider

	*	**	***	
AOR	195	5	8	kHz
IOR	190	10	13	kHz
POR	185	15	18	kHz

Fig. 1 Four different methods of acquisition and demodulation

5.5 Comparison and Comments

The above system alternatives can be summarized as follows:

Method	5.1.	5.2	5.3	5.4
Modulation	PSK	PSK	PSK	FSK
Acquisition time (seconds)	10	1	1	1
Information bitrate (b/s)	32	32	32	16
Transmission time for a message of 500 bits, incl. acquisition time (sec)	26	17	17	32

Methods 5.2 and 5.3 appear attractive as far as transmission time is concerned. Method 5.2 requires a more elaborate satellite design (the beacon) whereas method 5.3 requires a more elaborate receiver design (the filter bank). Method 5.4 would be ideal if it should turn out that reliable operation of phase shift keying in the presence of multipath effects is severely hampered.

6. SELECTIVE BROADCAST

It would certainly be unnecessary to broadcast an alert, intended for only a restricted area, to the entire hemisphere in view of the satellite. It would rather be appropriate to subdivide the satellite view coverages into suitable SAR areas. IMCO is presently planning for such SAR areas which will presumably be oriented along the existing ICAO-plans. In order to enable a quantitive evaluation at this time it is proposed to simply subdivide the earth into square areas about 120 nautical miles apart. Dividing the satellite view coverage of 216.6 million square km into square areas of side lengths of 120 nautical miles yields some 5000 such areas, about 3000 of which are covered by water. These 3000 areas could be sequentially numbered and the numbers coded in binary, leading to area code numbers with 12 bits each.

On the high seas a ROSES should be able to receive alerts addressing an area of

- 360 NM x 360 NM or 9 square areas normally for SAR
- 600 NM x 600 NM or 25 square areas during hurricanes, weather or ice warnings.

With the addition of two more bits, this variable receiving zone could be effected through the coast earth station. The ROSES would receive all SAR transmissions but print out only those as dictated by the coast earth station, unless programmed otherwise.

To ensure access for the various coast earth stations the 60 minutes in each hour could be, e.g., discretely allocated to the individual coast earth stations, reserving, of course, the even-numbered minutes for the pressing maritime distress alerting, in a random access scheme, with the meteorological warnings being configured during the odd-numbered minutes in every hour.

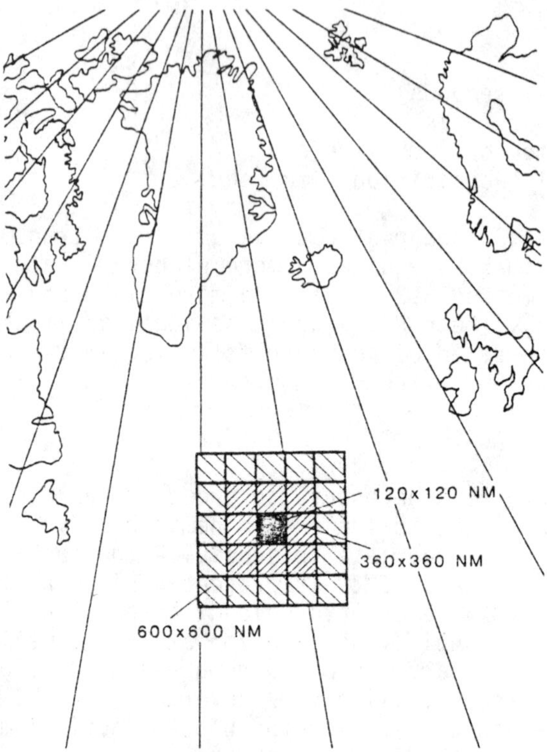

Area Call to Selected Ocean Areas with Extension Option

REFERENCES

(1) W. Goebel, G. Landauer, E. Messerschmidt. "Maritime Distress Radio Call Systems Employing Space-borne Technology", XXXI IAF Congress, Tokyo, September, 1980.

(2) D. Lipke, D. Swearingen, J. Parker, E. Steinbrecher, T. Calvit, H. Dodel. MARISAT - A Maritime Satellite Communication System, Comsat, (*Technical Review*, Vol. 7/No. 2, Fall 1977)

IAA 81-262

SARSAT: A SATELLITE-AIDED SEARCH AND RESCUE SYSTEM FOR LOCATION OF DISTRESS RADIO BEACONS*

D. Ludwig, D. Levesque, Ph. Goudy[+]

After a short description of present needs and methods for search and rescue, the paper presents the international cooperation between the United States, Canada and France on the Search and Rescue Satellite-Aided Tracking program.

The 406 system, which is more specifically described, uses low orbiting polar satellites and a network of ground stations in order to provide fast detection and location of new 406 MHz distress radio beacons.

System performances are described and in particular the mean waiting time for distress location is analysed. A study shows the reduction of this time that should be introduced by adding a specific memory for search and rescue data.

An international demonstration and evaluation phase scheduled to begin in 1982 will use the SARSAT system as well as the COSPAS system developed by USSR, the two systems being compatible.

Keywords: Search and Rescue; COSPAS/SARSAT program; Emergency Locator Transmitter (ELT); Local User Terminal (LUT); Search And Rescue Processor (SARP); Search And Rescue Memory (SARM); Demonstration and Evaluation Phase.

COMMENTS ABOUT PRESENT SEARCH AND RESCUE AIDS

Initiation of a Search and Rescue mission after occurrence of an air or maritime distress situation is dependent on a first action: alerting the responsible services.

Air traffic regulations define a strict three phase procedure: uncertainty phase, alert phase and distress phase. They can occur very quickly in the event of a controlled flight and also very slowly in the event of VFR flight without flight plan. U.S. statistics** show that the average alert delay increases from 45 minutes in the first case to 32 h 30 minutes in the second case (these figures do not concern accidents occurring in the immediate vicinity of airports).

---

* *This paper also appears in Volume 41, AAS Microfiche Series.*

+ *Centre National d'Etudes Spatiales, (ESA), 8-10, rue Mario-Nikis, 75738 Paris, Cedex 15, France*

** *A.F.R.C.C.: Summary of 1978 SAR activity.*

At sea, the same situation exists and the alert can be transmitted quickly by radio equipment or eventually by optical facilities near the coastline, but can also require extensive delays. The number of capsized vessels without the crew being able to establish radio contact is substantial. A study made in 1978 by the French Center CROSS-A (Centre Régional Opérationnel de Surveillance et de Sauvetage - Atlantique) of 9 cases of fishing boats distress occured not far from the coast, showed an alert delay varying from 10 hours to 168 hours. With regard to the same shipwrecks, the estimate of the maximum survival durations of a man at sea varied from 3 to 7 hours. The alarm being transmitted, the success of the rescue mission is largely dependent on precise localization of the distress. This is obtained rapidly in the case of controlled flights, which are in Europe and U.S.A., almost all the time carried out under radar surveillance. The localization is more hazardous for non-controlled flights due to the absence or non use of the radar transponder. Moreover, a lot of these flights are at low altitude, below the radar coverage. Here again, american statistics show average localization delays from nine hours for controlled flights to 3 days and 22 hours for VFR flights without flight plan. In France, smaller extension of the national territory, high population density and absence of large desert areas, considerably reduces these figures and, in their large majority, the searches are less than 24 hours. Cases of unfound fallen aircraft are rare.

From the maritime standpoint, the situation is somewhat different. If he is able to establish a radio contact for transmitting the alert, the navigator in distress generally has the time to transmit his reckoned position. Unfortunately, the experience gained in transatlantic races has largely shown that the reckoned position is often wrong.

Starting 1970, this situation led the authorities in the U.S.A. and Canada to impose on particular vessels and aircrafts the carrying of radio distress beacons transmitting on aeronautic distress frequencies of 121.5 and 243 MHz. In France, automatic distress radio beacons were made mandatory for aircraft in 1979. Altogether, approximately 190 000 civilian aircarfts and 2 000 boats are now equipped with such beacons in the U.S.A., Canada and France. In Norway, 600 aircrafts and 6 000 boats are equipped with the same type of beacon.

The system based on the only 121.5 and 243 MHz existing beacons unfortunately entails some negative points :

- For aircraft, the on-board installation deficiencies, the insufficient reliability of the equipment, the inadapted specifications of the automatic activation system (G. switch) and faulty knowledge of the equipment on the part of private pilots are the origin of numerous cases of non-activation. The very high rate of false alarms (95 %) and cases of non-operation on impact have considerably altered the confidence of users.

- The second point concerns detection of the distress transmission. Listening to the 121.5 MHz frequency is ensured by aircraft only on transoceanic flights, and the 243 MHz frequency is reserved for military aircrafts. The distress signals are also heard by land stations, but reception is dependent on a low power VHF transmission (about 100 MW) and the ground stations are not all equipped with automatic direction finding equipment. This yields a sometimes long and difficult localization, and a heavy workload for the SAR organizations.

In spite of these problems, the already long experience of the USA and Canada, fully demonstrates the utility of the system (Chart 1). Furthermore, a substantial effort has been made in the framework of the studies carried out by the RTCA[2] for

2 Radio Technical Commission for Aeronautics

eliminating the deficiencies of the aeronautical beacon.

	ELT* FUNCTIONED	ELT DID NOT FUNCTION
AVERAGE ELAPSED TIME FROM LAST KNOWN POSITION TO LOCATION	23 HOURS	4 DAYS, 18 HOURS
SURVIVAL RATE	33 %	23 %
AVERAGE TOTAL HOURS FLOWN TO LOCATE OBJECTIVE	19 HOURS	130 HOURS

* EMERGENCY LOCATOR TRANSMITTER

CHART 1 Statistics of the Air Force Rescue Coordination Center (USA) for 1978

The problems encountered with beacons used by boats are apparently less acute with regard to the reliability of the equipment and false alarms. However , the problem concerning detection remains and it is to be noticed that, in the framework of the present system, the extension of the use of aeronautical frequencies to the maritime alert service would place on the aeronautical S.A.R. services an important extra load.

The major insufficiency of the present system therefore concerns distress detection and localization. The purpose of the SARSAT/COSPAS program is to provide this service.

THE SARSAT/COSPAS PROGRAM

The SARSAT program results from a process initiated by the USA and Canada beginning 1976, the main objective of which was to obtain Doppler localization of the 121.5 and 243 MHz transmissions from present distress beacons.

In 1977, the program was extended to France, with the proposal of a concept developped by CNES (Centre National d'Etudes Spatiales) using new 406 MHz beacons.

The program as a whole became official in 1979 with the signing of a formal agreement between NASA (National Aeronautics and Space Administration), DOC (Canadian Department of Communications) and CNES.

The SARSAT program comprises two aspects : firstly, it will improve the search and rescue system based on the use of the presently existing 121.5 and 243 MHz distress beacons, and secondly it will demonstrate the advantage offered by a new concept using 406 MHz beacons.

This project will lead to put on-board NOAA[3] meteorological satellites of the TIROS-N series repeaters for the 121.5, 243 and 406 MHz frequency bands, as well as a 406 MHz processor.

3 NOAA : National Oceanic Atmospheric Administration (USA)

Equipment of three spacecrafts (NOAA-E, F and G), the first of which should be launched in 1982, is planned in the present framework of the SARSAT program. The extension of these series with NOAA-H, I and J satellites is now under study. The satellites of the TIROS-N series are low polar orbiting satellites. Each satellite covers entirely the earth in less than 12 hours.

The responsibility for equipment development between the SARSAT participating agencies is summarized by Chart 2.

	NASA	DOC	CNES
SPACEBORNE	TIROS-N SPACECRAFT TIROS-N MODIFICATIONS SPACECRAFT SAR ANTENNAS	3-BAND REPEATER	406 MHz RECEIVER & PROCESSOR
NOAA GROUND SYSTEM	MODIFICATION OF GROUND STATION AND CONTROL CENTER		
LOCAL USER TERMINALS (LUT)	U.S. LUTs – 2 USCG – 1 USAF	CANADIAN LUT	FRENCH LUT
MISSION CONTROL CENTERS (MCCs)	U.S. MCC (USAF)	CMCC	FMCC
406 MHz ELT/EPIRB	DEVELOP TEST UNITS	WILL PARTICIPATE	DEVELOP TEST UNITS

CHART 2 Project Responsibilities of the parties

While the SARSAT partners designed their system, the Soviet Merchant Marine Ministry, MORFLOT, expressed its interest in the 406 MHz system and its intention to design a system meeting the same overall goal. The objective of interoperability of both systems is maintained throughout the development phase and will lead to a demonstration phase using simultaneously satellites and ground stations of both systems.

The agreement signed in 1980 by the four partners formalizes the MORFLOT participation which, in term of development, will mean in particular two more spacecrafts. The coordinated launches will ensure the availability of two satellites at the beginning of the demonstration phase and the availability of four satellites later.

THE 121.5 / 243 MHz REPEATER SYSTEM

The spaceborne repeaters relay the 121.5 / 243 MHz distress signals to local user terminals (L.U.T.) which provide detection and position location of the distress. Localization is obtained from measurements of the doppler shift affecting the uplink frequency. These measurements are carried out by the LUT and the position location is obtained on a single satellite pass.

The technique associated the 121.5 and 243 MHz beacons and the spaceborne repeaters entails several limitations :

- The beacons which continuously transmit a low power signal (approximately 100 mW) are not fitted to space techniques. Continuous transmission limits the capacity of the system. The objective is to process 10 simultaneous transmissions in the satellite field of view, the radius of which is approximately 2 500 Km.

- The high noise factor of VHF frequency bands, the instability of the oscillator and the poor spectral characteristics of the transmission, substantially complicate the doppler measurement. The aimed accuracy of localization is 20 Km.

- Lastly, the simultaneous beacon/satellite and satellite/terminal visibility requirement in this repeater system limits the coverage area obtained with the Local User Terminals.

A priori designed as a comprehensive space system, the principle of the 406 MHz digital system does not include the limitations mentioned above.

THE SARSAT 406 MHz SYSTEM

Principle

The objective of the system is to demonstrate the superiority of the integrated 406 MHz system over the 121.5 and 243 MHz system in the following areas :

- detection probability,
- position determination accuracy,
- worldwide coverage,
- operational interest and quality of information received (coded message),
- capacity of system for processing numerous simulteneous distress situations.

The experimental beacons of the 406 MHz system, put into operation in the event of a distress, periodically transmit every 50 seconds a high power (about 5 watts) burst signal (440 milliseconds). This signal consists of a pure carrier transmission followed by a digital message including data useful to the rescuers.

When the satellite is in view of the transmitting beacon, the on-board processor receives the signal, measures the doppler shifted frequency and process the message data.

The 406 MHz informations, consisting of the time tagged doppler measurements and the message data, are transmitted in real time on a 1544,5 MHz downlink(Fig. 1). It is simultaneously stored aboard the satellite to be transmitted with the whole stored telemetry of the satellite at 1 700 MHz to the NOAA data acquisition stations Gillmore (Alaska),Wallops Island (Maryland).

The real time telemetry is received at 1544.5 MHz by the Local User Terminals. This affords a local coverage, limited from a geographical stand point by the relative positions of the L.U.T., the beacon and the satellite.

On the other hand, the stored telemetry affords a full orbit/worlwide coverage.

On the ground, for both local and worldwide coverage, the doppler measurements carried out on each message received by the satellite and the time tags are used to compute the location. For each distress situation, the computed location and the informations contained in the ELT message (Fig. 2) are transmitted to the search and rescue coordination center in charge of the SAR operations.

This concept which uses a simple beacon operating independently from the satellite and from the other beacons is based upon random access techniques. The transmission frequencies of the beacons are spread around the 406.025 MHz central frequency. Dispersion on both the time of emission and the frequency (to which is added the

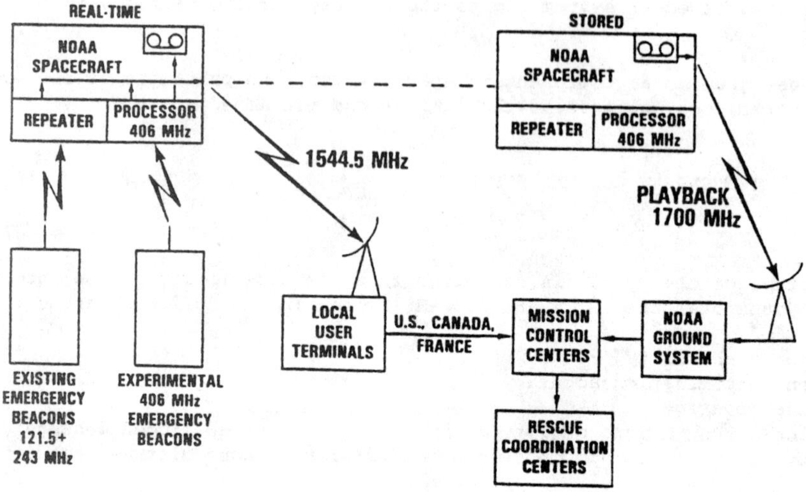

Fig. 1 Sarsat system

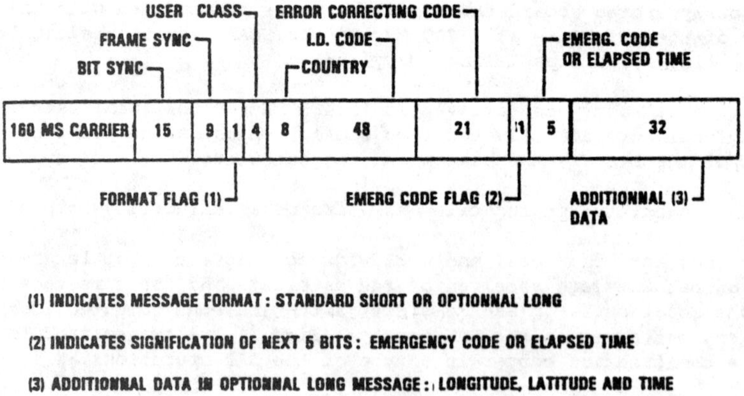

Fig. 2 406 MHz Message format

doppler shift) characterises the random access. The absence of interrogation of the beacon by the satellite (no receiving function) makes possible a simple design of the beacon which will allow low cost production.

Computation of the location is carried out on a single pass, with a minimum of 4 doppler measurements. It uses an adjustement by a least squares technique. The ambiguity due to the two possible positions yielded by the computation is automatically solved taking into account the doppler effect induced by the rotation of the earth.

To achieve this result, the LUT must carry out satellite ephemeris computations on a daily basis. Then, after each pass, it is necessary to adjust the orbit, to ensure the required precision in finding out the satellite positions used in the localization computation.

In the SARSAT L.U.T., the orbit adjustement has the original feature of being carried out from doppler measurements performed on the transmission of a single orbitography beacon associated to the L.U.T.

Performances

The performances of the 406 MHz system can be characterized by its coverage, its capacity for processing simultaneous distress situations, the probability of detection and the quality of the localization.

Coverage : Polar low orbiting satellites cover all the earth surface but the visibility of a given location is not continuous and the important parameter is the waiting time for a distress to be seen and transmitted to a ground station. This delay is essentially random since it depends on the position of the satellites at the time the beacon is activated. The statistics on the waiting time depend on the number of satellites, the phasing of their orbital planes and the latitude of the distress.

In the full orbit mode, the waiting time corresponds to the time for a satellite to come in view of the beacon plus the time needed for it to come in view of a Data Acquisition Station where the information is dumped.

In the local mode, the waiting time corresponds to the time for a satellite to come in view of both the beacon and the Local User Terminal. It increases with the distance between the distress and the L.U.T. Figure 3 shows the regions corresponding to mean waiting time in the local mode less than 2 and 4 hours for L.U.T.s in Toulouse and Tromsoe.

Figure 4 presents the enviewed location for the SARSAT / COSPAS Local User Terminals and Fig. 5 and Fig. 6 show the mean waiting time for alert (taking into account the full orbit mode and the local mode) with two or four satellites. These simulated times do not include ground segment processing and dispatching delays.

Capacity : The choice of random access allows high capacity. The present configuration of the on-board processor allows the processing of 90 beacons in the field of view of the satellite which is a circle of radius approximately 2 500 Km. The design of the space segment makes possible an extension of this figure.

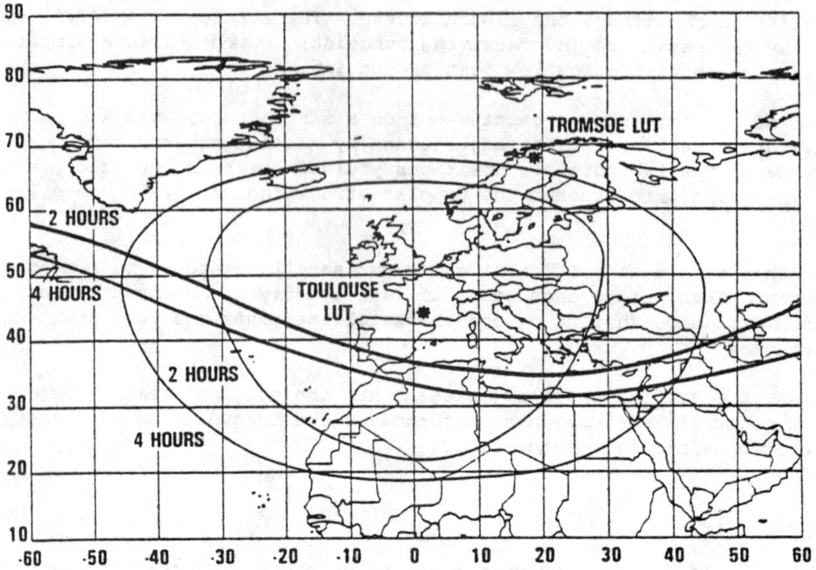

Fig. 3 Toulouse and Tromsoe L.U.T.s : Local coverage with 4 satellites.
Geographical limits for mean waiting times less than 2 and 4 hours

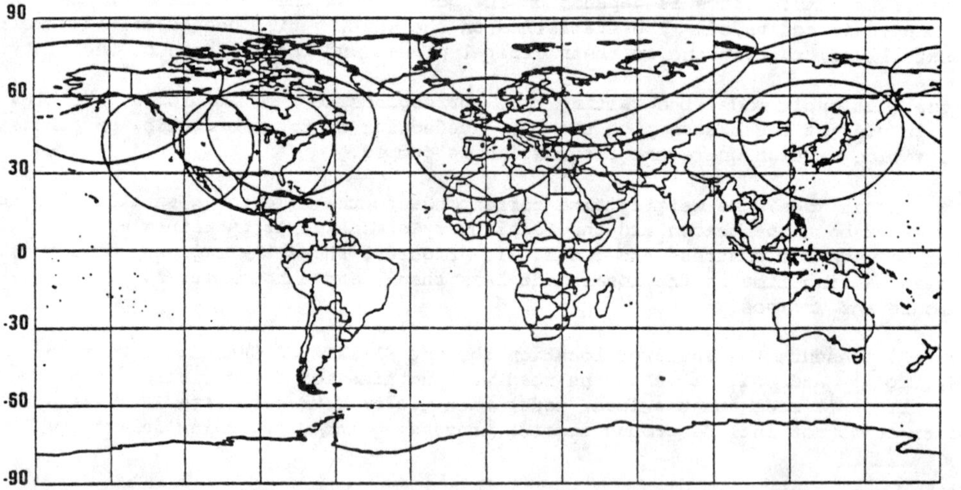

Fig. 4 SARSAT / COSPAS Local User Terminals
Visibility circles

A 1/2 H < T < 1 H
B 1 H < T < 1 H 30
C 1 H 30 < T < 2 H
D 2 H < T < 2 H 30

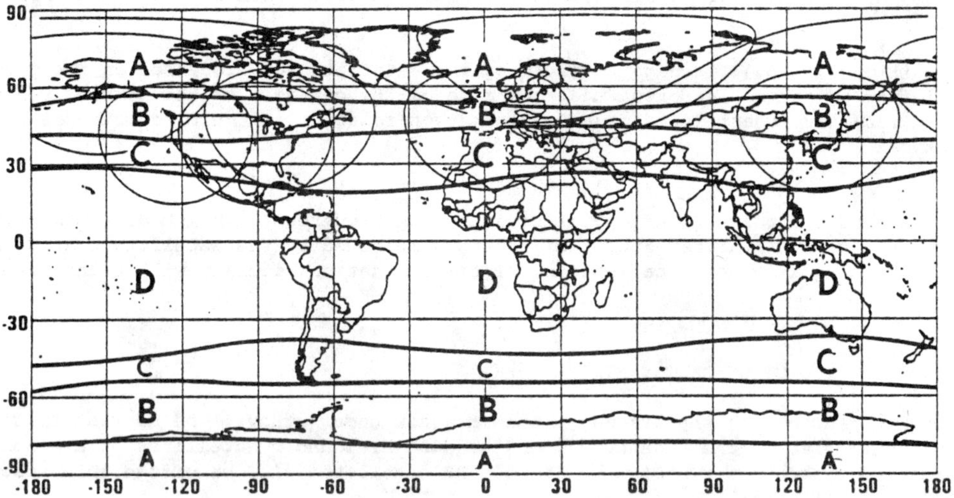

Fig. 5 SARSAT system present configuration :
- Mean waiting time with 2 satellites
- 8 L.U.T.s and 2 Data Acquisition Stations

A T < 1/2 H
B 1/2 H < T < 1 H
C 1 H < T < 1 H 30
D 1 H 30 < T < 2 H

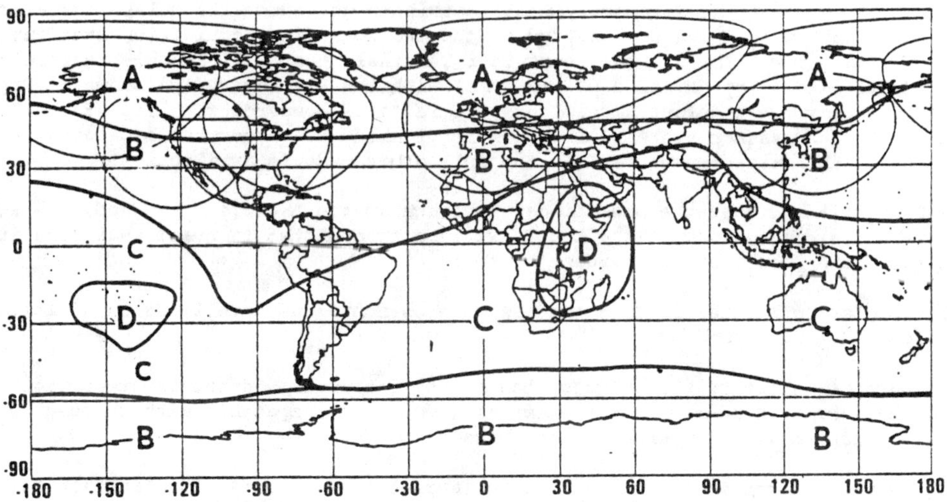

Fig. 6 SARSAT system present configuration :
- Mean waiting time with 4 satellites
- 8 L.U.T.s and 2 Data Acquisition Stations

Detection probability : The power of the 406 MHz transmission (5 W) gives sufficient margins in the link budget and should lead to a detection probability in an operational environment, definitely superior to that expected from a 121.5 / 243 MHz system.

In addition, a single message received by the satellite and processed on the ground, apart from detection, ensures the knowledge of the category of the carrier vehicle, its nationality, identification and information relative to the circumstances of the distress.

Localization accuracy : The doppler localization requires a minimum of 4 measurements during the pass. A higher number of messages received by the satellite considerably improves the quality of localization. The localization accuracy will be better than 5 Km.

EVOLUTION OF THE SYSTEM

Several extensions to the present system are now under study. The nearest in time to be considered is the adjunction, on the NOAA-H, I and J satellites of a specific search and rescue memory (SARM). The COSPAS satellites will be gifted this feature from the first satellite.

The storage of the search and rescue informations in the full orbit mode would then no longer require the use of the general purpose spacecraft memory.

The reading will be initiated by telecommand. The interesting feature as regarded to the present configuration is that the stored data will be dumped on the 1 544.5 MHz SAR downlink, thus allowing any SARSAT / COSPAS L.U.T. to receive both the local mode and the full orbit mode data.

The capacity enviewed for this memory will allow the storage of up to 2 300 messages. The introduction of the SAR memory will contribute to reduce the delays for alert in the full orbit mode. Indeed, instead of the two NOAA data acquisition stations presently used, it will be all the Local User Terminals that will be able to receive the stored data. The expected waiting time with the SAR Memory are shown on Fig. 7 and Fig. 8 for two and four satellites. It is to be compared with Fig. 5 and Fig. 6 related to the present system. The gain is of about 40 % over the equatorial regions and the south hemisphere where only the full orbit mode is available.

The SARM will increase the operational coverage of the L.U.T.s . Indeed, the alert near the visibility area of the L.U.T. will be transmitted in quasi real time after the satellite pass over the beacon.

Other evolution studies include the use of geosynchronous satellites in addition to the low orbiting satellites.

The characteristics of the beacons, mostly the 5 W power and the error correcting code, have been chosen so that the emission would be compatible with the use of a repeater aboard a geosynchronous satellite.

By this mean, alert would be transmitted in real time, precise localization being given afterwards by the present system. The additional bits in the optional long message format (Fig. 2) will then be very useful in providing a first rough estimate of the location of the distress.

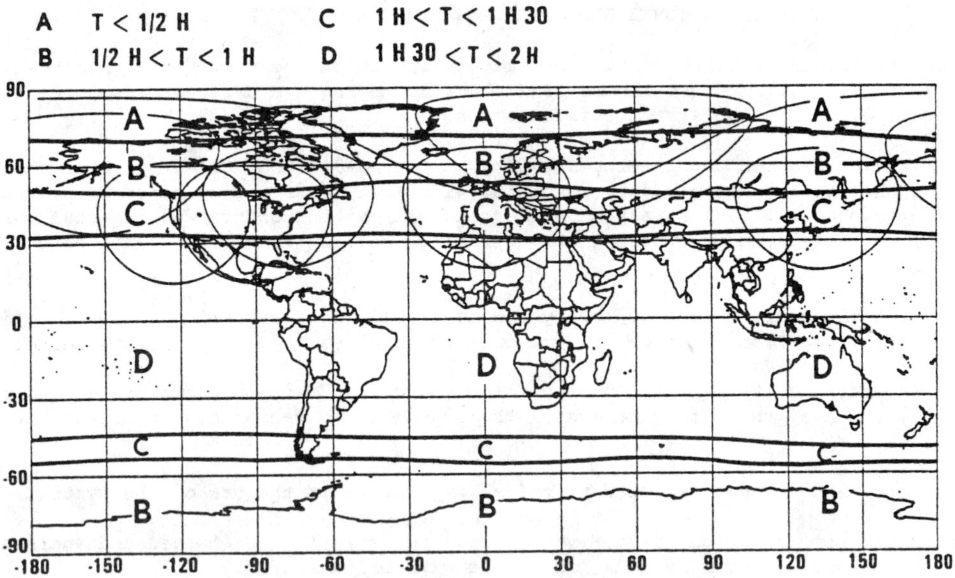

Fig. 7 SARSAT system with SARM
- Mean waiting time with 2 satellites
- 8 L.U.T.s

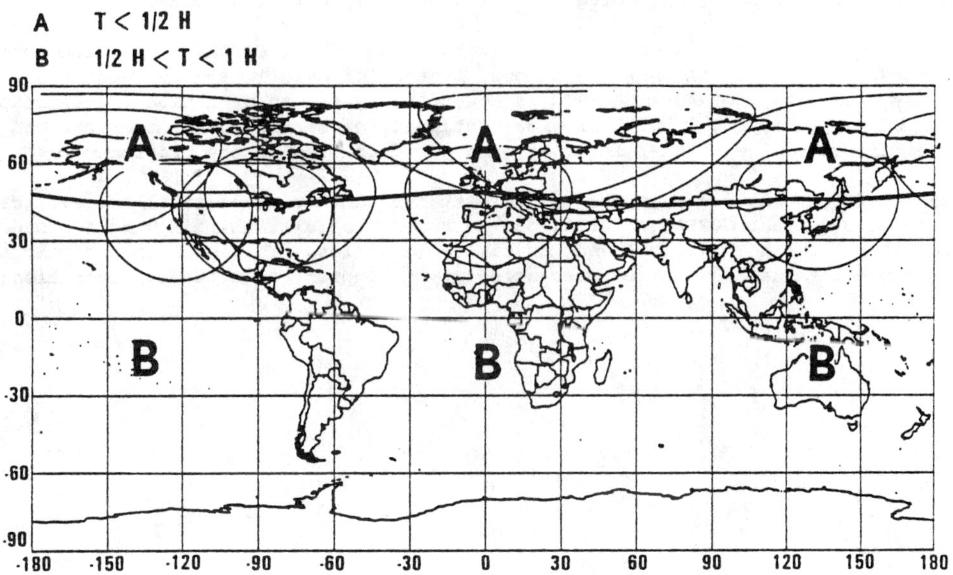

Fig. 8 SARSAT system with SARM
- Mean waiting time with 4 satellites
- 8 L.U.T.s

THE INTERNATIONAL DEMONSTRATION AND EVALUATION PHASE

It is anticipated that the first two spacecrafts in the series, NOAA E and then COSPAS 1, will be launched in mid 1982 allowing to begin a one year demonstration phase to determine the system effectiveness.

After having checked the interoperability of the SARSAT and COSPAS systems, the aim of the demonstration phase, conducted with active participation of user agencies having responsibility for search and rescue, will be to confirm the expected performances of the system and to show its operational effectiveness. The main objectives will be :

- To test the performances of the system in a real world situation : L.U.T. coverage, capacities, location accuracy, reduction in the delays, ...

- To demonstrate its utility in supporting search and rescue operations and to allow an evaluation of the effectiveness of the system as an operational tool for SAR activities,

- To provide user agencies with practical experience in the use of the system.

During the demonstration phase, evaluation results will be exchanged and analysed and at the end, a common evaluation report will be published.

Participation in the demonstration of additional countries and international organizations is also planned. Stated ground rules insure technical compatibility and contribution to the general objectives of the program.

At the present time, participation of Norway is under final formal approval. The participation of this country in investigating a system which has a universal vocation will increase the data collected during the demonstration phase.

If the demonstration is successful, it is the goal of the COSPAS / SARSAT partners to establish an international operational system for satellite aided search and rescue. To this end, COSPAS / SARSAT is now planning for an operational evaluation phase which, if approved, will provide continuity of service until near the end of the decade and allow greater international participation in the program.

A successful demonstration phase, along with the continuity of satellite service are conditions that COSPAS / SARSAT partners hope to fullfill, allowing a broader agreement by the aeronautical and maritime communities. This will then permit to set up the international structures necessary to support an effective operational system.

IAA 81-265

SPACE TECHNOLOGY CONTRIBUTIONS TO EMERGENCY AND DISASTER MANAGEMENT*

Paul B. Richards[+]

Each year, natural and man-made disasters take hundreds of thousands of lives, cause millions of dollars worth of property damage, and create tremendous social upheaval. The U.S. Government alone spends billions of dollars each year on foreign and domestic disaster relief operations. Space technology has, during the past ten years, demonstrated its great potential for contributing to disaster management activities. Space emergency communication systems provided a valuable communications link during the Johnstown, Pennsylvania, flood in July 1977, and their role in disaster relief operations has expanded greatly since then. Satellite remote sensing systems, originally designed for geological and meteorological purposes, have demonstrated their value in disaster mitigation, preparedness and warning, and in damage assessment. Possibly one of the largest contributions of space technology to disaster management will be that of satellite-based data collection systems, which will collect, process and relay data from earth-based data collection platforms to earth-based receiving stations. These three areas of space technology - communications, remote sensing, and data collection systems - will find additional use in disaster management through recently developed analytical methods which include models of predicting certain natural disasters and models which evaluate concepts for disaster response.

Keywords: Disasters, disaster relief operations, disaster management, emergency medical services.

---

* *This paper appears in full in Volume 41, AAS Microfiche Series.*

+ *Dr. Richards is associated with the Naval Research Laboratory, Washington, D.C. 20375*

IAA 81-267/268

MARITIME DISTRESS ALERTING SYSTEM*

U. Hammerschmidt[+] and K.P. Galligan[++]

The current status of the system is presented by U. Hammerschmidt and the satellite aspects of the system by K.P. Galligan.

The provision of an international global maritime mobile satellite communications service by the INMARSAT Organization furnished partly by the use of European satellites provides the prospect for greatly improved communications capability.

At an early stage, in the discussion of the founding of INMARSAT, in anticipation of an enhanced communications capability, it was realized that a considerable improvement should also be effected in the transmission of distress, safety and urgency messages.

The Inter-Governmental Maritime Consultative Organization (IMCO) has therefore been developing for three years in its subcommittee "Radiocommunication" the "Future Global Maritime Distress and Safety System" (FGMDSS), the implementation of which is planned for around 1990.

The discussion of this system includes distress alerting, identifying, positioning, SAR coordinating communications, on-scene communications, locating, and preventive actions. Requirements of the system and recommendations are given.

---

* *This paper appears in full in Volume 41, AAS Microfiche Series.*

+ *Deutsches Hydrographisches Institut, Hamburg, West Germany*

++ *European Space Agency (ESA), 8-10, rue Mario-Nikis, 75738 Paris, Cedex 15, France*

SPACE SAFETY AND RESCUE
1980

The following papers also appear in full in Volume 41, AAS Microfiche Series, supplement to this volume:

IAA 80-12 and -14

The following papers, abstracted in this section, appear in full in Volume 41, AAS Microfiche Series:

IAA 80-13 and -16

SAFER WAY WITH EVA

R.E. Breeding, H.R. Griswold*

The Space Transportation System (STS) and Spacelab represent international resources that will soon provide important new capability for manned space research and work. The capability to perform Extravehicular Activity (EVA) has been developed as part of STS and will be available on each flight to support space operations. This paper discusses the requirement for EVA capability in future STS missions. Essential features of EVA capability are described together with potential applications which directly support STS mission safety.

Keywords: Space safety, Extravehicular activity (EVA), EVA Equipment, Extravehicular Mobility Unit (EMU), Manned Maneuvering Unit (MMU).

INTRODUCTION

A notable space achievement occurred in the year 1965 when cosmonaut Alexer Leonov, outfitted in a spacesuit, ventured beyond the protective confines of his spacecraft to take that uncertain first step into the vacuum of space. Later the same year, a second successful spacewalk was taken by astronaut Edward White, who, in performing a set of prescribed procedures outside of his spacecraft, gave birth to a new space science to become known as Extravehicular Activity (EVA). The spacewalks of Leonov and White served to indicate that man, properly equipped, would be able to perform useful work outside of spacecraft. In manned space flight programs that followed, such as Apollo and Skylab, a trend of increased dependence placed on the use of EVA in support of both planned mission objectives and contingency operations was established. So successful was the use of EVA in these programs that, today, it stands qualified as an operational capability for performing mission tasks both in earth orbit and in deep space.

The Space Transportation System (STS) and Spacelab will soon be operational and mark the beginning of a new era in space. These reusable systems will support a new spectrum of manned space research and work activity, with STS providing the capability for conducting EVA. STS has baselined EVA as an operational capability, available on each flight, to support mission operations. In planning for STS, National Aeronautics and Space Administration (NASA) has imposed stringent space safety standards aimed at providing wide margin for mission safety. Part of STS safety margin will be provided in the form of EVA capability available for use in contingency operations.

* *Hamilton Standard, Division of United Technologies Corporation, Windsor Locks, Connecticut, USA 06096*

This paper discusses the requirement for EVA capability in support of future STS mission objectives. Essential features of EVA capability are described, together with potential applications which directly support STS mission safety.

REQUIREMENT FOR EVA

Planned and projected STS space operations, to be conducted from the Shuttle Orbiter in earth orbit, include:

- Research and Observation Using Spacelab
- Satellite Service
- Structure Assembly and Construction

To support these operations, NASA has developed a space work system which includes capability to perform EVA, the designation for STS crew orbital activity to be conducted outside of the Orbiter cabin. In contrast to the shirt-sleeve environment of the cabin, EVA will require the crewmember to don an Extravehicular Mobility Unit (EMU), a self-contained life support spacesuit system, permitting safe egress into the vacuum environment of space. The Shuttle EMU is, itself, a reusable anthropomorphic spacecraft comprising the essential hardware portion of a hybrid man/machine system. This hybrid system combines the important capabilities of man with those of hardware designed to provide for life support and protect against the space environmental extremes in pressure, temperature, radiation and illumination. Capabilities that man brings to this system include dexterity, tactility, mobility, ability to assess the state of an evolving process by sensory feedback, and adaptive decision making utilizing past experience. The wide spectrum of work tasks, projected for STS mission operations listed above, underscores the combinational advantages of the EVA hybrid man/machine system developed for STS.

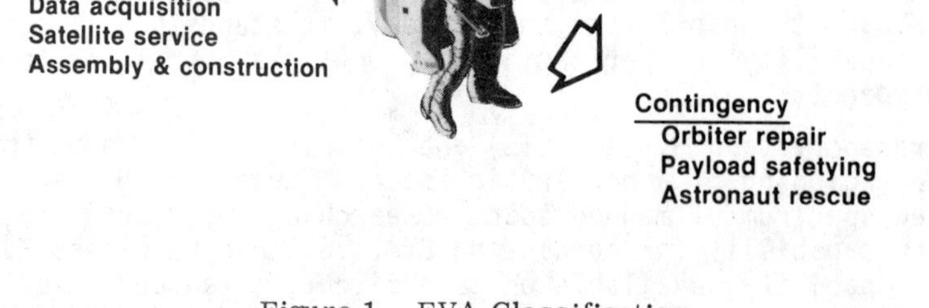

Figure 1. EVA Classification

As defined by NASA, EVA falls into one of three categories indicated in Figure 1. Planned EVA, as the term suggests, is planned prior to launch. This type of EVA is intended to provide a reliable, versatile, and cost-effective means for achieving orbital mission objectives. Planned EVA could include work activity such as replace/refurbish/repair satellite hardware, or conduct research experiments in the Orbiter payload bay. Unscheduled EVA, on the other hand, will not be planned prior to flight but may be required to achieve payload mission objectives or perform maintenance on non-critical Orbiter equipment. Unscheduled EVA could include activities such as diagnosing a payload problem followed by repair or cleaning optical sensor surfaces that have become contaminated. Finally, contingency EVA, not planned prior to flight, represents EVA required to maintain crew safety. This type of EVA could be required should a critical Orbiter system fail or be placed in jeopardy. Critical Orbiter equipment has been defined as equipment which could jeopardize crew safety should it fail. For example, contingency EVA could be used to render a payload safe that might otherwise endanger a critical Orbiter system, render rescue assistance to a fellow crewmember, or perform Orbiter thermal tile inspection/repair as depicted in Figure 2.

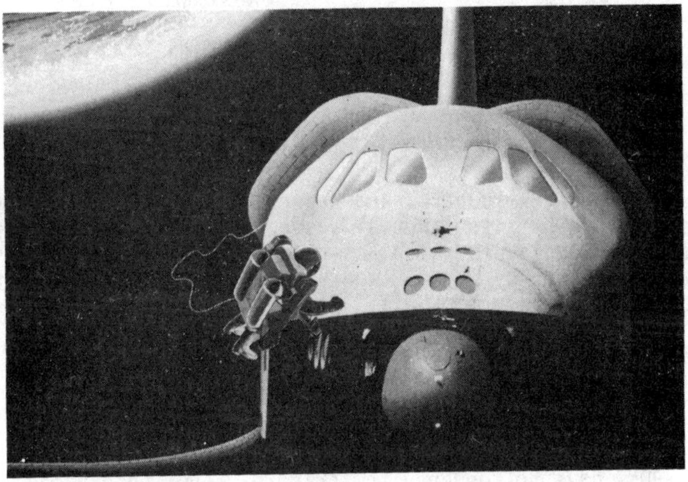

Figure 2. Orbiter Tile Inspection

NASA will provide EVA capability on each STS flight including support equipment and consummables necessary for three two-man EVA operations, each lasting a maximum of six hours. Two of the EVA's will be available for payload operations with the third retained to support any contingency. STS user requirement for additional payload EVA will be satisfied by provisioning additional EVA consummables on that flight. Safe operational procedures for conducting planned, unscheduled and contingency EVA will be established for each STS flight, having been developed as part of preflight operations.

EVA EQUIPMENT

Fundamental to EVA is the equipment designed to provide for life support, environmental protection and space work. The following discussion reviews important safety features of EVA equipment developed for STS, listed in Figure 3, which serve to promote safe EVA space operations.

EMU - The Shuttle Extravehicular Mobility Unit (EMU) represents the most advanced spacesuit and life support system developed to date by the USA. The EMU design provides for pressure retention, oxygen for breathing, atmosphere

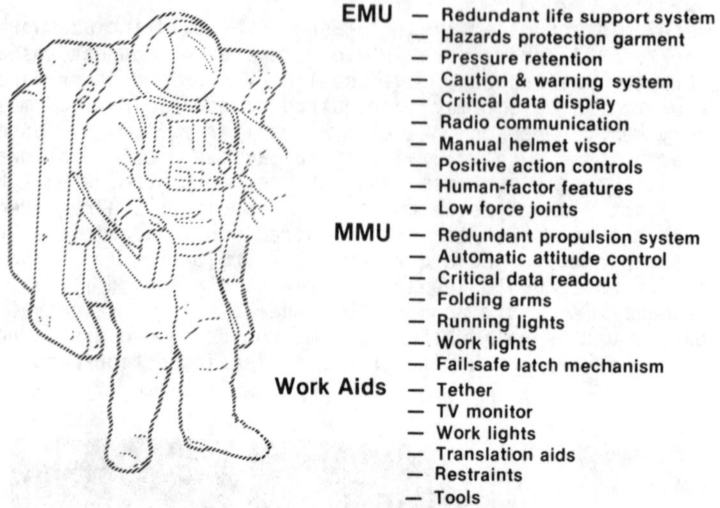

Figure 3. EVA Equipment Features

revitalization, temperature control and hazards protection. The thermal hazards protection is provided by a low thermal absorption outer spacesuit garment, radiation protection by layers which act to decelerate radiation electrons, and micrometeorite protection by way of layers that absorb impacting particles. The tough outer spacesuit garment also affords abrasion and snag protection.

Important human factors engineering is evident in the EMU design. The use of new joint designs and bearings provides pressurized spacesuit mobility 20% better than that of the Apollo suit and nearly 85% of nude range capability. Spacesuit joint torques have been lowered by an order of magnitude so that the EMU equipped crewmember can move easily. In addition, joints such as an elbow or knee, bent 90°, have no tendency to spring back to the extended position. This improved spacesuit mobility means that crewmember reach envelope and work capacity have been substantially improved. In addition, the size of the EMU has been reduced from that of the Apollo spacesuit system by a full 10 inches front to back and 8 inches in width. In packaging the EMU, care has been taken to round corners and contour the life support backpack. These features, coupled with elimination of hoses and external wiring, yields a clean, compact external configuration that fits into smaller work spaces with reduced tendency to snag.

The modular EMU spacesuit arms, legs and boots are interchangeable with various sizes of the fiberglass hard upper torso element to allow assembly of a properly fitting garment for different size crewmembers, avoiding costly custom constructions. Work gloves feature insulation padding for both palms and fingers, capable of withstanding contact temperatures of -180°F to +200°F, and are available in several sizes. The gloves provide ventilation cooling to finger and palm areas for comfort and have an outer abrasion resistant covering. The helmet is a rugged one-piece hemisphere fabricated of ultraviolet stabilized polycarbonate, capable of providing protection against micrometeorite particles. A series of sun visors, with thermal and optical coatings applied on inner surfaces, are contained in an assembly which is integrated with the helmet. The visors can be adjusted by the crewmember as required to protect eyes during varying EVA light conditions.

An EMU Caution and Warning System (CWS) monitors EMU instrumentation and provides the crewmember information by way of audio tones and visual display on the chest mounted Displays and Controls Module (DCM). The CWS provides the following information: (1) startup, checkout, egress, and ingress procedures to be executed, (2) quantitative readout of system parameters such as temperature, pressure, including oxygen, water and battery amp-hour expendables, (3) notification of failure or out-of-limit condition, the specific parameter value and necessary corrective action. The CWS measures critical parameters every 1/5 of a second providing essentially continuous EMU system interrogation. The EMU communication system provides an audible alarm for 10 seconds in the event of an unsafe condition detected by the CWS, uninterrupted voice communications between EVA crewmembers, and continuous biomedical telemetry from EVA crewmember to the Orbiter vehicle.

The DCM contains all displays and controls required for one-man operation of the EMU. The DCM houses the fan switch, water pump switch, communications selector switch and audio volume control, oxygen quantity indicator, suit pressure gauge, cooling control valve, purge valve and status indicators. The location and size of the DCM is such that the crewmember can read displays and reach manual controls without undue compromise of work visibility and mobility. The individual DCM controls have been engineered to prevent inadvertent actuation by either the crewmember or external equipment contact.

An EMU Secondary Oxygen Pack (SOP) provides spacesuit backup pressure regulation and emergency oxgen at a rate sufficient to support the crewmember for 30 minutes. Two separate SOP bottles are used to store the high pressure oxygen. If the primary EMU oxygen supply is depleted, SOP automatically begins to supply oxygen. A pressure sensor in the ventilation circuit would then alert the crewmember, by way of the CWS, that the SOP is in use.

MMU - The Shuttle Manned Maneuvering Unit (MMU) has been designed to provide EVA crewmembers with free-flight mobility for use in the vicinity of the Orbiter. Two MMU's will be stored in Flight Support Stations (FSS) bolted to the Orbiter forward bulkhead in the payload bay. The EVA crewmember backs into the MMU and, using foot and hand restraints, latches the EMU life support system backpack to the MMU. The MMU is then unlatched from the FSS for free-flight. Should the latching interface between the EMU and MMU fail to release, a safety feature is included which shears the latching mechanism effecting crewmember release from the MMU.

The MMU features a high pressure nitrogen cold gas propellant system consisting of 24 thrusters divided into two redundant systems of 12 thrusters each. The MMU includes redundant batteries and electronic controller, running lights, and work lights. Also included is an automatic stabilization control mode which can be used to null unwanted EVA crewmember translations or rotations. A 200-foot tether with reel has been developed for initial use with the MMU, providing an additional level of redundancy for crewmember return to the Orbiter.

Work Aids - Hand rails on the Orbiter and on payloads will be provided for use in EVA crewmember translation to payload bay worksites. In addition, tethers will be available for securing crewmembers to worksites and translation rails. Payload bay lights and TV monitors are included as part of Orbiter equipment for use in EVA work. The Remote Manipulator System (RMS), a powered mechanical arm operated by a crewmember in the Orbiter cabin, can be used in conjunction with

EVA and incorporates hand holds so that it can be used as a translation path. Work lights and TV monitor, mounted near the end of the RMS, can be used to support EVA work. A variety of hand tools and restraints, including those developed for the Apollo and Skylab programs, will be available to support STS mission EVA.

Figure 4. Spacelab Payload Support

Using versatile EVA capability, with equipment features described, a variety of orbital operations can be supported. An example of EVA is shown in Figure 4 where one crewmember in the Orbiter payload bay is conducting a Spacelab experiment. The second crewmember, free-flying with an MMU, is deploying an experiment which avoids development costs and weight penalty associated with a self-deployment mechanism.

ESSENTIALS OF EVA

Essential aspects of STS mission EVA, depicted in Figure 5, are now described. The individual frames shown in Figure 5 represent operational phases of the mission. As indicated by arrows, several factors contribute directly to EVA preparation for an STS flight, required to support safe EVA space operations.

Operations Definition - STS preflight preparation will include dialogue between STS user and NASA personnel to define those space operations that will require the use of EVA. Once requirements for EVA have been established, the specifics of planned operations, outlined in Table I, will be determined. For example, EVA tasks to be performed may impose certain design requirements on the payload with regard to features such as geometry, safety interlocks, caution and warning labels and lights, crewmember foot and hand restraints and worksite access, volume and lighting. The payload design will be made to accommodate certified tools and work aids included in the EVA inventory, if practical, thereby avoiding unnecessary development of new hardware.

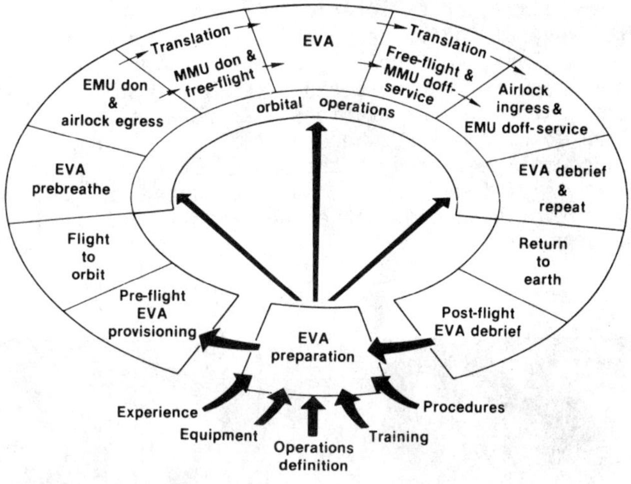

Figure 5. Essentials of EVA

Utilizing past experience, task time estimates will be made for use in planning individual EVA sorties. Preflight training requirements will be defined together with program, utilizing available NASA training facilities. For certain payloads, it may be determined that full-scale payload mock-ups will be required for use in preflight EVA training. Finally, the total number of EVA's projected to support planned payload mission objectives will determine EVA provisioning requirements for the STS mission, where consummables include water, oxygen and LiOH-canisters for the EMU and nitrogen for the MMU.

TABLE-1
EVA OPERATIONS DEFINITION

- **Task definition**
- **Payload design requirements**
- **Crewmember-payload interface requirements**
- **Work access, volume and lighting requirements**
- **Tool and equipment requirements**
- **Task time estimates**
- **Training and mock-up requirements**

Experience - The USA's manned space program has established an excellent safety record having benefited from over 50 years of aviation industry experience. Aviation guidelines for design of flight equipment and operational procedures have provided useful prototypes for evolving space program safety standards. In addition, cumulative space program experience comprises a valuable data base for use in preparing for future STS missions. EVA experience gained in the Apollo program, depicted in Figure 6a, clearly demonstrated that EVA can effectively support work activities in the vacuum environment of space. In the Apollo program, there were six lunar exploration missions which accumulated over 150 hours of EVA with several individual sorties extending beyond seven hours. As part of the program, lunar surface samples were collected, thousands of pictures taken including movies, and numerous experiments set up for making on-site as well as long term scientific observations. Astronauts, in support of these activities, walked some 37 miles, where virtually all lunar surface mission objectives were accomplished through the use of EVA. In addition to lunar surface EVA, 20 hours of free-space EVA was conducted in the Apollo program. To support free-space EVA, handhold, tether and translation aid hardware was developed and used with success.

(a) Apollo EVA

(b) Skylab EVA

Figure 6. Capability Builds on Experience

The Skylab EVA experience, depicted in Figure 6b, is of particular value since this was acquired in the zero-gravity orbital environment where STS space mission activity will take place. Both planned and contingency EVA experience gained in Skylab provide valuable examples of how EVA can be used to support mission safety and success. At the time of the Skylab launch, only 29 hours of EVA was planned and, that, primarily for the retrieval and replacement of film cartridges and particle collection devices. As a result of the loss of a heat shield, solar panel and the jamming of the remaining solar panel on launch, the success of the Skylab program depended on the ability of an EVA crew to repair and reconfigure Skylab. Fifty-three hours of contingency EVA were conducted to accomplish 13 repairs required to return Skylab to operational status. EVA activities included freeing the docking latch, erecting a sun shield, freeing a solar panel, and servicing a voltage regulator.

Between May, 1973 and February, 1974, three, three-man crews spent 171 days in Skylab, and conducted a total of 82 hours of EVA. The spacesuits worn by Skylab astronauts were the same as the Apollo suits, with the exception of an improved sun visor assembly. Life support and cooling was umbilical supplied, with a pressure and flow control unit mounted on the front of the suit. An important lesson learned from Skylab was that umbilicals limit EVA effectiveness by restricting movements, access and return paths and by limiting travel distance. This experience was valuable in evolving new concepts for the present Shuttle EMU. In Skylab EVA's, a number of tools were used by astronauts in accomplishing work objectives, including pliers, cable cutter, vice grips, dental saw, prybar, box wrench, ball peen hammer, screwdriver and brushes. Also, within the large Skylab, astronauts experimented with a prototype maneuvering unit, the forerunner of the present day MMU. These experiments proved viability of the MMU concept and provided valuable design data for the present MMU.

The EVA experience data base will continue to grow in the future as a result of STS missions. Information obtained from post-mission debriefings and analysis of mission data from telemetry, on-board recordings, and flight crew reports is expected to lead to further improvements in EVA equipment and procedures.

Equipment - EVA preparedness depends, to a large extent, on the equipment available to support planned space work. In the previous section, a description was given of important features of the EMU, MMU and work aids that will be available to support STS mission EVA. As STS matures to support a wider range of space operations than will initially be possible, there will be requirement for new EVA equipment to accomplish the new tasks effectively, safely and with minimum need for special skills. Analysis of new tasks to be performed, EVA crew performance limits, and time-line studies will play an important role in defining optimal designs for new equipment. Human factors considerations with regard to packaging design, crew/equipment interfaces, control locations, mobility and visibility will continue to be important in evolving new EVA equipment.

Already, planning has begun on future improvements to the EVA capability that will be available when STS first becomes operational. In Figure 7, a NASA advanced concept is shown for a Manned Remote Work Station (MRWS), which attaches to the end of the RMS and includes RMS controls for use by the EVA crewmember. The MRWS includes work lights, tool storage area and provision for astronaut foot restraint. This work platform would be maneuvered by the EVA astronaut to provide work access and restraint at select work sites in the Orbiter payload bay area.

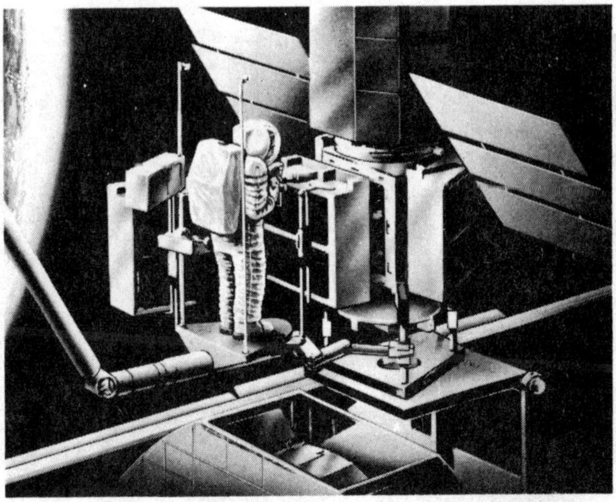

Figure 7.

TABLE-2
EVA OPERATIONS TRAINING

- **Payload familiarization**
- **Tool, equipment and restraint familiarization**
- **Verify work access and volume requirements**
- **Task and procedures familiarization**
- **Establish task times**
- **Residual hazards familiarization**
- **Zero-gravity familiarization**
- **Evaluate new tools, equipment and restraints**
- **Evaluate new work techniques**
- **Evaluate human engineering features**
- **Determine solutions to human error-producing situations**

Training and Procedures - The Orbiter spaceflight crew will consist of the commander, responsible for overall operations, the pilot, and mission specialists who will be the principle EVA crewmembers, proficient in payload and experiment operations. Training of EVA personnel will be required to achieve objectives, listed in Table 2, which serve to promote safe EVA operations. NASA's crew training program is aimed at developing crewmember expertise in using EVA related STS systems and specific payload EVA operational objectives. EVA procedures will be developed for performing planned, unscheduled and contingency EVA associated with the mission. Training will consist of classroom sessions, training in various simulators, and for some missions use of full-scale payload mock-ups in neutral buoyancy facilities to provide EVA crewmembers thorough familiarization with orbital task procedures to be performed. Special Orbiter mock-up facilities will be utilized so that familiarity with airlock ingress and egress can be acquired, along with practice in donning, doffing and servicing EMU equipment. In addition, training in the altitude chamber will provide familiarization with actual operation of the EMU and Orbiter airlock under simulated flight conditions in a vacuum environment.

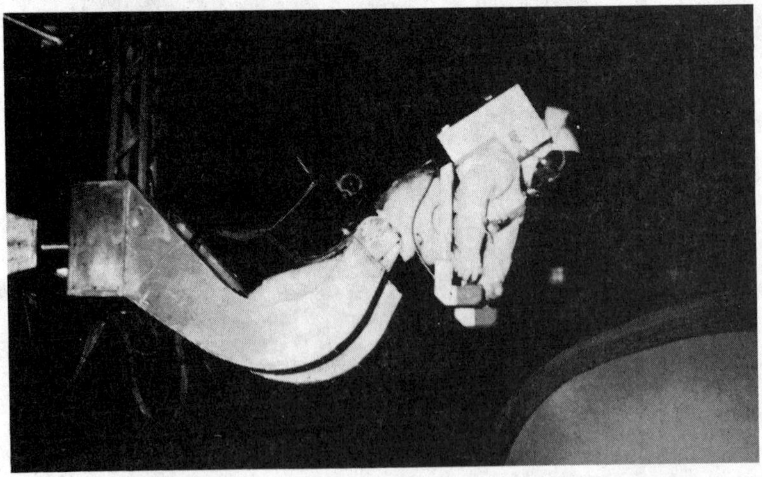

Figure 8. MMU Simulator Training

In Figure 8, the MMU training simulator is shown with astronaut in simulated free-flight approaching a mockup of the Orbiter vehicle with nose in the foreground. This facility simulates MMU orbital flight dynamics, includes visible MMU controls and instruments and responds in six-degrees of freedom to hand control inputs which are computer processed. The facility consists of a tower mounted on a rack-driven platform with range of motion in two horizontal directions. A sled rides the face of the tower to provide vertical motion. A roll motion adapter is attached to the sled to provide the remaining three-degrees of freedom.

Figure 9. Zero-Gravity Simulator Training

Astronauts, shown in Figure 9, are in NASA's Water Immersion Facility (WIF), which was used extensively for Skylab crew training. The astronauts are engaged in simulated space structure construction operations. WIF, a neutral buoyancy facility, provides a high fidelity zero-gravity simulation that can be sustained for periods of up to three hours. The large size of the water tank accommodates full-scale structures to provide a realistic simulation of the work environment to be encountered in orbit. Individuals with WIF simulation experience will be able to perform more effectively than if launched without such training. This will be especially beneficial for those payload specialists having no prior orbital experience.

<u>EVA Orbital Operations</u> - The EVA orbital operations sequence is indicated in Figure 5. The STS crewmember preparing for EVA first dons a Liquid Cooling and Ventilation Garment (LCVG) while in the Orbiter cabin. The LCVG provides cooling during EVA and is worn inside of the EMU spacesuit. The crewmember then, using a Portable Oxygen System (POS), breathes pure oxygen to de-nitrogenate the body. It is noted that NASA has recently initiated studies aimed at eliminating requirement for EVA prebreathe. During prebreathe activity, the astronaut enters the airlock with the POS and begins EMU spacesuit donning and checkout. Following this activity, the crewmember egresses from the airlock into the vacuum environment of space. For work activities in the payload bay, Orbiter

hand rails are provided enabling the crewmember to translate to EVA worksites. For free-flight work activities remote from the Orbiter, the astronaut dons the MMU propulsion unit located in the Orbiter payload bay. Upon completion of EVA work, the astronaut translates back to the airlock or free-flies using the MMU back to the Orbiter where the MMU is doffed and serviced in preparation for subsequent EVA. Upon entering the airlock, the EMU is doffed and serviced in preparation for subsequent EVA. Finally, the crewmember enters the Orbiter cabin completing the sequence. Post-EVA debrief may be required as well as subsequent EVA's to complete the STS orbital mission plan. The execution of EVA orbital operations described is the culmination of EVA preparation. Utilization of past experience, specially designed tools, restraints, equipment and payloads, experience gained in pre-flight training, and established operational procedures all support the safe and successful completion of the EVA.

EVA APPLICATIONS

Using versatile EVA capability described above, a wide range of STS work operations can be supported. Already, use of EVA has been planned for satellite service and space research operations. Also of importance, is the role that EVA can play in resolving STS contingencies that directly involve mission safety. Several examples of this latter application of EVA are highlighted below.

Critical Systems - A potential use of EVA is in contingency operations associated with critical Orbiter systems. Figure 2 shows an EVA crewmember engaged in an inspection of the Orbiter's Thermal Protection System tiles. A tile repair kit is currently being developed that will permit EVA tile repair in orbit. Should Orbiter payload doors fail to close properly, EVA will be available as a backup capability to manually effect closure. Finally, EVA could be used to resolve contingency problems involving payloads such as obstruction of payload door closure or, as shown in Figure 7, a satellite which has failed to self-deploy from the FSS platform.

Figure 10. Rescue Operation

Personnel Rescue - EVA has been selected as the prime method for effecting STS crew rescue in orbit. Rescue equipment, shown in Figure 10, includes the Personnel Rescue System (PRS) for use by personnel being transferred between disabled and rescue Orbiter. The PRS is a 34-inch diameter, inflatable pressure enclosure that includes thermal and micrometeoroid protection, and is used with the previously described rebreathing POS. Should an emergency require STS personnel transfer to a rescue Orbiter, the crewmember to be transferred would don the POS and enter the PRS through a zipper opening. The sphere would then be pressurized. During EVA transfer, the POS with face mask would provide the PRS occupant oxygen for breathing and maintain pressure inside of the sphere. This rescue system is capable of providing life support for one hour.

Three modes of PRS transfer from disabled to rescue Orbiter spacecraft are being studied, all of which utilize EVA. One method is for an EVA astronaut equipped with an MMU to free-fly the rescue ball between vehicles as shown in Figure 10. The second is to rig a line between the two vehicles for transfer of the ball. The third is to use the RMS in the cargo bay of the rescue Orbiter to retrieve the ball with its passenger out of the disabled spacecraft. This STS rescue capability is sufficiently flexibile to support rescue operations associated with manned space programs, other than STS, should a contingency arise.

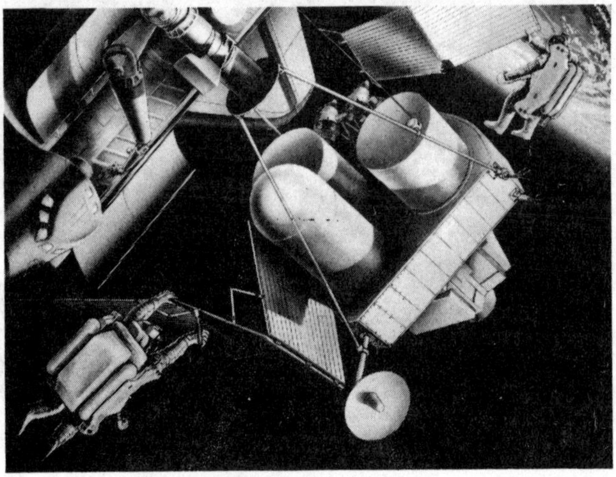

Figure 11. Satellite Safetying

Payload Safetying - With further additions to baseline EVA capability, additional STS operations will be supportable in the future. Figures 11 through 13, depict operations which would utilize an expanded EVA capability. These operations have been concepted to indicate the range of potential EVA usage in support of STS mission safety. In Figure 11 crewmembers are shown engaged in a satellite safetying operation prior to RMS transport of the satellite into the Orbiter payload bay for service or return-to-earth. The EVA safetying activity might include engaging safety interlocks, disconnecting electrical connectors, placing baffles over thrusters, deactivating antennas and removal or folding of solar arrays and radiator panels. An improved high tactility glove, TV monitor mounted to the helmet for use by Orbiter personnel, automatic sun visor system and voice control of the MMU are features of an expanded EVA capability[5] work system that could prove useful in activities shown.

Payload Stabilization - With present EVA capability, an MMU equipped crewmember could be used during payload berthing operations to assist in alignments and remove unwanted payload dynamics. Figure 12 depicts a concept in which crewmembers are stabilizing an inoperative spinning/tumbling satellite, remote from the Orbiter, required prior to approach for service or retrieval. The crewmember in the background is shown using a spring loaded device with which a line is cast across the path of the satellite for capture. The astronaut in the lower left has set reel-drag and is removing angular momentum from the satellite through energy exchange with the MMU discharge gas stream. When the satellite has been despun to a safe level of low energy dynamics, crewmembers can safely approach the satellite and complete stabilization using the MMU operated in the automatic stabilization control mode. A laser ranging device is concepted for this activity which mounts to the helmet and provides signals to the EMU computer for calculation of differential orbital mechanics between satellite and astronaut. The EMU computer, in turn, generates control signals for the MMU propulsion unit.

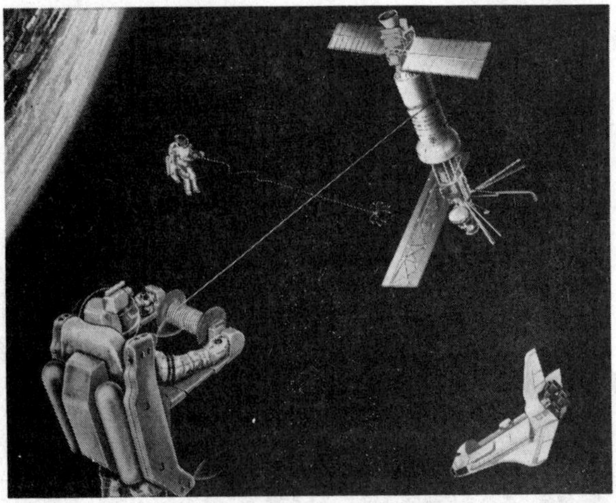

Figure 12. Satellite Stabilization

Debris Collection - There are many thousands of objects currently in earth orbit with orbital lifetimes, dependent on altitude and inclination, measured in years. Conventional space delivery systems have sometimes generated as many as 10 pieces of space debris for each usable payload delivered to orbit. With the planned frequency of payload launches, the issue of space debris is a concern especially at the intersection of polar orbits.

Present capability EVA could be used in berthing operations involving retrieval of large debris by offering assistance with alignments and damping of unwanted debris dynamics. The collection of small debris, however, represents a more difficult problem since objects orbiting at different altitudes, travel at different velocities and therefore tend to drift apart. Figure 13, an oversimplified depiction of the space debris problem, is used to illustrate concepts for collecting small debris. The EVA crewmember shown on the left, with expanded EVA capability features described in the previous examples, is using the MMU for free-flight retrieval of debris. A collection device, mounted to the EMU, is used to stow pieces of debris gathered by hand.

Figure 13. Debris Collection

The crewmember on the right has reeled out a line, attached to the satellite, and is moving about retrieving debris. Debris is attached to the line using straps where the line with debris would subsequently be reeled aboard the Orbiter for stowage.

SUMMARY

The Space Transportation System (STS) and Spacelab are ushering in a new era in space. These reusable systems will provide cost effective means and important new capability for space exploration, research, and utilization. Included as part of STS capability will be provision for conducting Extravehicular Activity (EVA), available for use in a wide spectrum of orbital operations including contingencies involving critical Shuttle Orbiter systems. This EVA capability includes features designed to promote safe STS space operations. As STS matures, to support a wider range of space operations, further improvements to EVA capability can be expected where a prime requirement for EVA will continue to be, support of STS mission safety.

ACKNOWLEDGEMENTS

The Authors wish to acknowledge valuable input from NASA information sources and Martin-Marietta Corporation.

REFERENCES

1) Space Transportation System, NASA Report, June 1977.

2) A Compendium of Future Space Activities, NASA Report, JSC-11532 Rev. C, October 1979.

3) Shuttle EVA Description and Design Criteria, NASA Report, JSC-10614, July 1977.

4) Apollo Experience Report: Safety Activities, C. Rice, NASA Report, TN-D-7950, April 1975.

5) EVA Equipment for Satellite Service, H. Griswold, R. Wilde, ASME Paper, 80-ENAs-48, July 1980.

EVALUATION OF RADIOACTIVE HAZARDS FOLLOWING NUCLEAR-POWERED SATELLITE REENTRY

E.A. Hutchinson-Benson*
J. Svoboda*
H.W. Taylor+

The reentry of failing spacecraft, especially those powered by a nuclear reactor, poses many problems and a serious hazard. This paper documents the experience with the Cosmos 954 reentry following the event and underlines the public concern over the problem of dealing with radioactive debris resulting from the impact in a frozen, remote and inhospitable environment. It also gives recommendations on how to cope with similar situations in the future.

INTRODUCTION

The crash of the Soviet satellite, Cosmos 954, over northwest Canada brought to world-wide attention the fact that nuclear power is being used in spacecraft orbiting the Earth. Cosmos 954, which was powered by a compact, Uranium 235-enriched reactor, was launched by the U.S.S.R. on September 18, 1977, and reentered the earth's atmosphere on January 24, 1978. It scattered its radioactive debris over an area of approximately 124,000 km^2 (Galloway, 1979). In addition to numerous identifiable engineering parts which reached the ground surface, a large portion of the debris consisted of particles ranging in size from 200 to 1000 μm that could be trapped in clothing or inadvertently ingested by grazing animals. In addition, several thousands of pieces of contaminated material with the dimensions greater than 1000 μm were recovered by search teams (Gummer et al, 1980).

* *University of Toronto, Department of Botany, Toronto, Canada*

\+ *University of Toronto, Department of Physics, Toronto, Canada*

Since the government of the United States and Canada knew about the presence of a nuclear reactor on the failing satellite, a large scale search for satellite debris in the crash area was undertaken immediately. Many Federal Government agencies were involved in the coordinated efforts to determine the degree and extent of the possible radioactive contamination. The Department of National Defence (DND) was given the primary responsibility since it had experience and the necessary equipment for working under the climatic conditions of the harsh northern Canadian winter. The Canadian Nuclear Accident Search Team (NAST), together with the U.S.A. Nuclear Emergency Search Team (NEST), worked with the Canadian Forces unit to promptly and safely locate any radioactive remnants that had been scattered over the impact area. The Atomic Energy Control Board (ECB), the Canadian Federal Governmental agency which regulates nuclear power in Canada, was responsible for the isolation and recovery of any debris located. The AECB was also in charge of formulating replies to questions concerning threats to public health and safety arising from the Cosmos 954 incident.

REVIEW OF THE COSMOS 954 INCIDENT

The DND and the AECB were involved in the initial search "Phase One", from January 24 to April 20, 1978 (Gummer, 1979). This operation was suspended when the deteriorating ice conditions on the northern lakes made aircraft maneuvers hazardous. By the end of Phase One the search corridor extended from Yellowknife eastward to Baker Lake, near Hudson Bay, some 1,000 km distant. The southern distribution of radioactive particles was believed to extend to at least 60°N, near the northern boundary between the prairie provinces of Alberta, Saskatchewan and the Northwest Territories of Canada (Bristow, 1978).

"Phase Two" of the search, from July 14 to October 14, 1978, included a second examination of the impact zone now free of snow, particularly areas in and around the human communities where an initial cleanup was conducted under conditions of snow and ice. "Phase Two" also included a search for radioactive debris in northern Alberta and Saskatchewan where fine particulate matter could have been carried and deposited by the NW prevailing winds at the time of the satellite reentry. By the end of Phase Two approximately 65 kg of material ranging in size from control rods to minute particles had been detected and recovered (Gummer, 1979). It was believed that the estimated 3500 particles retrieved from the areas searched thoroughly during both search operations were the remnants of the fuel of the nuclear-powered satellite which was 90% enriched ^{235}U.

In addition to the recovery of the satellite debris, two other Federal agencies were involved in the analyses of potential hazards resulting from radioactive contamination of the northern environment. The Department of National Health and Welfare (DHW) monitored radiation levels in the air near the ground in several northern communities where particular debris was recovered. Studies were also conducted soon after the Cosmos accident to determine the solubility of the radioactive particles in a medium stimulating human gastric juices in order to predict the likelihood of these particles being metabolized in the body if accidental ingestion occurred. The results of these tests, carried out in the early stages of the search, indicated that ingested particles would pass through the intestinal tract without being assimilated by the body. Nevertheless the human or animal organism affected would be exposed to intensive irradiation from the material for many hours. However, the Federal Department of the Environment continued to monitor fish taken from Great Slave Lake and caribou meat from grazing herds. In this way, it was assumed that any radioactive contamination of the northern food chain would be detected and the necessary precautions taken to minimize consumption of potentially hazardous food sources. (National Health and Welfare, 1979; unpublished).

CONDITIONS SURROUNDING SEARCH AND RECOVERY OPERATIONS

Several difficulties were encountered in the massive search for radioactive debris. The remote location of the crash, although fortunate in many ways, made prompt transportation and logistical support of field personnel more difficult due to the intense cold (about -40°C) and snow-covered terrain. In some cases, power batteries froze and instrument cables snapped so that equipment maintenance required advice from experts trained in survival techniques in arctic conditions. According to Bristow (1978) the major difficulty in the satellite search was in aircraft navigation under the conditions of the arctic winter when the snow covered ground appears to blend into the white overcast sky. In some areas, the search for radioactive debris was made even more difficult by the occurrence of natural deposits of radioactive elements in outcropping bedrock. Bristow (1978) reported that a modified version of methods designed by the U.S. Department of Energy to discriminate between natural and manmade sources of radiation were used to circumvent this problem. Fortunately, most, although not all, of the satellite debris was deposited on the ice of the frozen Great Slave Lake where it could be more readily detected since the background radiation was shielded by the column of lake water. Figure 1, due to Grasty (1978), shows the approximate distribution of debris on the lake along a number of transects.

The safe transportation of recovered debris from Yellowknife to Edmonton and finally to the AECB laboratory at Pinawa, Manitoba, required the use of steel containers well protected with lead sheeting so that contamination of personnel and equipment was avoided. The close cooperation and coordination between the Canadian Government and the Government of the United States made the search and recovery of the satellite debris an efficient and noteworthy operation.

FOLLOW-UP SEARCH FOR POSSIBLE VAPORIZED FISSION PRODUCTS

The large-scale search program conducted by Canadian and American government agencies established that the Cosmos 954 reactor had partially disintegrated before reaching the ground and that a portion of the reactor material had impacted on Great Slave Lake (Bristow, 1978). It was considered possible by the present authors that a significant amount of fission products, vaporized by frictional heating in the earth's atmosphere, could have been released at high altitude and distributed over a wide area by the prevailing winds at the time of reentry. The search

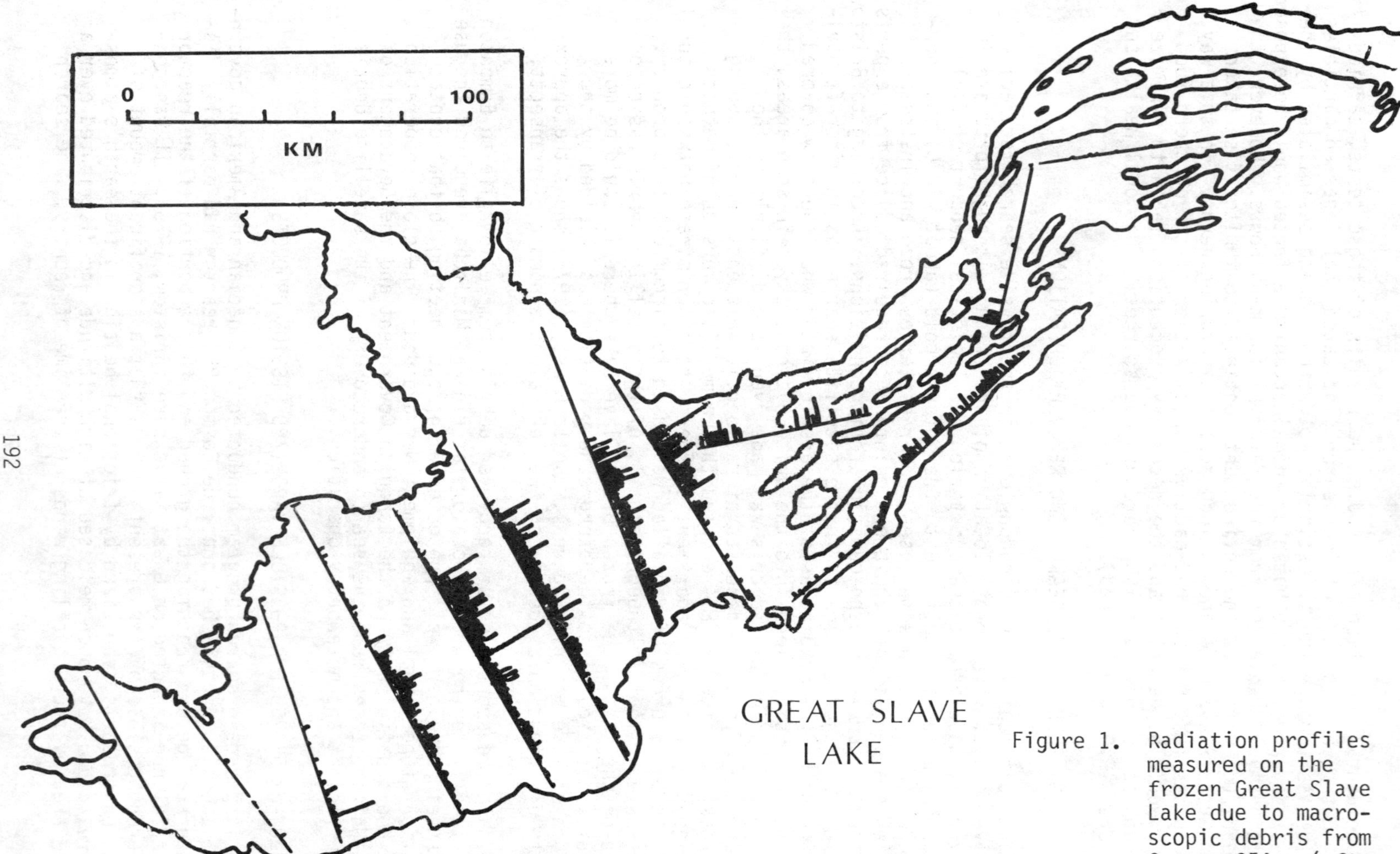

Figure 1. Radiation profiles measured on the frozen Great Slave Lake due to macroscopic debris from Cosmos 954. (after Grasty, 1978).

teams organized by the AECB recovered macroscopic pieces of debris. We, on the other hand, availed ourselves of an opportunity to investigate the distribution of airborne vaporized products which could contaminate vegetation in the same manner as nuclear fallout does. Unlike the large-scale airborne surveys supported by CAF aircraft, we used a helicopter and following the advice of DND personnel sampled vegetation rich with lichens point by point along 100 Km transects south and east of Great Slave Lake (Taylor et al., 1979). A total of 112 samples of lichens were collected at 10 km intervals along the transects in order to determine whether the region SE of Great Slave Lake had been contaminated by vaporized fallout. As controls we used lichen samples collected from arctic and temperate regions remote from the area of the satellite crash. All the samples were collected during the summer months of 1978 after the vanishing of the snow cover. Fallout from Chinese nuclear explosions carried out during late winter months (March 1978) was detectable on samples collected during the summer of that year, so that snow cover itself does not impede the detection of fallout on lichens by the methods described. An additional point may be useful concerning the removal of lichens through grazing. Only about 1% of northern vegetation is grazed by animals. Lichens grow extremely slowly and are selectively grazed by caribou. Clusters of lichen are normally pulled from the ground and completely consumed. It is therefore evident that the remaining lichen standing crop which we sampled was not influenced adversely by grazing. Figure 2 shows the impact area investigated by the DND and AECB and the sites from which we took lichen samples for counting. Of the eight sites compared in our original paper (Taylor et al., 1979) we present the most relevant gamma ray spectra of three sampling sites in Figure 3. The occurrence of ^{137}Cs ($T_{\frac{1}{2}}$=30.1 yr) is most apparent in all of the spectra. Short-lived isotopes ^{144}Ce (284.4 d), ^{125}Sb (2.73 yr), ^{7}Be (53.3d), ^{103}Ru (39.6 d), ^{106}Ru (369 d), ^{95}Nb (65.5d) and ^{95}Nb (35.1 d) were found in the lichen samples from the vicinity of the crash site to which we had access, as well as from those remote from the crash site. Contrary to our expectations, the line intensities were larger in the southern region of Niagara Escarpment, some 3000 km away from the crash site. There is no evidence from our data that would indicate that the vaporized core of the reactor on Cosmos 954 had made any detectable contribution to the fission product spectrum at sites near the impact zone. Since the gamma ray spectra were comparable for all the sites sampled, it was concluded that most of the short-lived radioactivity on the lichen samples was due to the Chinese nuclear explosions of September 1977 and March 1978, which masked low level contamination by Cosmos 954 if it occurred. A comparison of the spectra shown in Figure 3 suggests that the very fine airborne contamination from the Cosmos 954 did not substantially increase the level of fission product contamination at ground level at the sites sampled (Taylor et al., 1979).

UNITED NATIONS REVIEW OF THE PEACEFUL USES OF OUTER SPACE

Galloway (1979) has summarized several recommendations that address the problems of crashes arising from nuclear-powered spacecraft re-entry. The United States contributed a working paper to the United Nations Committee on the Peaceful Uses of Outer Space soon after the Cosmos 954 accident. The report suggested that the use of nuclear power sources

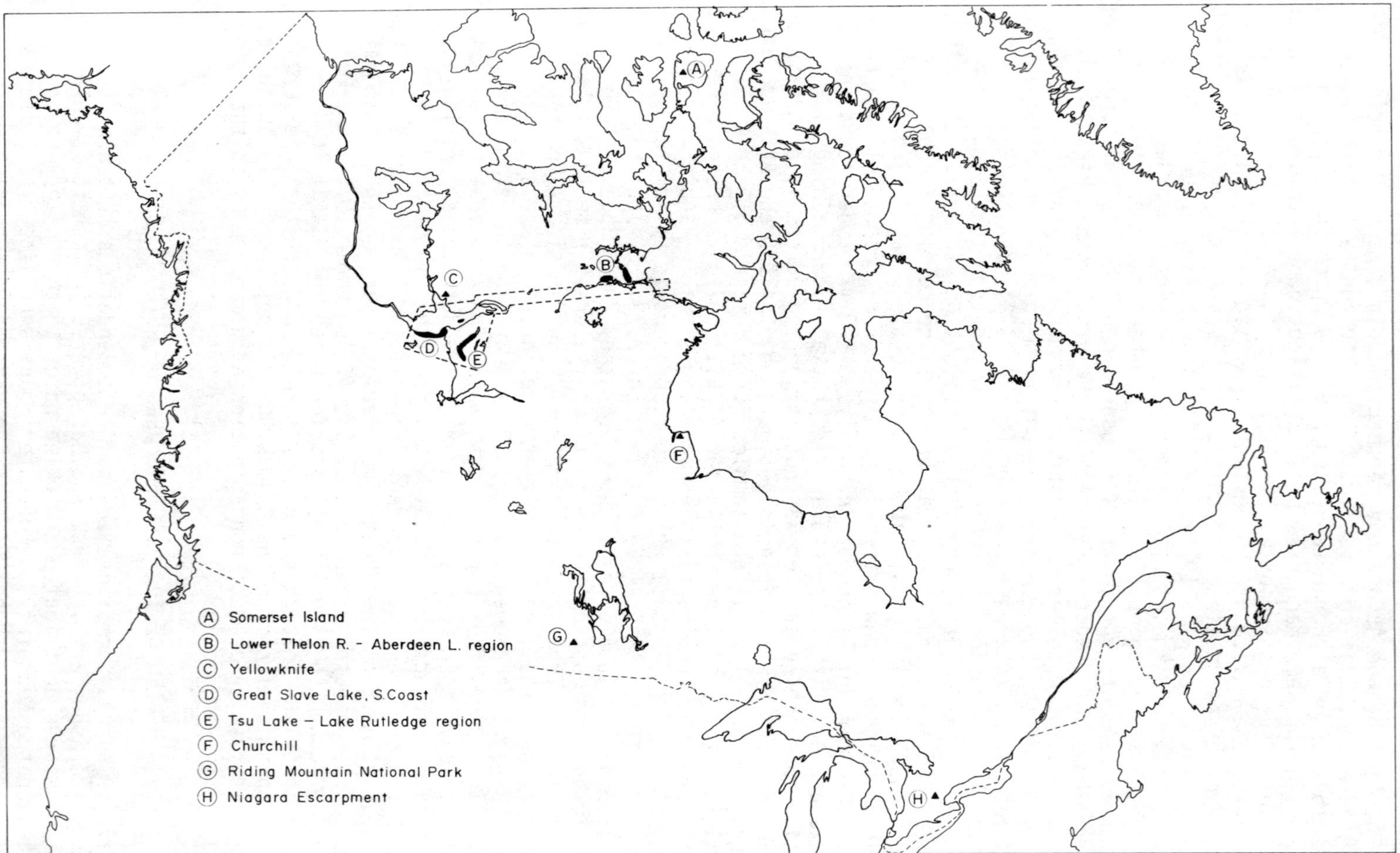

Figure 2. Simplified map of Canada showing the search corridor for Cosmos 954 debris (after Bristow, 1978) and the location of sites used for sampling lichens. (after Taylor et al., 1979).

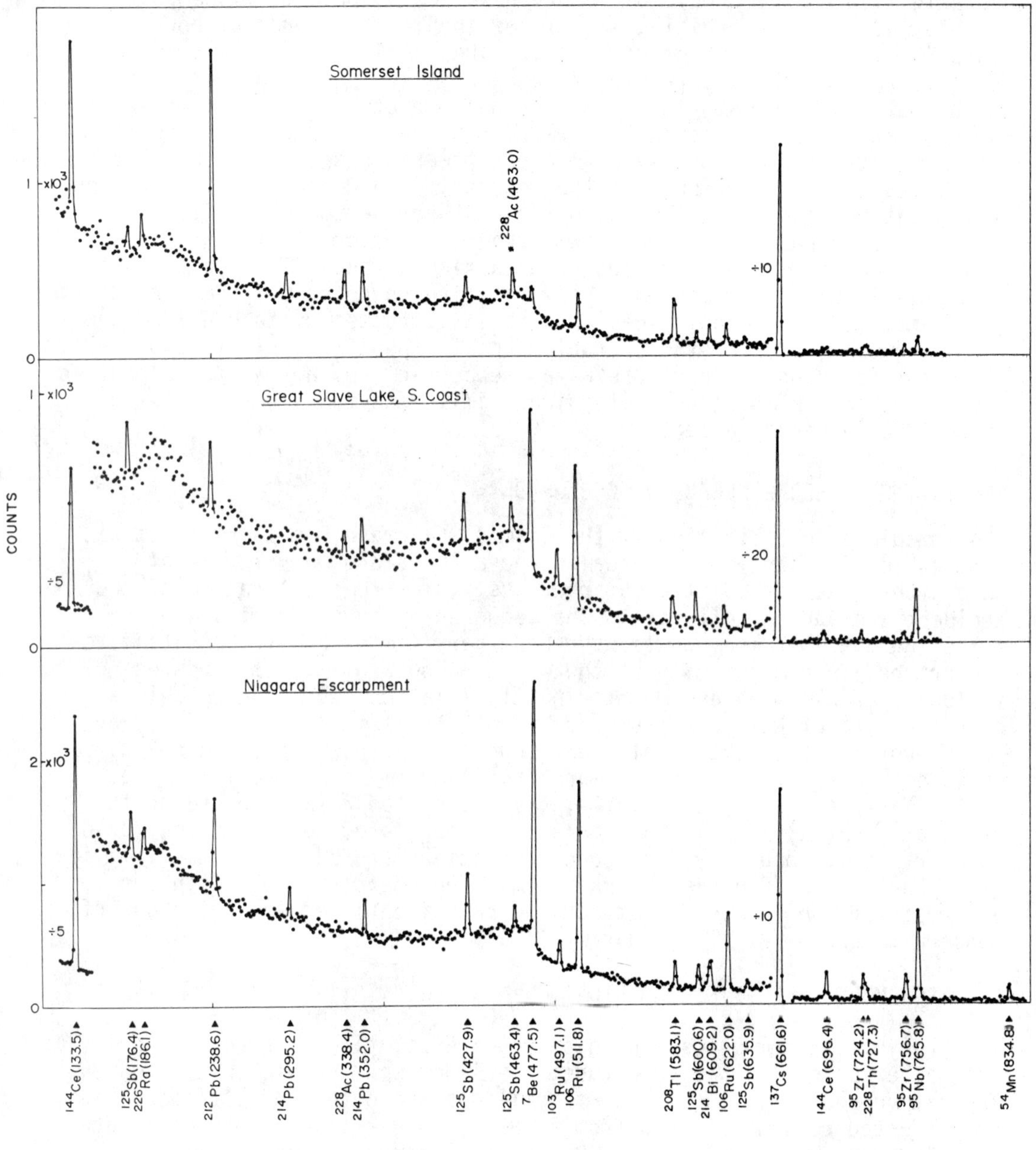

Figure 3. Gamma ray spectra obtained from lichen samples SE of Great Slave Lake (impact zone) and control sites. (after Taylor <u>et al</u>., 1979).

for satellite energy should be reserved for deep space missions only and all future earth orbiting satellites should be powered by non-nuclear means (U.N. Document, December, 1979). The complexity of the issue concerning the use of nuclear-powered spacecraft was underlined in recent United Nations debates. It is not the purpose of this paper to discuss recommendations already brought forward by others to deal with the problems arising from the use of nuclear power sources in earth-orbiting satellites. Rather, we suggest that scientific and technological refinements must be encouraged. We also expect that the current United Nations debate concerning the international legal regulation of the use of nuclear power sources in space will produce a lasting directive on the safe use of sophisticated technologies. We would like to point out that in addition to scientific and technological answers it is necessary to consider the psychological impact of these problems on the public. This is especially important for peoples in remote areas separated from normal communication channels when new emergency situations similar to the experience with Cosmos 954 in Canada arise.

PERSISTING CONCERNS REGARDING COSMOS 954

The immediate investigation of the Cosmos 954 crash was carried out with remarkable efficiency and unusual cooperation among many levels of government. As was pointed out previously, the remote location of the accident was indeed fortunate. The sequence of events during and following the emergency would have been very different if the satellite had crashed in a major city of the world. How would the procedures followed for clean-up and monitoring of radiation levels succeed in a city the size of New York, Rome or Tokyo? It is implicitly assumed by most countries that the development and use of a new space technology includes the risk of future failure, bodily harm and property damage. In the case of failing satellites, the calculated risk is based on the postulate that a direct impact on a major populated centre of the world has very low probability. As concerned scientists and observers familiar with the northern Canadian terrain and independently involved in the aftermath of the Cosmos 954 impact, we have identified several areas of concern to us.

NEED FOR MORE EFFECTIVE COMMUNICATION

The problems of communication which require solutions are not technological but rather psychological in nature: bewilderment, suspicion and fear. The native people living in remote villages and detached settlements in the affected region suffered from a lack of understanding of the nature of the danger they faced. The Inuktitut (Eskimo) language does not possess terms such as "radioactive hazard".

In order to minimize communication problems of this type in the future we would like to suggest that a universal code of Emergency Symbols be developed that would visually explain the nature of the imminent hazard, suggest emergency precautions and direct the distressed people to centres of aid. The system could be patterned after the international system of road signs which has been so successful in a number of countries.

Particular attention would have to be paid to the visualization of the more subtle hazards of radioactive contamination, spills of chemicals, toxic biological materials, etc. A comprehensive system of symbols should be designed by a group of social scientists working under the aegis of the United Nations. The authors have communicated this suggestion to the appropriate United Nations agency. As an example of an unconventional warning system used in the Canadian North, a sign informing the non-native population of dangers posed by the presence of polar bears is shown in Figure 4.

If the probability of a crash in an urban area is small then the probability of a crash in remote areas which are often populated with native peoples will be significantly larger. In such areas the hazards of modern technology are often unknown - perhaps even unrecognizable - to the inhabitants. In situations where the threat to human health is undetectable to the individual (e.g. nuclear radiations, biological infections, toxic chemicals) it is imperative that the authorities have at their disposal a clear method of communicating the nature of the hazard to native peoples. Anxiety, fear and panic are universal responses to unknown or unrecognized dangers. The Canadian experience with the Cosmos 954 affair illustrates clearly that the trust and confidence of the public can only be maintained by vigorous efforts to communicate effectively and be seen to communicate effectively.

The need for successful communication with the general public during an emergency situation does not end with the end of the emergency. In Canada the sense of frustration which literate, knowledgeable citizens have regarding the Cosmos incident is still in evidence. Although the official governmental agencies responsible for the search and recovery of radioactive debris did a creditable job, we suggest that the overall problem of communication with the Canadian public at large left something to be desired (Hutchison-Benson, 1981). The Government was berated publicly in the House of Commons for its apparent reluctance to issue a final report as quickly as possible (Broadbent, 1978).

During a relatively short time following the accident a number of technical papers dealing with various aspects of the Cosmos 954 incident were presented and published in the open literature (Bristow, 1978); Grasty, 1978; Galloway, 1979; Taylor _et al._, 1979; DND (Sentinel Publ., 1978); U.S. Department of Energy, 1978; Landingham and Casey, 1978. Most of these reports were published by experts directly involved in the search for radioactive debris, and their existence was known only to parts of the scientific community. In addition, a journalistic account of the accident was published in book form (Heaps, 1978). Also, several United Nations documents concerning the international implications of the use of nuclear power sources in outer space have been drafted and made available to the public. However, the official Canadian Government report which was designed to inform the general public, was delayed until the fall of 1980 (Gummer _et al_, 1980), more than two years after the event. The difficulty in obtaining official information tends to be compounded by the bureaucratic structure of most modern governments.

Figure 4. A sign used in the Canadian Arctic warning the non-native population of polar bear dangers. This sign is printed black on a fluorescent orange background to attract attention.

The tardy publication of official reports has only increased the uncertainty felt by the Canadian public over the Cosmos affair. Clearly the way around this problem is for official agencies to issue interim statements immediately and their final reports as rapidly as possible before they become objects of political pressure, or part of a forgotten history.

PARTICIPATION OF THE PUBLIC SECTOR IN EMERGENCY RESPONSE

One final recommendation that we make is that governments be encouraged to utilize private sources of pertinent expertise more effectively. By exploiting the non-government resources during and after the emergency situation and establishing an early liaison with the public sector, many communication problems could be eased. Teams of experts drawn from both governmental and private research and technical agencies would ensure that the results of crisis investigations reach the public domain with minimum delay. Furthermore, the presence of non-governmental experts would encourage the general public to constructively interact with the investigative agencies. Hopefully this liaison would also produce more ideas for solutions through the means of "cross-fertilization".

SUMMARY

The use of nuclear power as an internal source of energy for satellites is but one example of the modern perils for people on our planet. We believe that some improvements in our handling of accidents of the Cosmos 954 type could be realized through the acceptance of the recommendations made elsewhere (Galloway, 1979) and here. In summary our recommendations are: 1) the development of universal symbols to inform all affected people of the hazards confronting them, 2) the acceleration of the report-producing processes of investigative agencies and 3) the involvement of independent experts in investigative procedures as an interface between the general public and governmental bureaucracy.

We are likely to experience further major accidents during the years ahead. The problems arising from them will require solutions of ingenuity and courage. We should all unite in our efforts to preserve the healthy functioning of our newly conceived "Homosphere".

ACKNOWLEDGEMENTS

This work was supported by the NSERC grants to J. Svoboda and H.W. Taylor. Presentation of this paper by E.A. Hutchison-Benson at the 31st IAF Congress, Tokyo, was made possible by the financial assistance of the Science Advisory Board of the Northwest Territories, Emergency Planning Canada and the University of Toronto.

REFERENCES

Bristow, Q. 1978. *Geol. Surv. Can. Pap.* 78-113, pg. 151.

Broadbent, E. 1978. House of Commons Debates, November 15, Ottawa.

Department of National Defence. 1978. "Operation Morning Light. The Search for satellite Cosmos 954". *Sentinel* 14, 4-23.

Galloway, E. 1979. *Akron Law Review* Vol. 12 (3), Winter 1979.

Galloway, E. 1979. United Nations Consideration of Nuclear Power for Satellites 79-IISL-19. Preprint XXX Congress IAF. 1979.

Grasty, R.L. 1978. *Trans. Am. Nucl. Soc.* Vol. 29, 116.

Gummer, W.K. 1979. Summary of Cosmos 954 Search and Recovery Operation. AECB Report 1157 (unpublished).

Gummer, W.K., F.R. Campbell, G.B. Night and J.L. Richard. 1980: Cosmos 954. The occurrence and nature of recovered debris. Supply and Services Canada. pp. 60. Hull, Quebec.

Hanafee, J.E. 1978. "Analyses of Be Parts for Cosmos 954". *LLL Tech. Rep.* UCRL-52597.

Heaps, L. 1978. *Operation Morning Light: Cosmos 954.* pp. 208. Paddington Press Ltd. New York.

Hutchison-Benson E. 1981. "A Second Look at the Canadian Experience with Cosmos 954". *Emergency Planning Digest* 8(1).

Landingham, R.L. and Casey, W. 1978. Morning Light Clean-Up and Recovery Operation: Simulation Studies of Possible Reactor Fuels. *LLL Tech. Rep.* UCRL-52543

National Health & Welfare. 1979: Health significance of radioactive contamination from re-entry of Cosmos 954 over Northern Canada on January 24, 1978. Radiation Protection Bureau. Unpublished report.

Taylor, H.W., E.A. Hutchison, K.L. McInnes and J. Svoboda. 1979. "Cosmos 954: Search for Airborne Radioactivity in Lichens in the Crash Area, Northwest Territories, Canada". *Science* 205, 1383-1385.

United Nations Document. 1978. A/AC 105/L 102, March 15. U.S. Working Paper.

United States Dept. of Energy, 1978. Operation Morning Light, Northwest Territories, Canada, 1978: A Non-Technical Summary of U.S. Participation.

PROTECTIVE CLOTHING TEXTILE RESEARCH FOR SPACE ACTIVITIES IN THE 1980's*

M.I. Radnofsky+, Joseph J. Kosmo++

A survey is presented of the role played by protective clothing textile research for space activities in past years and its relationship to such activities in the 1980's.

Both intra-and extravehicular clothing is discussed together with its associated textile components.

Applications of both general and specifically designed fabrics, fibers and composites are discussed as they relate to military precursors of space attire plus Mercury, Gemini, Apollo, ASTP, Skylab and Shuttle clothing ensembles.

Newer concepts of protective clothing are related as they were adopted for the Shuttle program. Near-term future applications of textiles in space clothing are also discussed in terms of their incorporation into applicable potentially operational hardware.

Included in these future systems are such items as, construction workers space gear, a portable lunar survival shelter and an extravehicular tunnel suit.

It is concluded that for near-term applications in the 80's and perhaps early 90's international industrial textile companies are already in a position to fulfill requirements of textiles applications in space-related hardware. It is also concluded that no major breakthroughs will be required to supply this material in quantity or at a reasonable cost to support whatever program is embarked upon.

The development of an international data bank on textiles for space is recommended as one way to insure rapid and accurate dissemination of knowledge in support of these programs.

Keywords: Clothing, protective, intravehicular, extravehicular, functional garments, space suits, pressure suits.

---

* *This paper appears in full in Volume 41, AAS Microfiche Series.*

+ *Associate Professor, University of Houston, Houston, Texas 77004*

++ *Senior Project Engineer, Space Suits, Crew System Division, NASA, Johnson Space Center, Houston, Texas 77058*

IAA 80-16

ANNUAL SURVEY OF SPACEFLIGHT SAFETY SYSTEMS: 10th SUPPLEMENT SURVEY PERIOD: JULY 1979 -- JUNE 1980*

Nelson E. Brown, Jeri W. Brown+

Identifying and eliminating potential hazards, incorporating safety systems, performing systems analyses, and planning for future operations in space all serve to enhance the safety of crewmembers and ground personnel, to maximize the probability of mission success, to increase the integrity of payloads and experiment systems, and to provide the best possible equipment for current and next generation vehicles and space structures. As the maiden voyage of the Shuttle nears, efforts to ensure the safe return of the Orbiter and its crew are as prevalent as during initial design phases. And as the Spacelab design matures for a later Shuttle flight, analyses and testing continue to enhance crew safety and mission success.

In this paper, published documentation encompassing spaceflight safety developments for the July 1979 through June 1980 period is surveyed and typical reports summarized. Specific areas which are addressed include program guidelines and requirements, ground support and inflight crew operations, spacecraft hardware and components, and crew equipment and provisions.

Keywords: Safety; spaceflight; hazards; safety requirements; life support systems; thermal protection system; extravehicular activity; crew training.

---

* *This paper appears in full in Volume 41, AAS Microfiche Series.*

+ *National Aeronautics and Space Administration, Lyndon B. Johnson Space Center, Houston, Texas 77058*

WORLDWIDE DISASTER RESPONSE, RESCUE AND SAFETY EMPLOYING SPACE-BORNE SYSTEMS 1980

The following papers also appear in full in Volume 41, AAS Microfiche Series, supplement to this volume:

IAA 80-17, -18 and -22

The following papers, abstracted in this section, appear in full in Volume 41, AAS Microfiche Series:

IAA 80-19, -25, -26, -27, and -28

IAA 80-17

SATELLITES TO AID FLIGHT SAFETY

C. Rosetti*

The author discusses the problems of flight safety in relationship with economy of operations and with communication and navigation systems in use.

The prospective of a global approach using satellites both for communication and for navigation which might lead to a simplification of on-board equipment, more safety, and better economy is presented.

Keywords: Aeronautical communication, navigation, GPS, NAVSAT, maritime satellites, European Space Agency (ESA), anticollision, flight safety.

1. The increasing density of flights in terminal areas, the unavoidable mixture of slow and rapid traffic, and the steadily increasing direct operating cost, (mainly due to the increase of fuel prices), which calls for optimising flight profiles, have all a direct impact on flight safety.

In this paper, we examine the present situation and discuss the extent to which satellites could help to improve flight safety.

2. Flight safety requires:

- precise knowledge by the pilot and ATC (when available) of the position of the aircraft and its trajectory with respect to other traffic;
- good communications, in order to coordinate the traffic, and to acquire meteorological and other in-flight information.

We must therefore distinguish between areas covered by radar and possessing good radiocommunication infrastructures and those lacking such facilities (e.g., ocean and desert areas).

2.1 Areas covered by radar

Although in principle both ATC and pilots have at their disposal a reliable and precise technical infrastructure, a number of major mid-air collisions and

* *European Space Agency (ESA), 8-10, rue Mario-Nikis, 75738 Paris, Cedex 15, France*

the increasing number of near-miss reports call for a discussion of the situation, as illustrated by the following examples :

- The mid-air collision between a British Airways and a Yougoslav airliner, in the Zagreb Control Area, was due to an ATC mistake that gave the latter aircraft a descent clearance cutting the trajectory of the former. The British pilot was unaware of the conflict created by the ATC, because communication was in Serbocroat.

  One cause of the accident, therefore, was lack of communications

- The collision on the ground at Tenerife between 2 B-747s (KLM & Panam) occurred in low visibility (fog on runway) ; in addition to a procedure fault by the KLM captain, loss of communications (due to interference) prevented him from receiving a vital piece of information, namely, that the runway was occupied by the taxi-ing Panam aircraft.

  In this case again, one cause of the accident was lack of reliable communications

- The mid-air collision at San Diego between a 727 airliner proceeding to a visual approach and a small aircraft carrying out instrument flight practice could not be avoided, although both pilots were informed by ATC on the developing conflict, but, for whatever reason, they failed to identify each other visually.

  This example highlights the lack of objective information in the cockpit which often presents positive identification of other traffic.

  Many other examples could be given to show that, even in areas having the best available infrastructures for surveillance and radio communications, accidents and near-misses happen due to :

  - communication failures of all kinds, including language
  - lack of a cockpit display of traffic.

2.2 Areas outside radar coverage and lacking a suitable communication infrastructure

In some parts of the world, aircraft sometimes arrive at destination before their flight plan which could not be transmitted due to the poor communication network.

In desert and ocean areas, the lack of radar coverage and the consequent lack of precise information on aircraft positions (obtained only from crew reports) imposes large separation minima. The chances of achieving the optimum flight profile while remaining within the safety envelope are therefore severely reduced. Navigation errors, attributable to either equipment or crew failures, although rare, introduce a risk factor that cannot be disregarded, particularly if one attempts, for the sake of economy, to decrease the separation between aircraft.

In such areas, a fail-safe navigation and communication system could lead to more efficient use of the available airspace, and hence increased economy, while at the same time increasing flight safety.

3. Present system and ways to improve it

3.1 Navigation

For long-distance flights, inertial navigation systems are currently used, and, more recently, Very Low Frequency (VLF) ones.

Reliability is generally good, partly due to on-board redundancy, but neither system is error-safe.

The inertial system is only as accurate as the data fed into it by the crew. Its accuracy is further diminished by drift, and its confidence level limited by the possibility of failure.

In some regions, the VLF system is subject to ambiguities or errors, usually affecting all aircraft in a given area and thus increasing the risk factor.

The altimetric system, in use virtually since the dawn of aviation, is reliable but not very accurate at high altitudes.

Recently, the USA began to introduce the "Global Positioning System" (GPS-NAVSTAR). Such a system enables extremely accurate tri-dimensional determination of the position of a mobile. GPS-NAVSTAR, however, is a military development, and therefore not available for civil navigation, or at least not with the high accuracy that could make it attractive compared to the systems now in use. Moreover, GPS-NAVSTAR is expensive for the civil user, because it complies with military specifications (e.g. with respect to jamming resistance). The possibility of a civil satellite navigation system will be discussed below.

3.2 Communications

Two factors determine the ability of the available frequency spectrum to satisfy the needs in terms of both capacity and geographical coverage.

(a) The current use of voice communications, although indispensable for certain phases of flight when the real-time, conversational mode is required, is very wasteful of the frequency spectrum.

Certain flight conditions permit the use of "keyboard/electronic display" alphanumerical digital time-division communications, which offer :

- highly reliable, interference-free communications, without linguistic problems ;
- rapid access both to and from ATC as well as to data banks for meteorological or other in-flight information ;

(b) In areas where propagation conditions prevent the use of VHF (e.g. over deserts or oceans) the existing HF system, although generally satisfactory - particularly since the introduction of SSB techniques -, does not offer much scope for future development.

One other aspect deserves mention. Aircraft flying at high altitudes are potential VHF frequency polluters, because of their wide radio horizon. The reduction of channel spacing from 50 to 25 KHz has made the capacity of the 118-136 MHz band adequate, but increasing air-traffic density may mean that this will not always be the case.

Limiting the use of this band for voice communications up to a given flight level, at which aircraft usually pass from the take-off/climb phase to the cruising phase, and using at higher altitudes L-band digital alphanumerical communications via ground terminals or satellites, depending on the region overflown, could significantly decrease crowding in the VHF band and improve the speed of access and reliability of communications.

3.3 Proximity warning/anti-collision devices

As already mentioned, experience has shown that even in the most developed air traffic control system the lack of objective information on the traffic environment in the cockpit has led to accidents and creates dangerous situations every day.

Several proposals have already been made and various solutions are currently being studied. They rely either on data acquired via the radar surveillance system or else on the active, air-to-air transmission of proximity warnings and evasive manoeuvres.

Another possible approach would be to try to integrate several functions into a single black box.

Accordingly, instead of tackling the problems of navigation, communications, proximity warning and anti-collision measures separately, we should attempt to define an integrated system, based on terrestrial infrastructures over developed continental areas, and on satellites in ocean and desert areas for communication as well as for global navigation.

3.4 Equipment existing on board aircraft

The communication and navigation equipment on board aircraft today is a beautiful representation of the history of telecommunications. The frequencies used range from the lowest ones (10 kHz) to the X-Band. Most known types of modulation can be found as well. And as time went on, we have added black box after black box, and of course the corresponding indicators and antennas. Among the devices on board, we will find the altimeter (barometric) which has been used since the beginning of aviation. Its accuracy at high altitude is poor. Fig. 1 illustrates a typical equipment set up found today in most civil aircraft (in duplicate for sake of redundancy : duplicates not illustrated). Current trends suggest that we might add in the near future two more boxes : one for the microwave landing system (MLS) and another for collision avoidance (thick frames in fig. 1).

The problems to be considered when looking at this equipment stack are :

- capital investment
- maintenance cost
- cockpit space required
- mass
- antennas (in particular on high-speed aircraft)

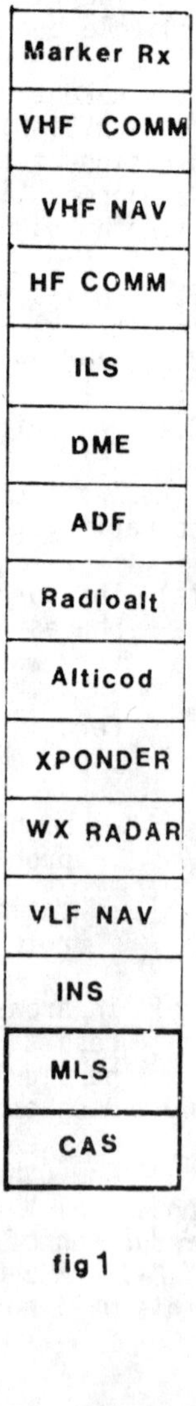

fig 1

3.5 Terrestrial infrastructures

For most of the boxes illustrated in fig. 1, there is a corresponding terrestrial network. The capital investment and in particular the high maintenance cost represent significant amounts, which will have to be taken into consideration when new systems are evaluated.

The impossibility of providing satisfactory (or indeed any) services in some parts of the world (e.g. ocean or desert areas) reduces the efficacity of most of the equipment on board. When it can be used, its accuracy and its expansion potential are both relatively limited.

4. First thoughts on a new approach

Mobile activities, be they aeronautical or maritime, are global. Therefore, when considering new options for navigation and communications, our approach must be global too. Furthermore, we must take account of the commonality between the aeronautical and maritime requirements.

In the following paragraphs, we shall suggest improvements that might make qualitative contributions both to navigation safety and to the economics involved.

4.1 Communications

As already mentioned, voice communications, although inefficient and frequency wasting, will have to be kept for the phases of the flight where instantaneous conversational mode is indispensable. That means retaining on board the aircraft the contemporary VHF communication equipment. For the other phases of the flight, alphanumerical, digital communications using terrestrial infrastructures where possible and geostationary satellites in the other regions might be a good solution.

The most suitable frequency range allocated to this service in the radio regulations is in the L-Band.

The considerations relevant to such an approach are :

- possibility of frequency re-use
- low-power transmitter, low aircraft-antenna gain
- low power on board the satellites

4.1.1 Link power budgets

The link budgets below are calculated for a transmission speed of 2400 bit/sec.

(a) satellite-aircraft link :

- satellite EIRP (per carrier)	34 dBw
- free space losses	- 188.5 dB
- margin	- 5 dB
- mobile terminal G/T	- 24 dB/°K
- up-link noise and intermodulation in the satellite	60 dBHz
- down-link c/No	45.1 dBHz
- resultant c/No	45 dBHz

(b) aircraft satellite link :

- mobile station EIRP	11.7 - (x) dBW
- margin (transmission stability)	- 1 dB
- propagation margin	- 5 dB
- free-space loss	- 188.5 dB
- satellite G/T	(x) dB/°K
- up-link c/No	45.5 dBHz
- down-link (C-Band) c/No	54 dBHz
- resultant c/No	45 dBHz

Notes on link budgets :

(a) Quality required for transmission

$$\frac{E}{No} = 10 \text{ dB where } P_E = 10^{-5}$$

$$\frac{C}{No} = \frac{E}{No} + R = 10 + 33.8 = 43.8 \text{ dBHz}$$

Equipment margin	1.2 dB
Quality required	45 dBHz

(b) A propagation margin of 5 dB has been allowed. Furthermore, protection against multipath effects could be improved by suitable coding. Taking together, these steps should be sufficient to ensure very good link quality.

4.1.2 Power required on board satellite and aircraft (per channel)

(a) Narrow, high-gain beams must be used, in order to : permit frequency re-use ; decrease power requirements on board the satellite ; and permit a low-G/T mobile terminal. The gain at the edge of a 2° satellite beam is about 35 dB. The power required by the satellite antenna from the power amplifier is therefore about - 1 dBW.

(b) Assuming a satellite G/T of 5 dB/°K, the transmitter on board the mobile (aircraft in this case) must deliver to a 0 dB gain antenna a power of 11.7 - 5 = 7 dBW.

These powers are extremely small compared with those envisaged in previous studies.

They offer sufficient margin for optimising the system (e.g. an increase of bit rate if required) while in principle remaining within an acceptable economic enveloppe.

4.1.3 Space-segment configuration

Since the system must serve all regions lacking suitable infrastructure (mainly ocean and desert), a space segment comprising 3 geostationary satellites is needed in order to achieve global coverage. While in this paper we only discuss the aeromobile system, we do not exclude the sharing of the space segment by several types of mobile user (e.g. maritime) who could benefit as well from the low G/T mobile terminal possibility. Furthermore, we shall bear in mind the fact that such a maritime system is currently being set up.

4.2 Navigation

In the main, terrestrial-based systems are currently used for aeronautical navigation, e.g. non-directional beacon (NDB) ; very high frequency "VOR" ; very low frequency (Omega type). When terrestrial systems are not available, on board inertial systems are used.

Auxiliary aids for poor-visibility landing, distance and speed measurements or identification (radar transponders, vertical beacons) complete the picture.

The U.S. Global Navigation System providing high-accuracy positioning and navigation is not and will probably not be available for civilian use, at least in its high-accuracy version.

The questions are then :

- Could a civilian satellite navigation system be envisaged and what would its impact be on safety of navigation and economy of operation ?

- Would the aeronautical community be the sole user to benefit from such a system ?

4.2.1 A possible system concept

Military global satellite navigation systems are expensive for a number of reasons, including :

- the requirement for resistance against jamming
- exclusive use of the system by one user, or a small group of users, for defence purposes only.

In the GPS/NAVSTAR system, 24 satellites (currently only 18) are spread over 3 orbits inclined at 60° to one another. Each satellite carries an atomic clock, a processor and a transmitter broadcasting (in the L-Band) high-accuracy time and satellite-position signals to passive mobile receivers. This information is, of course, encrypted. Using data from at least three satellites, the mobile receiver calculates the intersection point of three spheres, thus obtaining its position in latitude and longitude as well as altitude (with respect to the geoid). Each time a satellite passes through the visibility area of the ground tracking station, its clock and its position knowledge are corrected.

For civilian use, such a system is not required, since there is no need to protect the satellite receivers from intentional jamming. One might therefore withdraw the atomic clock and processor from the satellites and install them in a terrestrial facility. The time and position signals would there be suitably processed and corrected for ionospheric disturbances and trajectory, and then transmitted via the geostationary communication satellites to the navigation satellites, which would in turn broadcast them to the users.

Under this arrangement, the navigation satellite would become a very simple, cheap and reliable radio relay, while the very expensive equipment would be ground based. Since the latter do not need to be space qualified, its cost is considerably reduced. Moreover, a high degree of redundancy can be built in, thus giving the user improved reliability and availability at low cost.

Figs. 2 a and 2 b illustrate the space segment and links of such a system.

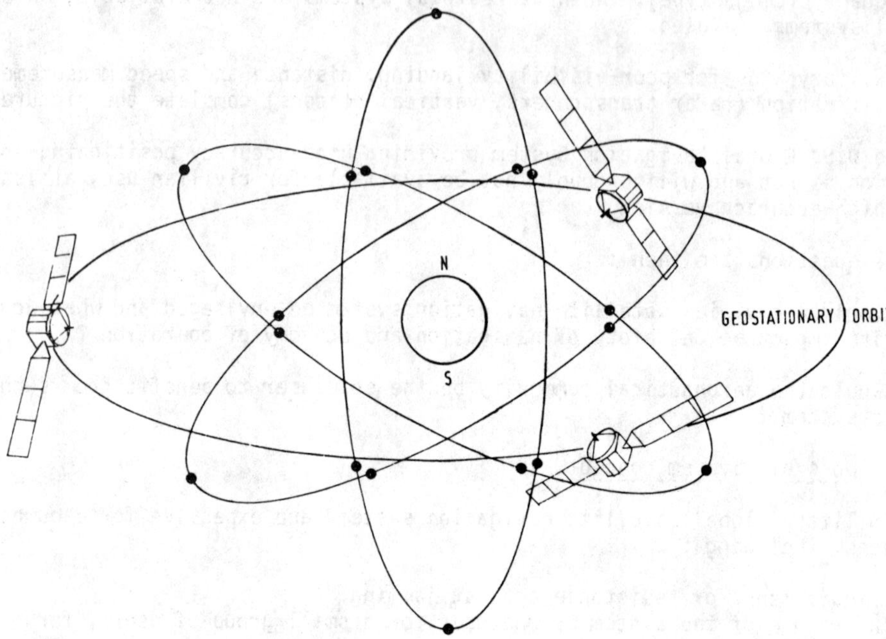

FIG. 2 a

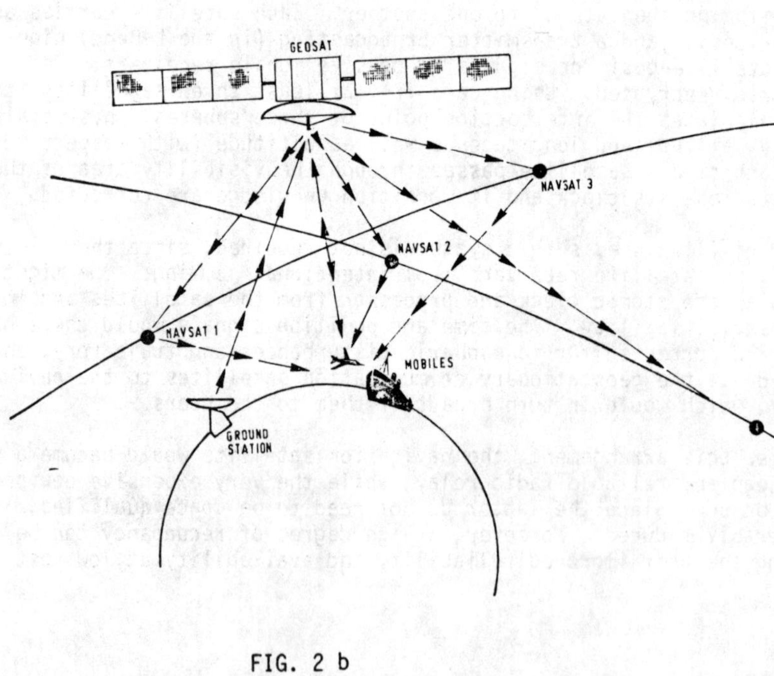

FIG. 2 b

5. Consequences of the adoption of a satellite communication and navigation system

The adoption of such a system could result in :

- improved functions offered to the crew with less complex equipment
- economy at system level
- economy of aircraft operations
- increased flight safety.

5.1 Functions that can be provided by an integrated L-Band satellite system

(a) Two-way radiocommunication with a high degree of reliability and availability

(b) Precise en-route navigation all over the world

(c) Precise altitude determination, even at high and very high altitudes

(d) Accurate speed, time and distance-to-station measurement

(e) Possibility of locating "ghost" beacons with high accuracy

(f) Blind approach and landing at any airport in the world, since no terrestrial infrastructure required

(g) Highly reliable dependent surveillance through fail/error-safe, precise, automatic altitude/position reporting

(h) Integrated traffic display - proximity warning function on board aircrafts.

5.2 Economy prospects opened by the system

In fig. 1, we showed the complexity of radio and navigation equipment on board aircraft that arose from the "box after box" introduction of functions over the years.

By adopting a L-Band satellite communication and navigation system, both functions might be served by one common equipment on board the aircraft.

Fig. 3 illustrates what equipment will be found on board the aircraft after such a system has been introduced.

Of course, the customary equipment redundancy must be provided.

As already explained, large terrestrial networks will then disappear as well.

Both simplifications will lead to significant savings, mainly in maintenance costs.

These will of course be offset (at least in part) by the capital and running costs of the new satellite system and its terrestrial infrastructures,

fig 3

but the functions provided as well as its flexibility will open other possibilities, such as :

- reduction of vertical separation from 2000 feet to 1000 feet, above flight level 290
- reduction of lateral and longitudinal separation in regions outside radar coverage
- more degrees of freedom in routing aircraft along the optimum flight profile and track
- more flexibility in handling traffic in the crowded terminal areas. All these have a direct impact on the cost of operations.

5.3 Safety aspects

Besides better communications, which was mentioned earlier, objective information on surrounding traffic can be displayed on board, with high accuracy since the position and altitude information is derived from a highly precise and fail-safe navigation system. One possible way of achieving such a display system is described below.

Principle of the anti-collision/proximity-warning system

In order to display the nearby traffic, each aircraft is surrounded by a "safety" volume, that moves with it. The volume is further divided into blocks numbered 1 to n (see fig. 4). The dimensions of the volume and of the blocks may be determined according to the nature of the traffic. They may also be altered, if need be, as time goes on, without changing the system. The block layout may likewise be altered according to the flight phase, but will always be the same for a given phase (see fig. 5 a).

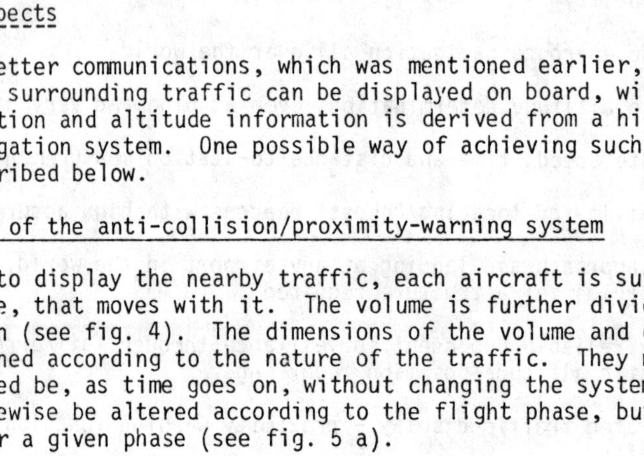

FIG. 4

Where there is radar coverage, the position and altitude of aircraft are known to the control centre. A reading device in this centre's computer explores the safety volume associated with each aircraft, and if another aircraft enters it the centre will inform the first one of the number of the block occupied by the second. The horizontal position of the second aircraft will then be directly displayed on board the first. Any traffic that is conflictual or simply in the neighbourhood of the first aircraft will therefore appear as a luminous spot giving its real direction and relative distance.

The vertical situation will be displayed in code, either by symbols (Λ, ✱, V) and/or in colour (blue = up, red = same level, green = down). This display will enable the positions and movement of traffic to be interpreted unambiguously (see fig. 5 b).

In areas without radar coverage, the ATC centre will receive information via satellite on the position and altitude of aircraft by means of an automatic reporting system. The ATC will read the position in either the inertial platform(s) or the radio navigation system (VLF, Navstar, etc.) on board the aircraft.

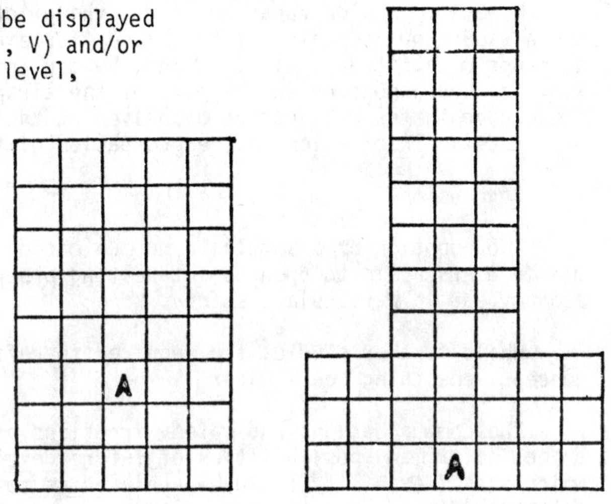

FIG. 5 a

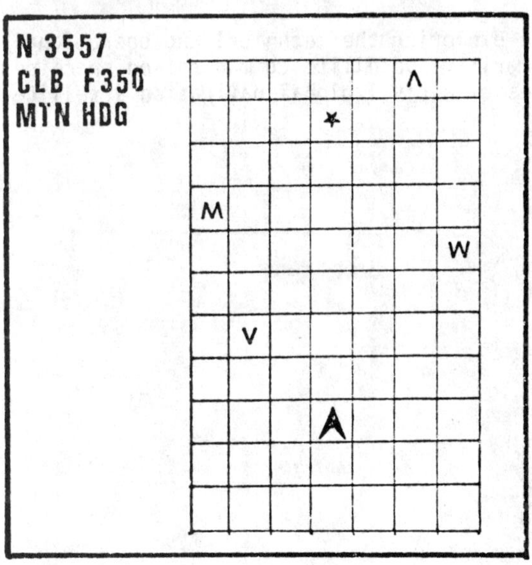

Own aircraft symbol -Own aircraft
✱ -Acft at same flight level
Λ V -Acft one level higher/lower
M W -Acft two level higher/lower

FIG. 5 b

The altitude will be read in the alticoder. This data will be transmitted to the ATC centre where they will be stored and displayed. In the centre, a reading device will explore the safety volume associated with each aircraft and, if another aircraft enters it, the number of the corresponding block will be immediately transmitted to the first aircraft, for display.

The arrangements for display will be the same as in regions with radar coverage.

However, outside radar coverage, the pilot will also be able to explore the airspace outside his safety volume (for example when he wishes to request a change of altitude). To this end, he can introduce fictitious coordinates, x, y, z, corresponding to the part of the airspace that he wishes to explore. These coordinates will not be displayed at the ATC centre, but the aircraft will receive information on the occupation of the space concerned.

6. Conclusions

The adoption of a satellite communication and navigation system serving mobile users seems to open up interesting prospects, in the fields of both economy and in particular, safety.

While further studies are required to refine the ideas and the system concept, one thing seems clear :

The communications and safety functions provided by systems in use today do not offer many possibilities of future development and the only way in which such flexibility could be achieved is to make use of the particular features offered by satellite systems.

The European Space Agency is exploring the technical and operational characteristics of a joint aero-maritime satellite communication space segment and is just about to begin studies of a civil global-navigation satellite system.

IAA 80-18

COMPUTER SIMULATION MODEL
ADJUNCT TO EMERGENCY SATELLITE COMMUNICATIONS SYSTEMS: APPLICATIONS IN CIVIL MASS CASUALTY DISTRESS RESPONSE

Paul B. Richards*

The background of a computer simulation model to determine resource requirements for wartime contingencies and to design medical support for such contingencies is presented. The first version, called NAMES (Navy Amphibious Medical Evacuation Simulation), was initiated in 1975 and has since been enhanced. The model completed in January 1980 is called the World Wide Military Medical Support System (WWMMSS). The basic concepts of this simulation model for worldwide disaster and rescue response are given.

Since 1975, medical planners in the U.S. Military Departments have used a computer simulation model developed by the author to determine resource requirements for wartime contingencies and to design medical systems to support such contingencies. Originally constructed for the U.S. Navy, it was called the NAMES (Navy Amphibious Medical Evacuation Simulation) Model. The first version of NAMES simulated the medical treatment, hospitalization and evacuation of combat casualties within the actual zone of combat. Two subsequent versions of NAMES expanded this simulation capability to include worldwide medical support and evacuation. During the past year, still more enhancements have been added to the model, and the latest version was completed in January, 1980. This model is named the World Wide Military Medical Support System (WWMMSS) simulation model, a title which is more indicative of this model's scope and capabilities than was the title originally given to the family of NAMES models which were developed during the period 1975-1979. The WWMMSS Model is written in the SIMSCRIPT II.5 computer programming language, which was especially designed for simulation. The model currently operates on the Control Data Corporation (CDC) 170 series computers, which are part of the worldwide CYBERNET system.

* *Naval Research Laboratory, Washington, D.C. 20375*

The basic concepts and features of the WWMMSS Model make it an appropriate tool for worldwide disaster and rescue response employing space-borne systems. First, the model concepts and the model itself are unclassified, and documentation of the model is available to the public. Second, the model is, generically, a tool for predisaster planning, predisaster training, and immediate postdisaster response for local or international disasters, either natural (e.g., earthquakes, floods, storms, volcanic eruptions) or man-made (e.g., aircraft accidents or nuclear power plant accidents). While the model was developed to consider primarily mass casualty disasters, its concepts apply as well to routine emergencies such as automobile accidents.

Basically, the model works as follows. First the area whose inhabitants would be affected by the disaster is defined, e.g., a river valley prone to floods or a community within a 20-mile radius of a nuclear facility. Within this area, predisaster conditions which will be affected by the disaster must be specified, for example: the population, including the elderly and the institutionalized; available modes of travel and access routes into and out of the area; locations of hospitals and other medical facilities, together with their normal staffs and average normal patient populations. In addition, external resources which will respond to the disaster are specified, both their location and their quantity; for example, rescue helicopters, rescue aircraft, and additional medical and rescue teams that may be brought into the area from other countries. The disaster scenario is next superimposed upon this predisaster situation. Depending on the type of disaster, there will be medical casualties, injured to be cared for, and people who must be evacuated. The current model identifies each casualty and follows that casualty throughout his/her medical treatment (if needed) and evacuation. Medical and evacuation requirements are recorded, as well as the utilization of available medical and rescue personnel and evacuation vehicles. The model simulates the actual movement of rescue and evacuation vehicles and records how long it takes to carry out the rescue operation, as well as the number of casualties who die. In this way the model predicts the effectiveness of the planned response system, its resources and its operation. If the planned disaster response system does not work efficiently when simulated, the model inputs can be modified until improved response systems are found.

Used in this way, the WWMMSS Model would be a significant aid in predisaster planning. Depending on the kind of disaster anticipated, space-borne remote sensing would provide valuable data needed for the model during this preparedness phase. Once an acceptable disaster response system is achieved with the assistance of the model, the simulated disaster response operation can be used as the basis for training exercises, using local inhabitants as well as international rescue teams and facilities. During this phase, space-borne communications systems would be used. In the event of an actual disaster, the WWMMSS Model, in conjunction with space-borne communications systems, could provide real-time information and direction to the disaster response operation. If this disaster had been anticipated and previously modeled, response to the disaster would be initiated according to earlier planning and training, predicated on the simulated disaster. As soon as information regarding the actual disaster would become available, this data could be used to refine the model's predisaster predictions and the current information would be immediately communicated to rescue teams at the site of the disaster, as well as to international groups involved with the disaster operation. If the disaster had not been previously simulated, the model could still accept data during the early postdisaster phase and within hours could provide valuable information for the disaster response operation. Space-borne communications systems would play a prominent role during this immediate postdisaster phase.

In summary, remote sensing satellites and communications satellites would play a significant role, in conjunction with a computer simulation model, in preparedness planning prior to a disaster and in response operations during a disaster. In the predisaster planning phase, satellites would employ remote sensing to locate potential natural disasters and thereby contribute to the construction of the model. Satellite communication channels would then be established to test proposed disaster response systems, and to assist in the training of disaster response personnel. During a disaster, satellites would perform two functions: (1) they would transmit emergency communications required in the actual disaster response operation, and (2) they would transmit on-site data to locations where the model would be used to simulate conditions corresponding to those existing in the actual disaster. Information obtained from the model would then be relayed back to the disaster response personnel to help direct the disaster response operation.

IAA 80-22

DISASTER WARNING USING THE GOES SATELLITE

J. Gordon Vaeth*

The islands and territories of the Pacific Basin frequently feel the fury of tsunamis, often called seismic sea waves or tidal waves. Unfortunately communications through the area and to its inhabited places are not always timely or reliable. The interrogation downlink (468 MHz) of the GOES Data Collection System holds promise for improving the ability of the peoples of the Basin to receive tsunami and other natural disaster warnings. This paper describes how this could be achieved using that link with a specially designed low cost (less than $1000 U.S.) receiving station that would issue an oral warning in the local language.

Twenty years ago---at 1:05 AM on May 23rd 1960---a wall of water 20 feet high roared ashore at Hilo on the island of Hawaii. It smashed into the downtown area, demolishing ten city blocks, destroying 537 buildings, and causing $23 million of damage. There were 62 people killed and 282 injured.

What struck Hilo that morning---and what had struck it in a like manner 14 years before leaving 173 dead---was a tsunami or seismic sea wave. It had originated off the coast of Chile and had travelled across the Southeast Pacific to Hawaii at an average speed of 440 miles per hour (710 km/hr).

Tsunamis are among the most dangerous and destructive of all Nature's forces. Less than a year ago, for example, on December 12, 1979, a tsunami created by a local seaquake took a toll of 200-800 dead in Western Colombia. So widespread is the tsuanmi danger throughout the Pacific Ocean Basin that some dozen nations, including Canada, Chile, Ecuador, France, Guatemala, Japan, Korea, New Zealand, Peru, the Philippines, Thailand, the USSR, and the United States have formed an international coordination group for tsunami warning in the area. The National Weather Service of the United States operates two tsunami warning centers for the Pacific, one in Palmer, Alaska, the other near Honolulu, Hawaii.

* *Director of Systems Engineering, National Environmental Satellite Service, National Oceanic and Atmospheric Administration, U.S. Department of Commerce, Washington, D.C. 20233*

The hardest part in giving warning of an approaching tsunami is not in detecting it---Nature sees to that---but in communicating the warning quickly and reliably enough. Conventional communications throughout the Pacific Basin, the main tsunami area, leave much to be desired. Tests of the tsunami warning network have revealed delays in warning reception incompatible with public safety.

How can the timeliness of these warnings---and the number of inhabited places in the Pacific that can receive them---be improved upon?

GOES, the Geostationary Operational Environmental Satellite of the United States, holds promise of making this possible by broadcasting them over the platform-interrogation downlink of its Data Collection System (DCS).

GOES is best known for the meteorological imagery provided by its Visible and Infrared Spin Scan radiometer. Audiences, even astronautics audiences, are far less familiar with its Data Collection System involving the observations made by buoys, river gages, tide gages, seismometers, aircraft, magnetometers, and other *in-situ* environment-sensing platforms. Radio sets that are part of these platforms transmit the observational data up to GOES which relays them to earth stations for centralized or individual distribution to users. Those radio sets operate in three modes: (1) self-timed, (2) adaptive random reporting, which is to say they come on the air only when they have some data of significance to report, and (3) interrogated. Interrogation is accomplished by a downlink (468 MHz) from the spacecraft which uses it to send addresses with accompanying interrogations and commands.

The National Environmental Satellite Service routinely keeps two GOES in operation over the equator at 75° West and 135° West. The westernmost has a communications field of view encompassing most of the Pacific Ocean Basin, the tsunami-prone area. The downlink over which it broadcasts interrogations and commands to DCS platforms in that region could also be used to broadcast warnings to islanders and coastal dwellers there.

We know that technically this can be done, because the technique that would be employed is already in use. GOES has used it for some years to broadcast National Bureau of Standards UTC (Coordinated Universal Time). It is a technique that transmits a message one piece at a time and then re-assembles it on the ground into the whole message. It is a technique that utilizes the first four bits of every sequential interrogation or command sent over that 468 MHz link.

Each of these interrogations or commands is 50 bits long and takes ½ second at the transmission rate of 100 bits/second. Sixty such transmissions---in other words 30 seconds---are used to send the complete time information consisting of year, day, hour, and minute. After the complete time message has been sent, received, and displayed, a 30-second pause could follow, after which the time transmissions would begin again.

Coverage Area of the GOES Satellites

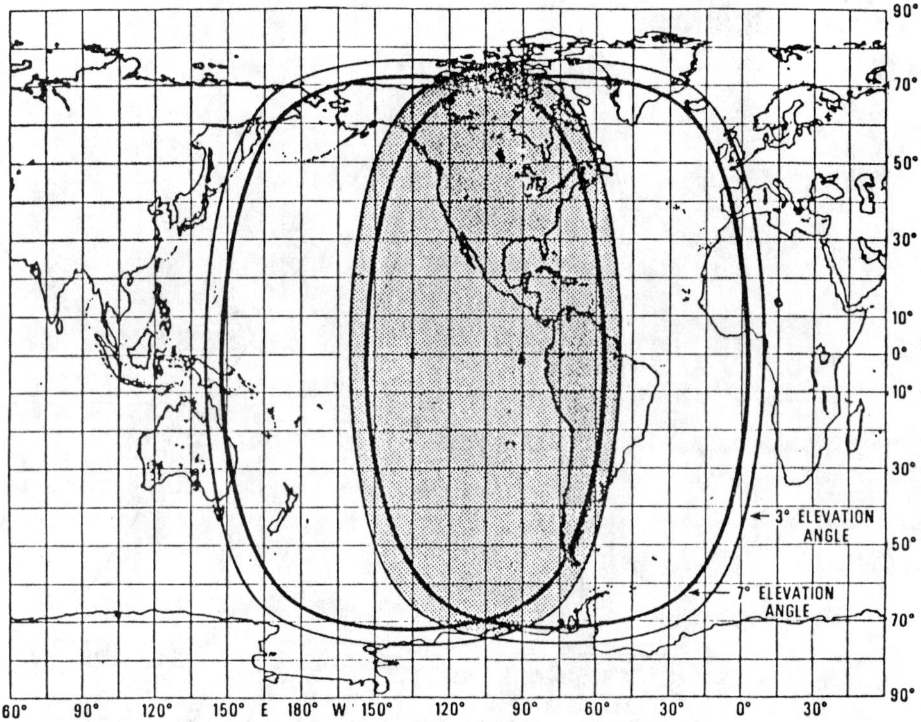

(AREA OF EAST AND WEST GOES RADIO OVERLAP IS SHADED GRAY.)

THE RADIO HORIZON OF THE WESTERN GOES (135° W) COVERS MOST OF THE PACIFIC BASIN

PACIFIC DISASTER WARNING

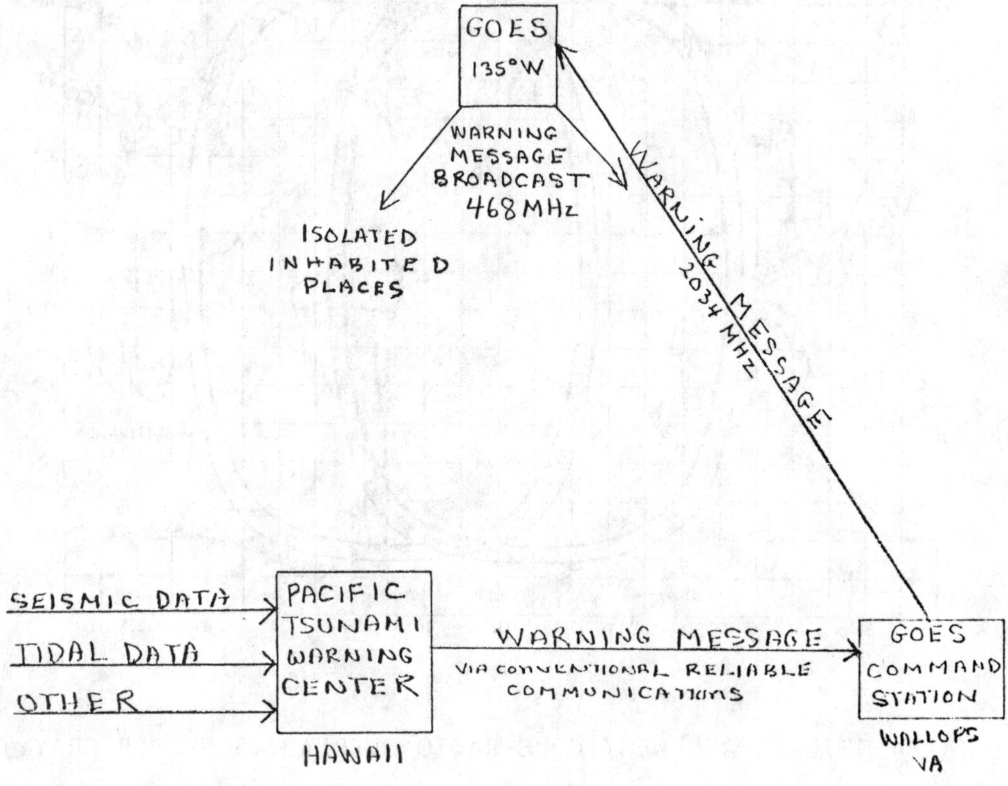

This half-minute interval between time transmissions is when the tsunami or other natural disaster warning messages could be sent. During this period, instead of 4 bits of time information, 4 bits of tsunami information would be transmitted every half-second. In the 30 seconds, a warning message comprising 240 bits could be sent. Then another 240 bits a half-minute later and so on...alternating tsunami data with time code data.

The warning message sent in this manner would be short, succinct, and standardized. The elements---wave arrival time, arrival direction, and the like---would be standardized as would be the format of the tsunami warning message and would be stored in memory. The transmission on the DCS downlink would provide the data to "fill in the blanks" in that standardized, pre-prepared, form message. At the warning receiving site, a microprocessor would translate the message with its filled-in blanks into a meaningful text and output it on a printer or display.

Instead of a printout or display, the receiving system could provide a voice alert, using a "talking chip." This type of alert would have more warning impact than a printer or Cathode Ray Tube. It could also be re-broadcast over the local radio to alert the immediate and nearby communities. That "voice chip", furthermore, could be programmed to give its oral warning in the particular language appropriate to the area. Thus, a multilingual oral disaster warning capability could be brought to reality.

What other features might that ground system have?

It could automatically activate warning sirens and alarms.

It could be powered by solar cells and storage batteries for use in the remotest of inhabited places where there may be no reliable, or perhaps no source of electricity.

The cost of the basic receiving system---(a) a flat plate antenna small enough to stand on an indoor window sill and with a beamwidth of 60° or more such that it can simply be pointed in the general direction of the satellite, (b) a radio receiver to do what a radio receiver normally does, (c) a microprocessor to store the framework of the standardized message (and fill in its blanks with the incoming data) also to store the vocabulary for the voice chip, (d) the chip itself, and, finally, (e) the audio or loudspeaker---is estimated at $1000 (U.S.) or less.

This remarkably low cost is anticipated by the fact that essentially this same receiving system is being developed jointly by the National Environmental Satellite Service and the National Weather Service of the United States to broadcast marine weather to ships at sea. The principle is exactly the same: a standardized form message with blanks in it stored in the receiver and incoming data, arriving 4 bits at a time, to fill those blanks. The message length is 300 words, meaning that the vocabulary stored for the benefit of the "talking chip" must be sized accordingly. While all this may sound complicated, the prospects of being able to develop a suitable shipboard receiving system that will sell in production for less than $1000 (U.S.) are excellent and are being pursued.

When such a receiving package, small enough to take up only a small portion of a table top or desk top, comes into being, so essentially will also come into being the tsunami warning receiver. They will, of course, be one and the same, except for changes in programming, vocabulary, and the like to enable tsunami warning messages addressed to it via the GOES Data Collection System's downlink to be received and to generate the warning voice output. A permanent or hard-copy output device, such as a printer, could be actuated, as well as, or instead of, the loudspeaker if desired, but of course at additional cost.

The missing data to fill the blanks in the standardized message stored in the receiving system would be inserted into the GOES Data Collection System interrogation sequence at Wallops, Virginia, by the GOES command and data acquisition facility operated there by the National Environmental Satellite Service. Of course, the data to fill in those blanks must come from an appropriate source. In the case of tsunami warnings affecting the overall Pacific Basin, it would logically be the Pacific Tsunami Warning Center on Oahu in Hawaii. The message content would be prepared at that center based on the various seismic and tidal information available to it. It would then be communicated directly to Wallops where it would be introduced into the interrogation sequences and broadcast 4 bits at a time by the western GOES to the Pacific Ocean area within its radio horizon.

This, then, is how the GOES satellite may someday soon convey tsunami warning messages to the peoples of the Pacific Basin, particularly those who live isolated and removed from conventional and reliable communications. Hopefully the low cost of the receiving system can be afforded by the nations and inhabited places there. The fact that the system will be usable as well to receive general area Pacific marine weather forecasts and is not limited just to tsunami messages will go a long way toward making it affordable and the concept practicable.

The westernmost GOES, as one can see from its radio horizons, covers a great deal---but still only a part---of the tsunami-susceptible Pacific Basin. Where its tsunami-warning broadcast cannot reach, the warnings will have to be received by other means, more than likely, in most cases involving remote and isolated areas, by HF (High Frequency) radio with its vagaries and uncertainties. If Japan's GMS (Geostationary Meterological Satellite) and <u>its</u> data collection system, however, could be applied like GOES to disaster warning broadcasts, then an environmental warning service for essentially the entire Pacific becomes possible.

The course of technological development takes many turns. The synchronous satellite's data collection capability, as it has evolved with GOES, is a case in point. When it was originally decided to make the DCS a part of GOES, its purpose was seen as conveying data <u>from</u> remote, isolated, and inaccessible regions. Now, with the operational feasibility and value of that aspect well demonstrated and proven, GOES DCS shows promise of taking on a new and equally important role in an <u>opposite</u> direction---the conveying of disaster warning information <u>to</u> remote, isolated, and inaccessible regions.

Will this promise be realized? I feel certain that papers delivered at IAF congresses three to five years from now will reflect that it has.

IAA 80-19

FREQUENCY SPECTRUM AVAILABILITY FOR SEARCH AND RESCUE SATELLITE SYSTEMS IN THE LIGHT OF THE RESULTS OF THE WARC 1979*

G.F. Block[+]

Based on the provisions of the ITU Radio Regulations, as revised by the World Administrative Radio Conference, Geneva, 1979 (WARC 1979), i.e., the definitions of relevant radiocommunication services, frequency allocations to these services, possible technical limitations imposed, etc., the frequency bands available for search and rescue satellite systems are compared with the projected needs of these systems. Where appropriate, recommendations for additional studies required in the preparation of future competent WARCs are formulated.

Keywords: International Telecommunication Union; Radio Regulations; search and rescue satellites; safety service; mobile satellite service; frequency allocation; emergency position-indicating beacon.

---

* *This paper appears in full in Volume 41, AAS Microfiche Series.*

+ *European Space Agency (ESA), 8-10, rue Mario-Nikis, 75738 Paris, Cedex 15, France*

DISCUSSION OF THE SAFETY CAPABILITY OF A MARITIME SPACE SEGMENT - A NEW PROPOSAL FOR A SATELLITE EPIRB SYSTEM*

R. Rogard+

This contribution presents a general discussion of the satellite EPIRB system, and defines its operational and technical requirements. Then a concept to be tested with the European Space Agency Maritime Satellite (MARECS) is described.

The convention on the Safety of Life at Sea which came into force in May 1980 states that every ship to which the convention applies shall carry a Free-Floating EPIRB (Emergency Position Indicating Radio Beacon), and every survival craft carried on ships to which the convention applies shall be fitted with a manually-activated EPIRB. The present systems, which operate on 2,182 KHz, 121.25 MHz and 243 MHz, do not entirely fulfil the operational requirement for such an important service: The successful reception of distress messages is dependent on the propagation characteristics at these frequencies. The problem can best be solved using satellites to relay the distress signals, and in particular the geostationary satellites of the Maritime Mobile Satellite Service.

The European Space Agency has been involved for a long time in this field. In cooperation with DFVLR.,++ experiments were performed between September 1974 and June 1975 with the NASA ATS.6 satellite. Starting with the experience gained through this experiment, an overall study of the EPIRB service was performed, resulting in:

- the definition of an EPIRB systems concept;
- the decision to equip ESA Maritime Satellites (MARECS) with a dedicated distress channel.

Keywords: Maritime Mobile Satellite Service; geostationary satellite; MARECS; EPIRB.

---

* *This paper appears in full in Volume 41, AAS Microfiche Series.*

+ *European Space Agency/ESTEC, Noordwijk, The Netherlands.*

++ *Deutsche Forschungs und Versuchsanstalt für Luft und Raumfahrt - F.R.G.*

IAA 80-26

MARITIME DISTRESS RADIO CALL SYSTEMS EMPLOYING SPACE-BORNE TECHNOLOGY*

W. Goebel, G. Landauer, E. Messerschmid+

Technical and operational requirements for a space-borne maritime alert system are considered and a brief survey of the presently proposed approaches and their basic parameters are briefly reviewed. The main objective of the paper is to compare various system elements such as position determination, capacity for simultaneous alerts, multipath resistivity, alerting time, return link, coverage, or link design, in relation to the requirements. Emphasis is given to the examination of system elements with respect to geostationary and low orbiting systems. The experimental activities in the past and in the future are summarized.

Keywords: Distress radio communication; alerting; multipath; position determination; super-position technique; downlink degradation; INMARSAT; EPIRB; Keyboard Sender.

---

* *This paper appears in full in Acta Astronautica, Vol. 9-1, pp 35-44, 1982 and also in Volume 41, AAS Microfiche Series.*

+ *DFVLR - German Aerospace Research Establishment, Institute for Communications Technology, D8031 Oberpfaffenhofen, West-Germany*

IAA 80-27

DEVELOPMENT AND TESTING OF A SATELLITE-AIDED MARITIME DISTRESS ALERTING SYSTEM*

J.J. Fee, Y. Kaminsky, W. Scales, R.E. Todd+

During recent years, interest in maritime distress signaling via satellite has developed on an international level. Distress signaling is currently limited to terrestrial communication channels except on ships equipped with satellite communication terminals. Conventional signaling frequently entails long delays in establishing contact; many ships have simply disappeared without radio contact. Commercial satellite communication terminals allow rapid distress alerting but are not economical for most commercial vessels and may not be functional in certain types of maritime accidents. A number of recent efforts have been aimed at the development of an inexpensive distress transmitter that would alert rescue authorities via satellite. At least three such efforts have reached the experimental stage. One of the proposed systems would use near-polar, low-orbiting satellites to provide Doppler-derived location fixes as well as distress alerts. The other two approaches would relay distress signals via the maritime geostationary satellites (MARISAT, INTELSAT V, or MARECS).

Location data would be entered manually or through a shipboard connection to an electronic navigation aid. In all three approaches, the distress transmitter would be powered by batteries and packaged as a watertight buoy, although permanent shipboard installations are also possible.

This paper describes a maritime distress alerting system that would signal via commercial (geostationary) maritime communication satellites. The approach described differs from other development efforts in that spread-spectrum signaling is used to provide multiple access and interference resistance.

* *The work described here was funded jointly by the U.S. Department of Commerce, Maritime Administration and the U.S. Coast Guard. This paper appears in full in Volume 41, AAS Microfiche Series.*

+ *The Mitre Corporation, Westgate Research Park, McLean, Virginia 22102*

POSSIBILITY OF CREATING COMBINED SEARCH AND RESCUE SATELLITE-AIDED SYSTEM*

Y.G. Zurabov, V.A. Bogdanov[+]
I.S. Bronitsky[++]

This paper considers a theoretical possibility of creating a combined search and rescue satellite-aided system based on both low-altitude and geostationary satellites. The advantages and shortcomings of both COSPAS-SARSAT and INMARSAT systems are considered to require some improvement. The 406.025 MHz EPIRB signal is supposed to be fully processed on board a low-altitude satellite to determine an EPIRB location and derive a distress message with these data being subsequently relayed via geostationary satellite to a ground station. Besides, a low-altitude system will be capable of operating in the modes which will be implemented within the framework of the COSPAS-SARSAT project (dumping stored data and real-time transmission).

Keywords: Search and rescue; COSPAS-SARSAT project; INMARSAT; Low-Altitude Satellite (LASAT); Geostationary Satellite (GEOSAT); Emergency Position Indicating Radio Beacon (EPIRB); Local User Terminal (LUT).

* *This paper appears in full in Volume 41, AAS Microfiche Series.*
\+ *V/O "Morsviazsputnik", Moscow, USSR*
\++ *Merchant Marine Central Research Institute, Lengingrad, USSR*

SPACE SAFETY AND RESCUE 1979

The following papers also appear in full in *Volume 39, AAS Microfiche Series, a supplement to this volume:*

IAA 79-A18, -A19, -A22, and -A24

The following papers appear in full in Volume 39, AAS Microfiche Series:

IAA 79-A20 and -A21

IAA 79-A24
(AAS 79-330)

PSYCHOLOGICAL TRAINING - AN IMPORTANT FACTOR OF INCREASING SPACE FLIGHT SAFETY

G.T. Beregovoy, I.V. Davydov,
N.V. Krylova, I.B. Solovyeva*

At present the pace of the outer space exploration by means of manned spacecraft is increasing. New spacecraft that provide for an extensive research program for the benefit of mankind are being developed and tested. The problem of greater crew safety becomes therefore of utmost importance. The task of enhancing space flight safety is a matter of highest priority.

While expansion of space research possibilities is ensured by the technological progress achievements, dependable accomplishment by man of ever more and more complicated missions on board the spacecraft is ensured by preflight professional and psychological training.

Equipment reliability is not limitless and an emergency situation can therefore crop up practically at any moment the crew is on board the spacecraft, starting from the moment of ingress into the craft before launch through all the mission phases and up to the egress after the landing.

Space flight safety depends not only on the reliability of the spacecraft and its organic systems and equipment but also on the crew training standard and their ability to counter unfavorable factors of a spaceflight, especially at the moment an emergency situation develops, with their intelligence and self-control.

Psychological training is among the major factors that help to solve the problem of space flight safety. The overall psychological training of a cosmonaut embraces the entire intricate dynamic complex of pedagogical, methodics, and psychological factors aimed at improving the preflight training standard proceeding from the program features and individual psychological characteristics of its participants.

Psychological training should take into rational and full account the individual psychological personality features of a cosmonaut as well as the regularities governing the origin and course of psychic processes that are a part of his professional training.

* *Institute of Psychology, Academy of Sciences, Moscow, USSR*

The basic component of psychological training is to mobilize the necessary psychic functions of a cosmonaut by means of pedagogical and psychological methods to help him actively form a conceptual model of the coming mission, to teach him to bring into effect his psychological reserves in the most efficient way.

Increasing complexity of spaceflight programs makes psychological training ever more universal and diverse. New requirements to the training are stipulated by such factors as increased flight duration, various cycles of scientific research and complex experiments included in the program, long-term stay of the crew in an ecologically closed, limited space under the conditions of a deformed sensor field.

In the process of a cosmonaut's professional training great attention should be paid to special psychological training that provides for a system of measures aimed at forming and developing the necessary psychological qualities that would ensure the efficiency of the cosmonaut's actions under complicated spaceflight conditions.

Crew operation efficiency in a complicated flight situation depends on the cosmonaut's ability to behave actively under emotionally stressed conditions. It is necessary therefore to develop in a cosmonaut a definite system of psychological qualities: emotional stability, ability of self-regulation, readiness for probable contingencies and stress effects.

All that substantiates the psychological approach to the organization of cosmonauts' professional activity, i.e., its concentration on modeling various psychic states characteristic of actual flight activity. In other words, achievement of training goals stretches beyond professional training based on available training facilities and implies the use of special psychological and experimental methods.

Efficiency of developing professionally significant psychological qualities is a direct function of purposeful use of emotiogenic effects in the course of training.

Various so-called exogenic stands and simulators employed in the system of cosmonauts' training (laboratory stands, surdochamber, etc.) make it possible for a man to familiarize himself with certain factors of a spaceflight and form definite professional skills. But training sessions on the basis of these facilities on the ground are remote from real-life activity. However complicated the emergency situation we simulate, it cannot cause the stress of a real-life situation when a man realizes that his mistakes might cost him his health and probably his life. Therefore we believe it necessary, in spite of organizational difficulties, to resort, in the course of cosmonauts' training, to real-life emotiogenic situations which cause a reaction to a real-life stress factor (centrifuge, underwater zero-G simulater, pressure chamber, parachute, etc.

The analysis of a cosmonaut's real flight activity professiogram allows singling out moments strenuous enough in terms of their psychological effect on the man. Primarily it is the period of psychological adaptation to unusual flight conditions.

An indispensable condition of maintaining a man's operationability during this period is his operational stability; hence the specific task of psychological training in this respect: to teach an operator to concentrate on his performance under unusual, emotionally stressed conditions.

Besides, conditions that result in any deviations from a nominal situation are to be regarded as complicated flight conditions. An emergency situation brings about two levels of back reactions: adaptive/protective (biological) and psychological ones that form the strategy of behavior and ensure the performance of new, unprogramed actions. This very reaction is to be formed in the process of a cosmonaut's training when modeling his psychic state in the course of training sessions.

Efficient means of achieving these goals are flying and parachute training.

An aircraft is considered as both an exogenic and a professional training facility providing a complex effect on man: a psychological sensation of unusualness, risk and responsibility as well as physiological stressors and operator's actions to control a moving object.

Employment of real stress conditions (both in parachuting and free falling) making it possible to reproduce to a certain extent the emotional background characteristic of a cosmonaut's activity (unusual conditions, time deficit, sensor stress, sensation of risk, responsibility and self-reliance) combined with the accomplishment of the task purpose - performance of professional activity elements - is a reason to regard parachute training as an indispensable means of cosmonauts' psychological training.

The fact that parachute and flying training provide for developing in a cosmonaut an ability to take optimum decisions against the background of rapidly changing processes adds to their credit.

Flight experience of flying vehicles, spacecraft included, shows that there might be cases when it is necessary to terminate a mission and perform an emergency landing outside the estimated landing area.

What makes such a situation difficult? It is the fact that the crew have been exposed to a stress factor as a result of an accident and have to perform an immediate landing, being unaware of the character of landing and conditions in the landing area.

The experience gained from a number of experiments has demonstrated that sometimes the stress of landing in the extreme conditions of a critical Earth zone (desert, high temperatures and wind; steppe, wintertime, low temperatures and wind; highland and hypoxia; stormy sea) is enough for the crew to fail to overcome it and therefore fail to preserve life.

There have been cases when the subjects who were not trained to face the possibility of exposure had to terminate the experiment ahead of time.

Such psychological factors as loss of nerve, passivity, lack of will, accompanied by fear, pain, loneliness, hunger, thirst, etc may lead to disability or death of the crew.

To attenuate unfavorable factors and remove the stress in case of an emergency landing in extreme conditions, cosmonauts are trained in various climatic and geographical areas. The basic aim of these training sessions is to prepare spacecraft crews psychologically for probable hazardous exposures and give them practical knowledge of their actions after the landing.

Training conditions are as close to real-life critical ones as possible so that the crews could have the feeling that through their knowledge, skill and self-control they are able to cope with a difficult situation and overcome it.

It should be noted that the accumulated training experience has borne fruit. At the final stage of their spaceflights the cosmonauts have more than once relied on this experience.

The most dramatic case was after cosmonauts Lazarev and Makarov emergency-landed on a steep slope of a snow covered mountain. The stress of the emergency landing at high G was not aggravated by the additional stress of uninhabited, difficult terrain because the cosmonauts had been trained in winter forest conditions. The crew showed courage and tenacity and coped successfully with a very difficult situation.

The crew of Zoudov and Rozhdestvensky needed just as much self-control and tenacity when they made an unplanned night landing on a half frozen salt lake. Due to psychological training they had been given in the course of their training sessions at sea, they were not caught unawares and could therefore act efficiently and correctly.

In this report we have touched upon the general principles of psychological training in the course of which a cosmonaut obtains and develops abilities, qualities and skills necessary for him to be confident in his preparedness for a spaceflight, which training guarantees his enhanced safety.

IAA 79-A19
(AAS 79-326)

MANNED REMOTE WORK STATION - SAFETY AND RESCUE CONSIDERATIONS

C.A. Nathan*

The exploration and utilization of space have witnessed a continuous growth in spacecraft size and weight. Many applications are now envisioned which require ultralarge space structures. Due to the restrictions of payload and volume limitations of current and projected launch systems, space construction of these ultralarge structures is essential. This paper discusses the concepts and application of a key piece of construction equipment needed to support assembly of these large structures. The Manned Remote Work Station (MRWS) is a universal crew cabin to be used as a construction cherry picker, space crane turret, work station on a construction base rail system, or a free flyer. Concepts and safety and rescue requirements for this spacecraft are delineated for early applications in support of Shuttle operations, applications in support of a mid- to late-1980's space construction base, and in support of constructing and maintaining a Solar Power Satellite system (SPS).

Keywords: Space work stations; EVA safety and rescue; satellite servicing; space construction.

INTRODUCTION

Concepts for an MRWS were developed for NASA Johnson Space Center (JSC) under contract NAS9-15507, with S.H. Nassiff as director. The first objective of this effort was to evaluate the MRWS flight article roles and associated design concepts for fundamental requirements, and to embody key technology developments into a ground simulation program. The second objective was to provide detail manufacturing drawings and plans for an MRWS simulator Development Test Article (DTA) to be used in the JSC Manipulator Development Facility. The current effort, under contract NAS9-15581, is fabricating a DTA for JSC. This paper emphasizes those aspects of the effort related to safety and rescue, and briefly discusses plans for simulating rescue operations.

A scenario of the MRWS program is presented in Fig. 1, in which is also indicated the approximate times for introducing elements of the system. Activity starts with introduction of a development test article in a ground base simulation program, where key system elements and safety procedures are tested and design parameters are defined. This simulation program will be performed at the Manipulator Development Facility, located at JSC Building 9A. This facility contains a 56-ft by 80-ft air-bearing floor and a 50-ft hydraulically actuated manipulator arm for simulating the Shuttle Remote Manipulator System (SRMS). The MRWS development test article will be mounted on one of various size air-bearing platforms that interface with the simulator manipulator.

* *Grumman Aerospace Corporation, Bethpage, New York 11714*

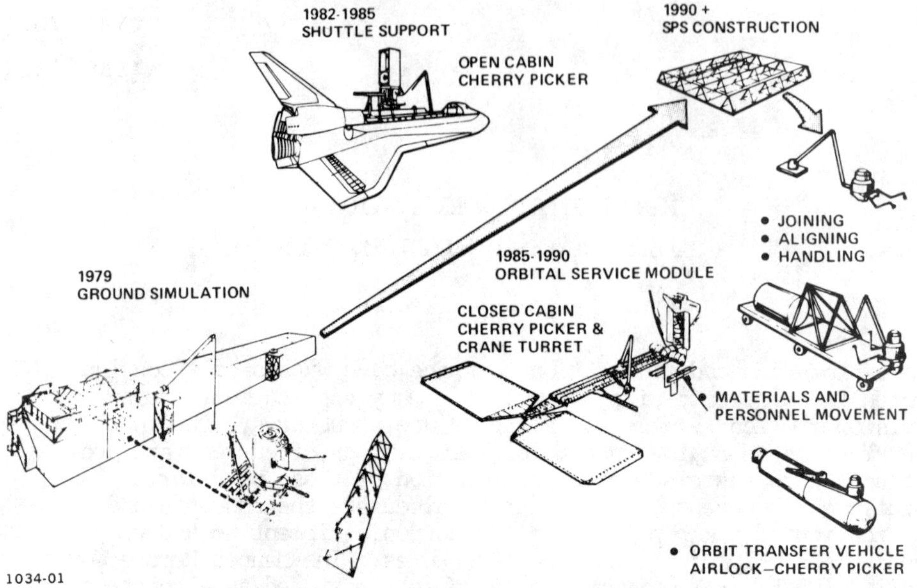

Fig. 1 Orbital construction support equipment development universal manned remote work station.

The near-term applications of the MRWS are in support of Shuttle operations as an open cherry picker mounted to the end of the SRMS. This MRWS provides a platform for EVA servicing of satellites, deployment and retraction of Spacelab experiments, and assembly of large space structures. A small open platform can transport the crew and tools to work sites within reach of the SRMS and provide a stable work station for detailed tasks best performed by man. This device will reduce the cost of EVA operations by bypassing the need to man-rate all hardware in the path of the astronaut during transport to the work site, and will minimize physical contact with the payload once the astronaut is at the work site. Hazard analyses have been performed for these operations, and key single point failures identified. Accurate prediction of entrance into the South Atlantic Anomaly radiation hazard, for example, is an essential safety consideration needing disciplined mission plans.

In the mid- to late-1980's, several planned programs will make use of MRWS's. A small space construction platform will use an MRWS as a crane turret and as a closed-cabin (shirtsleeve) cherry picker for assembly of large antenna and solar power development articles. The open-cabin (EVA-operated) cherry picker will also have applications in the early phases of this period. As crew stay time increases, the need for increased productivity and crew comfort and safety will dictate conversion of the open cherry picker to pressurized work stations that utilize highly dexterous manipulators for assembly and installation tasks. Because these crew cabins must be small to be cost effective, the system cannot accommodate a back-up space suit (EMU). Procedures must be developed that make use of the Personnel Rescue System (PRS), which is a small inflatable system that safely houses a man for several hours. Development of these rescue procedures will be an integral part of planned simulation.

The future application of the MRWS will be in constructing the Solar Power Satellite. The MRWS crane turret and cherry picker will be joined by free flyers, control center cabins, and railed work stations. The prime role of the MRWS free flyer is to rescue stranded workers in remote sections of the large (1 km by 1.5 km) space construction base needed to assemble the Solar Power Satellite.

An objective of the MRWS study is to evaluate the mission requirements for MRWS for the

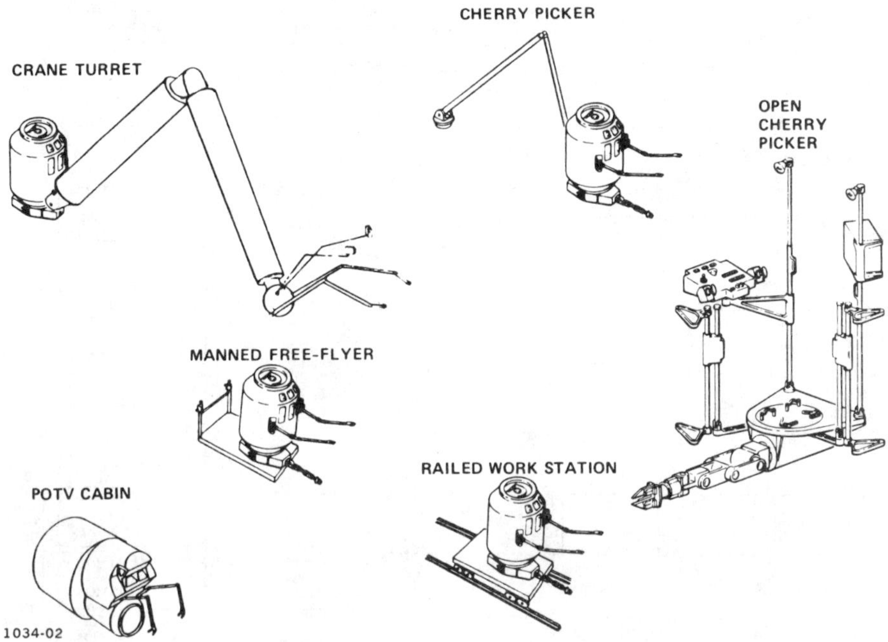

Fig. 2 MRWS multirole concept in support of large space systems.

time period between the years 1980 and 2000, and to determine the potential applicability of a single crew cabin providing the safe environment for performing the roles shown in Fig. 2. These potential MRWS roles include a manned cabin for large space crane operation, a cherry picker, free flyers, railed work stations, and an airlock for the Personnel Orbit Transfer Vehicle (POTV). The results of the MRWS studies indicate that a single-cabin design can accommodate these roles, but a near-term use of an open cherry picker for support of Shuttle operations would be a logical first step towards developing the ultimate closed-cabin MRWS.

MRWS MISSIONS AND ROLES

A survey of NASA's future plans was made using the results of several studies performed over the past few years[1-7]. Potential MRWS missions and roles logically divided into the three time periods and MRWS roles shown in Fig. 3. The near-term manned work station operations can be performed with an open-cabin version of the cherry picker. Requirements for this time phase have been derived from mission descriptions for automated payloads[2] and two studies[3,4] related to construction R&D operations directly from the Orbiter. The MRWS requirements for the midterm have been derived from two studies[5,6] related to the initial construction base operation. These studies identified the need for a closed-cabin cherry picker mounted on a railed manipulator arm and a cabin for a crane turret. The results of the OCSE studies[1] and the Solar Power Satellite System Definition Study[7] form the basis of the far-term MRWS requirements.

Figure 4 lists potential Open Cherry Picker (OCP) applications in support of Shuttle operations. A platform mounted on the end of the SRMS provides a means of conveniently transporting an EVA astronaut, tools and mission hardware about the Shuttle cargo bay. Similar in application to terrestrial cherry pickers used by power utilities, the OCP will enhance productivity during six-hour EVA periods.

The OCP will enhance Spacelab sortie mission operations by providing a convenient means

MRWS ROLE	NEAR TERM 1982–1985	MID TERM 1986–1990	FAR TERM 1991–2000+
• OPEN CHERRY PICKER–FIXED	SHUTTLE SUPPORT	SCB -STRONGBACK	
• CLOSED CHERRY PICKER–FIXED			
• RAILED CLOSED CHERRY PICKER		SCB -OCDA	SPS CONSTRUCTION (PHOTOVOLTAIC)
• FIXED CRANE		SCB -STRONGBACK	SPS CONSTRUCTION (THERMAL)
• RAILED CRANE			SPS CONSTRUCTION (PHOTOVOLTAIC)
• RAILED CABIN			
• FIXED CABIN			SPS CONSTRUCTION (THERMAL)
• FREE FLYER			SPS CONSTRUCTION
• OTV CABIN		GEO. SATELLITE SERVICING	SPS CONSTRUCTION

1034-03

Fig. 3 MRWS mission matrix.

of deploying and retracting pallet-mounted experiments. The cost of experiments needing mechanical extension desired to clear the cargo bay can be reduced by utilizing EVA assistance in the deployment/retraction operations. The OCP can also position an astronaut who can replace film or recording tapes, and thereby minimize the need for data interfaces between experiment and Shuttle.

The concept of in-orbit servicing of automated payloads can be enhanced with use of the OCP. The multimission spacecraft (MMS) with its replacement subsystem modules, is particularly suited for servicing by an astronaut in an OCP. The replacement module and the torque tools needed to withdraw the spent module and insert the new module can all be conveniently located on the OCP. Plans are also being prepared for in-orbit servicing of the Long Duration Exposure Facility (LDEF), as shown in Fig. 5. The LDEF experiment trays, slightly larger than the MMS subsystem modules, can conveniently be serviced by an OCP. Other automated payloads with potential need for an OCP service are the Space Telescope and LANDSAT.

The most extensive use of an OCP can be envisioned for support of construction R&D activities using the Shuttle as the construction platform. Many studies performed over the past few years call out the need for EVA crews to deploy assembly fixtures, fasten and align structure, and install subsystems. The OCP provides a convenient means of crew transport and provides a stable work platform, once the crew is at a work site.

An OCP can also have application in the checkout and in-flight repair of the Shuttle itself. All subsystems within reach of the SRMS, including the cargo bay doors, forward-mounted RCS and GN&C equipments, could be serviced with an OCP.

OCP safety and rescue issues are listed on Fig. 6 for Remote Manipulator System (RMS) Operations, Environment, Factors, and Servicing Operations.

The RMS standard end-effector has a potential failure mode that could release the OCP from the manipulator arm. To protect against this hazard, a safety lanyard can be secured be-

FUNCTIONS IN SUPPORT OF
SHUTTLE OPERATIONS

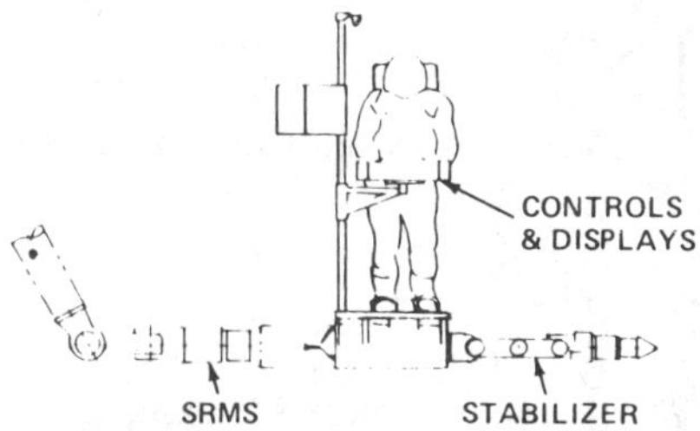

- SPACELAB
  - DEPLOY & RETRACT EXPERIMENTS
  - DATA RETRIEVAL & DATA STORAGE SYSTEM RESUPPLY
  - UNSCHEDULED MAINTENANCE
- AUTOMATED PAYLOADS
  - MMS SUBSYSTEM INTERCHANGE
  - LDEF SAMPLE TRAY INTERCHANGE
  - LST SERVICING
- CONSTRUCTION R&D
  - DEPLOY FIXTURES
  - RESUPPLY FAB MACHINES
  - JOINING & ALIGN OPERATIONS
  - DISASSEMBLE STRUCTURE
- SHUTTLE INSPECTION/REPAIR
  - FORWARD RCS
  - RADIATORS
  - EXPERIMENT TIE DOWN

1034-04

Fig. 4 Open cherry picker missions.

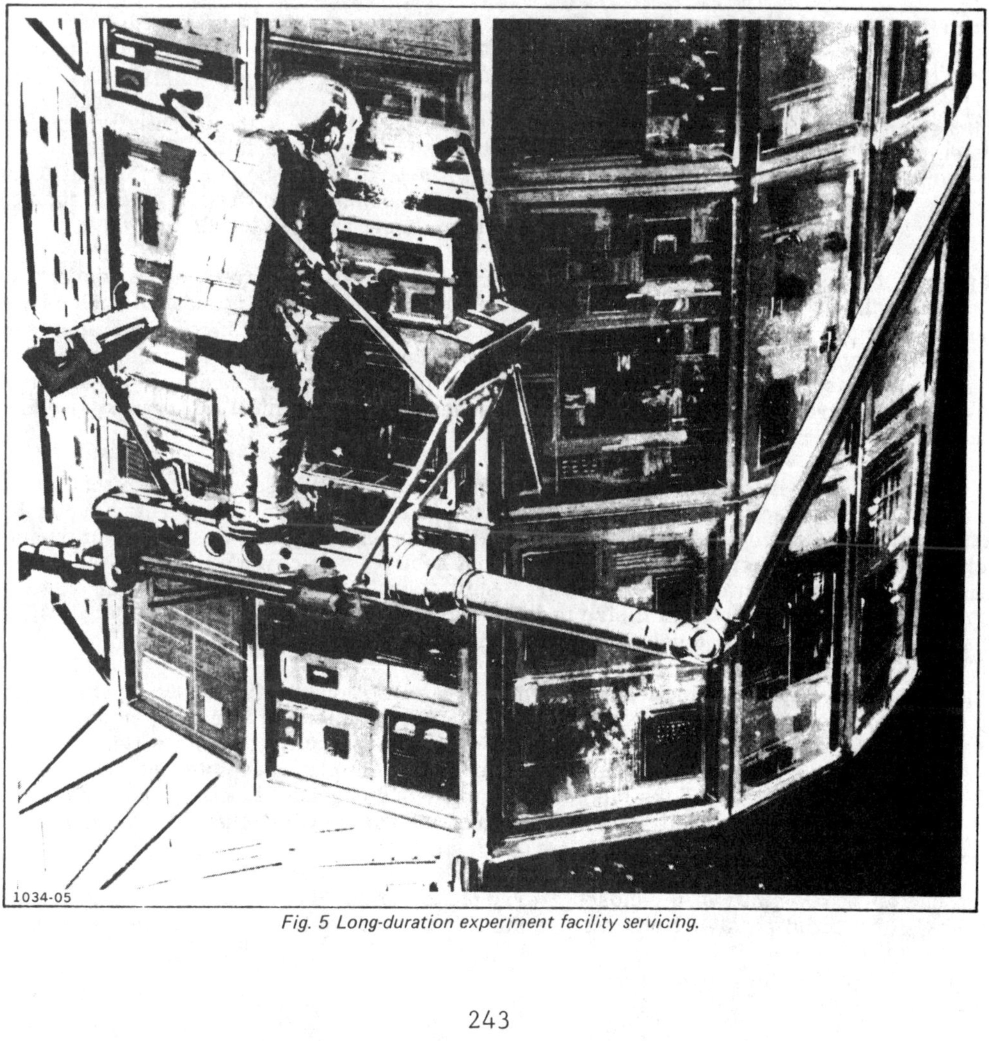

1034-05

Fig. 5 Long-duration experiment facility servicing.

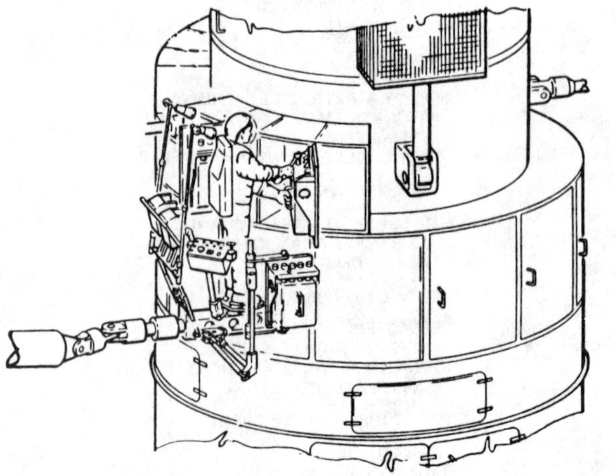

- RMS OPERATIONS
  - END-EFFECTOR RELEASE
  - HARD-OVER COMMANDS
  - STRANDED CREW
- ENVIRONMENT FACTORS
  - SOUTH ATLANTIC ANOMALY
  - THERMAL
- SERVICING OPERATIONS
  - EXPLOSIVE BOLTS
  - HIGH PRESSURE COOLANT TRANSFER
  - PROPELLANT HANDLING

1034-06

Fig. 6 OCP safety and rescue issues.

tween the OCP and RMS prior to actual operations. In the event of an RMS failure that strands the crew on an inoperative arm, a safety line and hand rail could readily be located along the full length of the RMS as a back-up means of retrieving the crewmen.

The key environmental issues associated with OCP operations are the potentially dangerous effects of the South Atlantic Anomaly and overheating of the space suits. Both hazards can be avoided through proper mission planning. EVA activities can be planned for the 16-hour periods of the day that the Shuttle orbit is outside the anomaly and Orbiter orientations can be selected that prevent the working crewmen from being exposed to full sun light.

Some tasks associated with actual satellite servicing operations may result in design constraints to the satellite manufacture. Techniques for disengaging appendages such as solar arrays, that use explosive bolts, may have to be replaced with fully mechanical approaches that take advantage of having an EVA crewman participant. Safety procedures, no different than ground-based approaches, must be developed for handling high pressure systems and potentially hazardous propellants. In addition, design attention must be given to avoiding sharp edges.

In the mid-1980's, it is envisioned that an initial construction base, similar to that shown in Fig. 7, will be deployed as an outgrowth of the Shuttle power module. This base will initially be in a Shuttle-tended mode, meaning that the Shuttle provides the habitat during construction operations. Eventually, permanent habitats will be added, as well as construction scenarios that could require three-shift operations of 30 to 90 days per Shuttle launch. Because of the potential size and modular design of these permanent habitats, design attention should be given to providing multiple escape paths.

Figure 8 summarizes the functions performed by a closed-cabin cherry picker and the types of structures expected to be assembled by the MRWS. Studies performed over the past few years indicate that the likely candidates for in-space construction are large radiometers, multibeam communications antennas, and Solar Power Satellite Development Articles. The MRWS will perform the handling, joining, aligning, and checkout functions associated with in-space construction, as well as those functions associated with general maintenance and repair of the construction base itself.

Because of the proximity of the closed cherry picker to the Shuttle, the safety and rescue issues at the initial construction base are similar to those of the OCP. If the crewman is

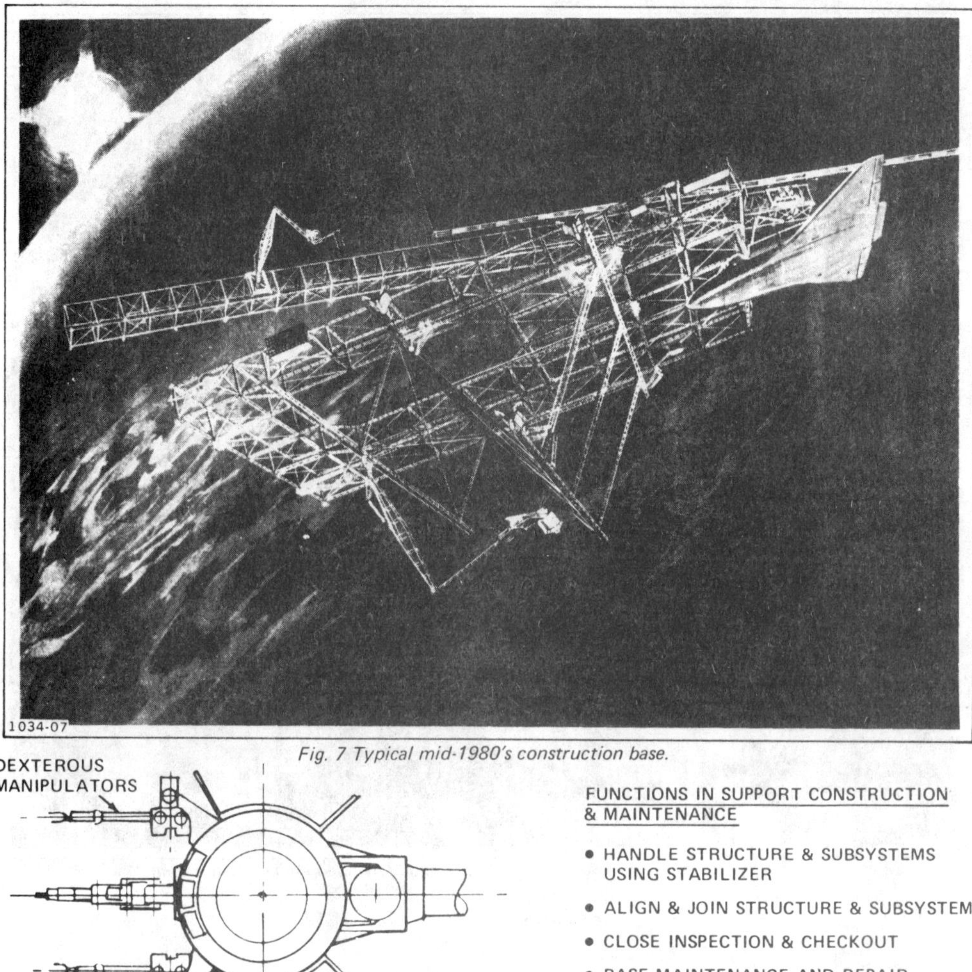

Fig. 7 Typical mid-1980's construction base.

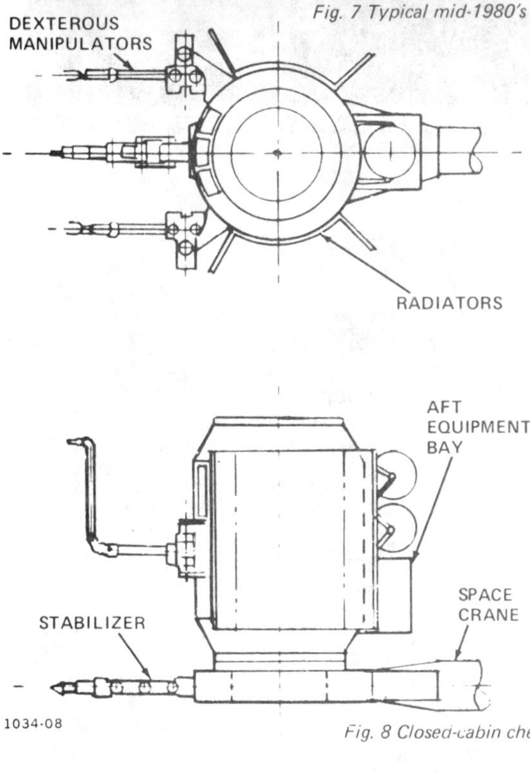

FUNCTIONS IN SUPPORT CONSTRUCTION & MAINTENANCE

- HANDLE STRUCTURE & SUBSYSTEMS USING STABILIZER
- ALIGN & JOIN STRUCTURE & SUBSYSTEMS
- CLOSE INSPECTION & CHECKOUT
- BASE MAINTENANCE AND REPAIR

REPRESENTATIVE CONSTRUCTION ARTICLE IN 1985 TO 1990 TIME FRAME

- SOLAR POWER DEVELOPMENT ARTICLES
- RADIOMETERS
- MULTI–BEAM COMMUNICATIONS ANTENNA

REPRESENTATIVE TASKS FOR FULL–SCALE SPS CONSTRUCTION

- CONTROL OF BASE TRANSPORTER
- STRUCTURAL FASTENING
- RESUPPLY AUTOMATED EQUIP
- BUS/SWITCH GEAR INSTALLATION
- MONITOR REPAIR SOLAR BLANKET INSTALLER & MW SUBARRAY INSTALLER
- GENERAL BASE MAINTENANCE

1034-08

Fig. 8 Closed-cabin cherry picker missions.

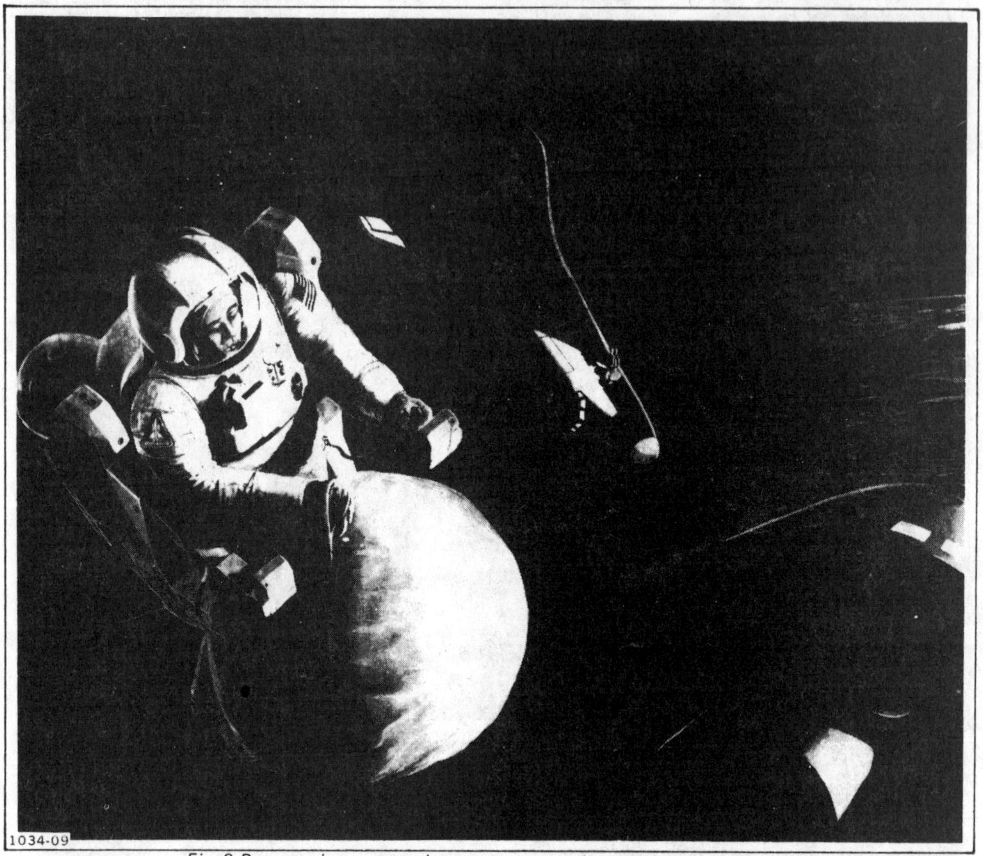

1034-09

Fig. 9 Rescue using personnel rescue system and manned maneuverability unit.

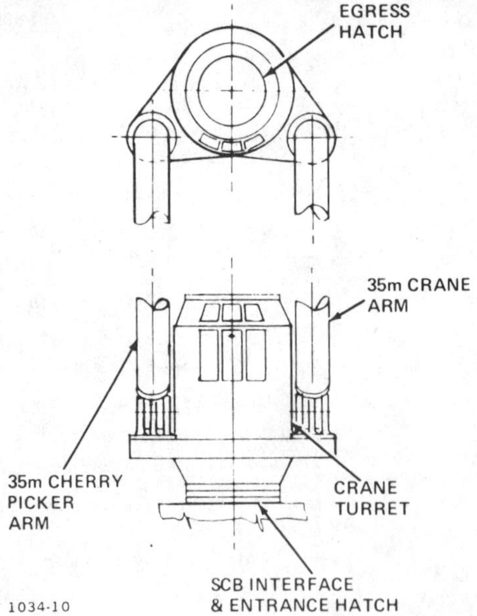

MRWS FUNCTIONS IN SUPPORT OF SCB OPERATIONS

- OPERATOR STATION FOR LARGE SPACE CRANE
  - TRANSPORT & HANDLE STRUCTURE & SUBSYSTEMS
- BACKUP OPERATOR STATION FOR CHERRY PICKER CRANE ARM.
- AIRLOCK FOR TRANSFER OF CREW FROM SPACE CONSTRUCTION MODULE TO CHERRY PICKER

Fig. 10 Crane turret missions.

provided an EMU, he could maneuver, hand-over-hand, the 75 m to the safety of the Orbiter, provided proper hand rails and tethers are designed into the basic construction base. The EMU, however, takes up considerable storage room and would drive the cabin size to larger volume than is needed. An alternate rescue approach would be to use the Personnel Rescue System (PRS)[9], which is a small inflatable ball that a crewman climbs into in an emergency. This system, however, requires rescue from an external source using the Manned Maneuvering Unit (MMU) in an operation similar to that shown in Fig. 9.

The space crane, which is a seven-degrees-of-freedom manipulator that is 35 m long, is also introduced at initiation of the construction base. This mechanism requires a manned work station located at the base of two space crane arms, one of which supports a cherry picker version of the MRWS. Figure 10 presents an outline drawing of the MRWS crane turret that houses the controls and TV monitors required to operate the two crane arms. The safety and rescue issues of this concept are minimal because the cabin would be mounted directly to a large space station habitat.

The future MRWS application is in the construction of the Solar Power Satellite (SPS). Small pressurized cabins have application for each major construction function identified in Boeing's Solar Power Satellite Systems Definition Study[7].

Cherry Picker MRWS's are required to position 816-m long beams for alignment and joining, in an operation similar to that shown in Fig. 11. Crew cabins are required to monitor and control beam fabrication machines, to install power buses and associated switch gear, to monitor and control solar blanket deployment machines, to install microwave system elements, and to assemble the antenna gimbal system.

The SPS construction base, Fig. 12, is extremely large, with dimensions measured in kilometers. The distance between Manned Remote Work Stations and the safety of the 50- to 100-man habitats is too large for simple EVA rescue. A free flyer concept that utilizes the basic MRWS cabin could be used for this rescue function. Figure 13 is a rendering of the MRWS free flyer that adds two structural rings which house thrusters, propellants, and fuel cell reactants, to the basic cabin arrangement. A concept for rescue is presented in

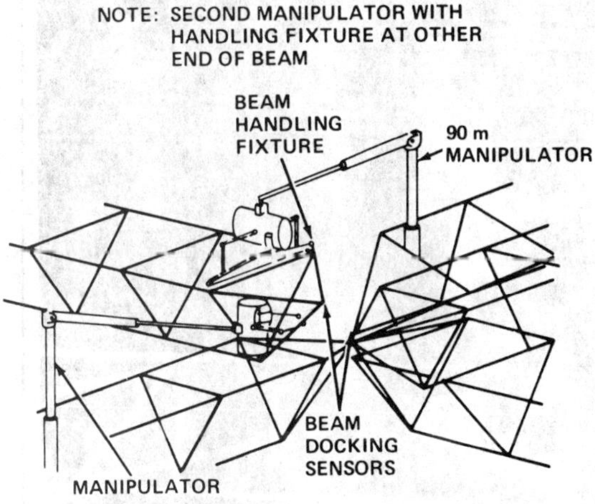

MRWS REQUIREMENTS

- HANDLE LARGE BEAMS
- POSITION FOR JOINING
- PERFORM ALIGNMENT
- FASTEN
- VERIFY JOINT INTEGRITY

TASK FREQUENCY = 3536 JOINING OPERATIONS PER SATELLITE
TASK TIME = 1 HR PER JOINT
NO. OF MRWS UNITS = 8

SOLAR ARRAY BEAM CHARACTERISTICS

DEPTH, m	LENGTH, m	MASS, kg
7.5	816	4513
7.5	660	3656
7.5	470	1603

1034-11

Fig. 11 Joint machines.

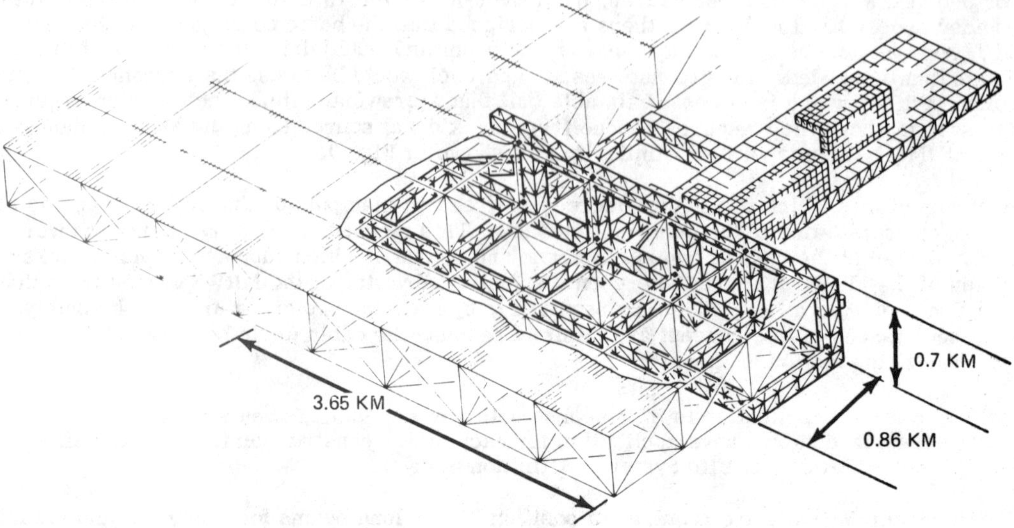

Fig. 12 Solar power satellite construction base.

1034-12

Fig. 13 MRWS free flyer.

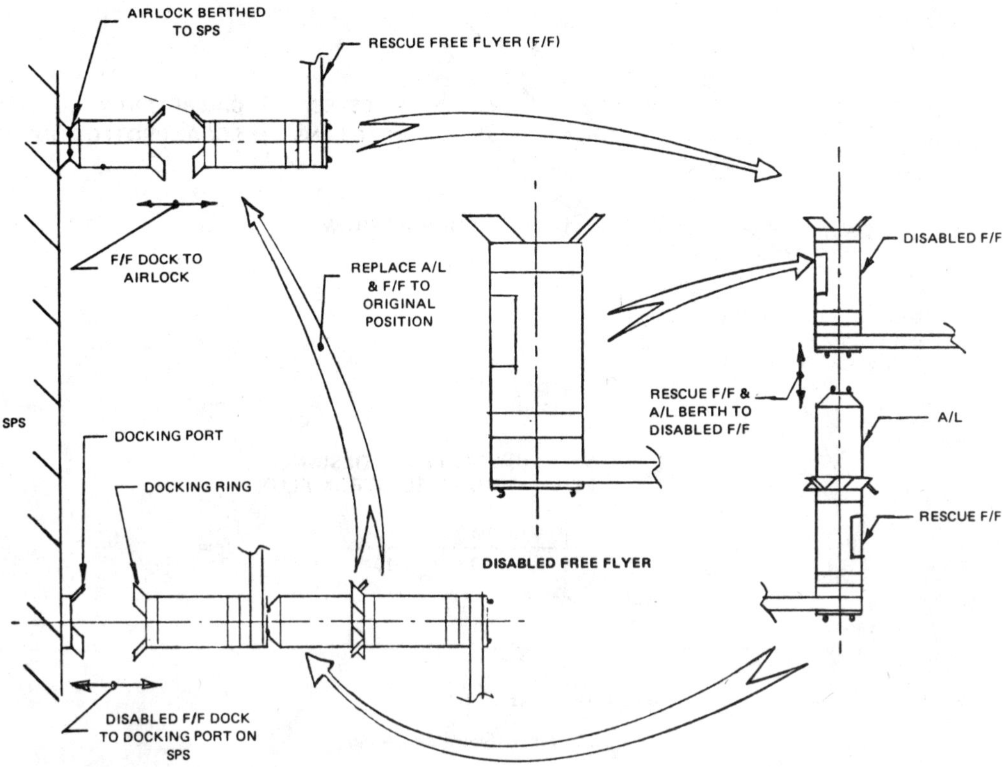

1034-14

Fig. 14 Rescue of disabled free flyer using two-astronaut air lock.

Fig. 14, in which a small air lock is berthed to the basic free flyer, providing as a result the extra room needed for the disabled crewman. Note that in this operation an exit path is needed at both the top and bottom of the spacecraft.

A major safety concern at the SPS construction base is that of solar radiation. Because the construction base is located at a geosynchronous orbit, the crew will be subject to up to 10 times the radiation levels experienced by astronauts working in low earth orbit. The impact of radiation on vehicle design is presented in Fig. 15. An analysis by Wilson and Denn[10] was used to show the effects of time from onset of a major solar storm on various parts of the body as functions of vehicle skin thickness. The MRWS cabin designed for "normal" electron flux with a shield thickness of $4 g/cm^2$ will provide only seven hours of protection in a major solar storm. Crew transfer to a storm shelter within the seven-hour period has many transport and auxiliary man habitat implications on the design of the SPS construction base, including selection of locations for storm shelters and the number of emergency free flyers.

The mission applications of a small pressurized cabin have been found to be broad in scope, with uses in the mid-1980's through the year 2000. The first step in the evolution of this pressurized multipurpose spacecraft is an open-platform cherry picker that operates from the end of the Shuttle RMS. Key subsystem elements, such as the RMS control interface and manipulators, are initially addressed in this simplified support tool and then refined for ultimate use in the pressurized cabin.

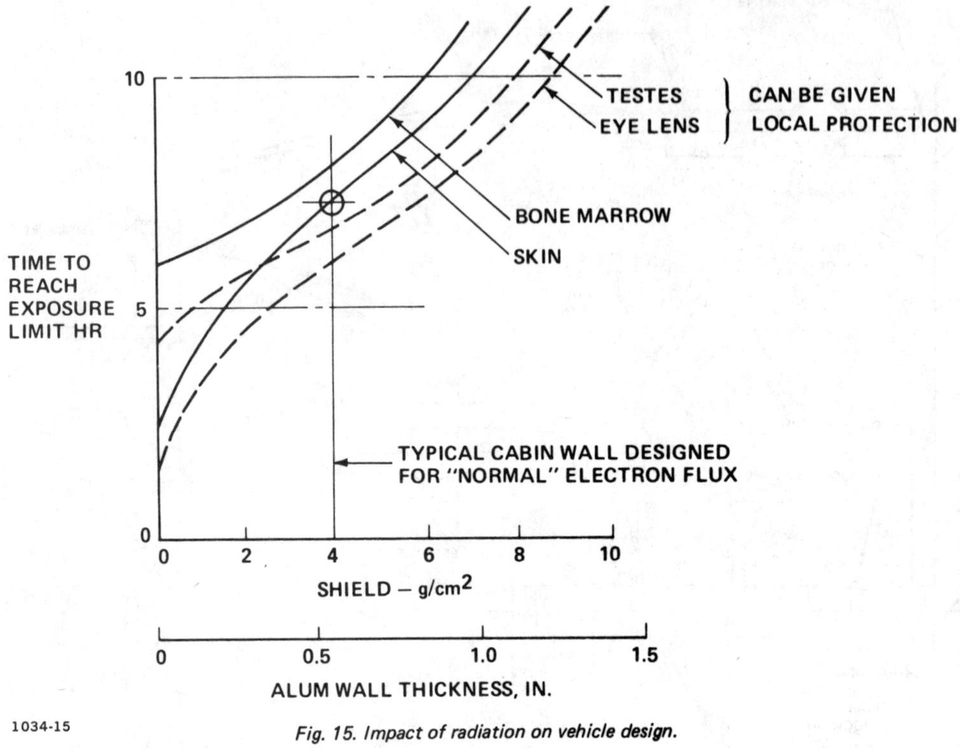

Fig. 15. Impact of radiation on vehicle design.

CONCEPT DESCRIPTION

Open Cherry Picker

Figure 5 is a rendering of the Open Cherry Picker (OCP) servicing the LDEF. The platform is 91 cm (36 in.) wide and houses a foot restraint assembly on a rotary bearing for full 360-degree rotation by the astronaut. A three-DOF stabilizer is mounted to the platform and is used to snare a worksite to minimize RMS motions during crew detailed work periods. A control console, which includes hand controllers and essential RMS controls and displays, is mounted to a swivel. This feature allows the astronaut to rotate the control panel to the rear while working out the front of the cherry picker. Mission-peculiar handling devices similar to the clamp mechanism, shown on the LDEF experiment tray are also mounted to the platform via a rotary joint. Two light stanchions are mounted to the rear of the platform and provide the astronaut with 30 footcandles of illumination at the worksite.

The concept shown is designed with mechanisms so that it can be packaged into a 106 x 152 x 91 cm stowed volume. The OCP is stowed in the forward section of the cargo bay at the structural frames established for the MMU. The overall mass of the OCP is 197 kg, including 25% contingency.

Mechanical and electrical interface with the SRMS is through the standard snare-type end effector. The 250 watts of power provided as a payload service at the end effector is adequate for OCP operations. A data bus system currently available in the SRMS up to the end effector can be utilized for signal interfaces with the OCP. An additional 12 signal lines are also available for dedicated analog signals.

The overall impact of introducing OCP into the Shuttle system is minimal. Development of the OCP also needs no special facilities. The equipments and capabilities already available at such JSC facilities as the Manipulator Development Facilities would be sufficient to flight-qualify the system as well as provide a training facility for the operational system.

Closed Cherry Picker

Figure 16 is a layout of a closed cherry picker (CCP). The CCP is configured for one-man operation from a single control console utilizing bilateral force-reflecting manipulators. The manipulator slaves are 2 m long, while the masters located in the cabin are 50 cm long. Test data indicate that men operating Bilateral Force Reflecting (BFR) manipulators have the same productivity as an astronaut in a EVA suit, but they benefit from the convenience of a shirtsleeve environment and can probably work longer shifts without fatigue.

The CCP has a 170-cm diameter dictated by the safety need for a 1-m egress hatch at the top and bottom. The cabin height is 200 cm, resulting in an internal volume of 4.76 m^3. The atmosphere of the cabin uses a two-gas system operated at 14.7 psi. A console area of 2.1 m^2 is available for the subsystem controls and displays. All subsystems are located in an aft equipment bay, with a separate heat rejection system. The cabin heat rejection system that operates at 75° to 85°F requires 13.4 m^2 of radiator area. The small size of the cabin requires installation of four deployable two-sided radiator panels. The overall mass of the CCP is 2244 kg, including a 25% contingency. The structure (549 kg) and mechanical system (555 kg) make up the major portion of the total mass.

A key issue needing resolution through simulation is the optimum crew size for the cherry picker. A brief cost study indicates that a cabin configuration that can house two astronauts, with one astronaut operating the manipulator systems and the second in a standby role, may be cost effective. Though a smaller single astronaut crew station is lower in cost to develop

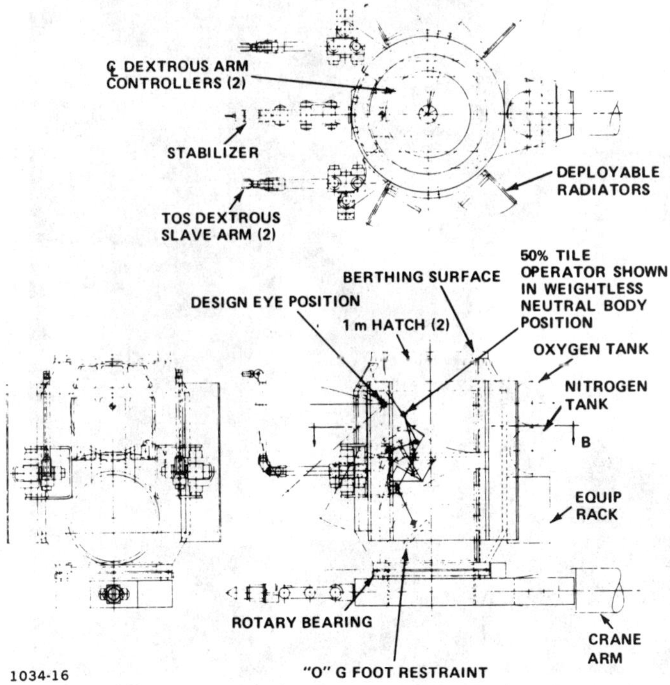

CABIN I.D.	170(67)–cm(in.)
CABIN HEIGHT	249(98)
OVERALL HEIGHT	290(114)
VOLUME TO IML	4.8(168)–m^3(ft^3)
PRESSURE	1.03(14.7)–kg/cm^2(psi)
WINDOW AREA	m^2(ft^2)
• FWD	0.65(7.0)
• UPPER	0.23(2.5)
TOTAL	0.88(9.5)
DISPLAY AREA	m^2(ft^2)
• FWD CONSOLE	0.17(1.8)
• L/R SIDE CONS	0.40(4.2)
• L/R VERT	1.53(16.4)
TOTAL	2.10(22.4)
RADIATOR AREA	13.38(142)–m^2(ft^2)
HEAT LOAD	WATTS
• WINDOW	225
• METABOLIC	350
• CO_2 REMOVAL	50
• INTERNAL ELEC	822
• EXTERNAL ELEC	3113
• AFT EQPT BAY	228
TOTAL	4788

Fig. 16 Closed-cabin cherry picker one-astronaut operation.

and build, the operational cost advantage of the two astronaut station indicates selection of this approach. Data on operating BFR manipulators indicate that a two-hour work period is reasonable with a period of rest required between periods. In a two-astronaut station, astronauts can interchange roles, thereby significantly increasing the utility factor of the two-astronaut station relative to the one-astronaut station. This fatigue factor must be determined through simulation before configuration selection is made.

Free Flyer

A rendering of an MRWS free flyer repairing a damaged structural element is shown in Fig. 13. The MRWS cabin is mated to two structural rings that contain the needed propulsion and electrical subsystems, maintaining a clear transition for the cabin from the roles of cherry picker and crane turret to that of a free flyer. The subsystem elements[8] that must be added to the cabin itself are the GN&C system, including a rendezvous sensor and inertial system, as well as added displays of key flight parameters such range, range rate, and line-of-sight rates.

Multirole Applications

The closed cherry picker configuration (Fig. 17) can be broken down into the six structural/functional elements shown in Fig. 18: namely, (1) berthing port; (2) cabin core; (3) rotary bearing; (4) stabilizer base; (5) lower berthing interface; and (6) crane arm interface. By adding an interface module for (7) a docking ring and (8) a crane turret, the basic MRWS

Fig. 17 Closed cherry picker constructing an antenna.

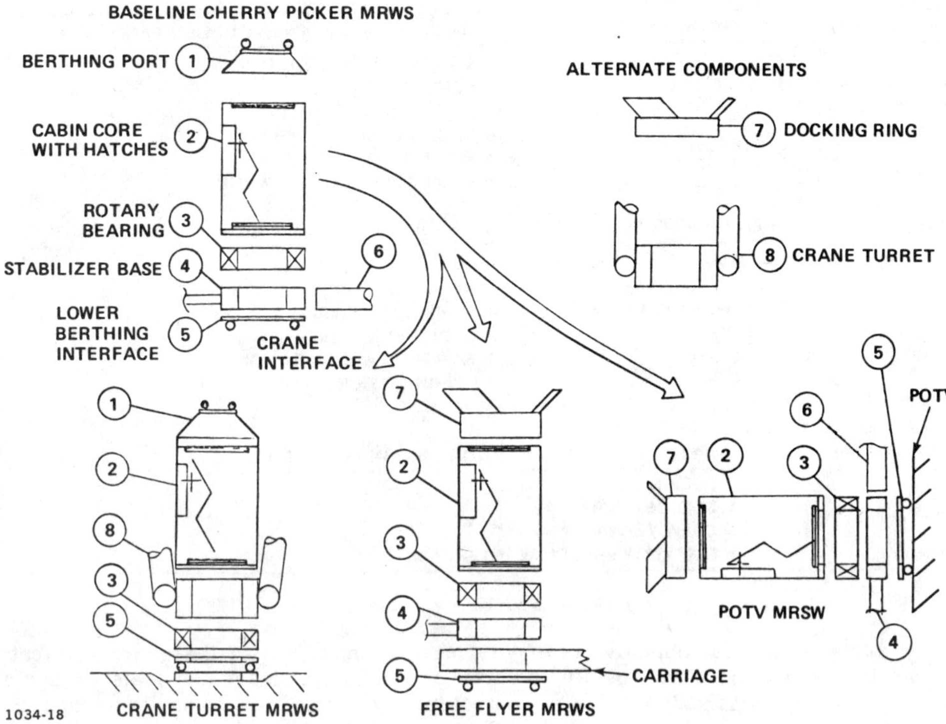

Fig. 18 MRWS future growth trade.

configured for a cherry picker can be used as a free flyer, crane turret, or Personnel Orbit Transfer Vehicle (POTV) air lock/cherry picker.

The cabin core provides the structural elements common to all four MRWS vehicles, such as pressure shell, windows, hatches, console support structure, flooring, restraint system platform, and mounting provisions for external equipment. The CCP and crane turret MRWS have a berthing port interface added to the top of the cabin core, while the free flyer and POTV MRWS vehicle have a docking ring interface. All four vehicles have identical rotary bearing and lower berthing interfaces added to the bottom of the cab core. The CCP free flyer, and POTV MRWS vehicles have the stabilizer base installed between the rotary bearing and berthing interface, while the crane turret MRWS has a crane turret added between the cabin core and rotary bearing. The controls and displays can be tailored to meet the requirements of each MRWS with minimum impact to the common cabin core.

SIMULATION REQUIREMENTS

The broad applications of the MRWS and the strong possibility of using one pressurized cabin for all mission roles suggest the need for early simulation to determine key design factors to verify this potential. During analysis of MRWS flight vehicle mission, several issues were identified that could benefit from simulations. These issues can be collected as simulation objectives under the five categories outlined in Fig. 19.

A typical cabin design objective (category 1) is to determine through simulation the relationship between work duration and crew size of the MRWS. As discussed previously, a two astronaut crew station, though more costly to develop and build, may be more cost effective than a one-astronaut station, if fatigue operating the BFR manipulator is severe.

CATEGORY 1 – CABIN DESIGN FACTORS

- WORK DURATION
- METABOLIC LOAD
- DISPLAYS
- VIEWING
- LIGHTING (INTERNAL)
- MEN & CONSOLES

CATEGORY 2 – EQUIPMENT DESIGN FACTORS

- MANIPULATOR REQUIREMENTS
- CONTROLLERS
- CCTV UTILIZATION
- GRAPPLER REQUIREMENTS
- END EFFECTOR DESIGN
- EQUIPMENT/TOOL STOWAGE

CATEGORY 3 – PROCEDURES DEVELOPMENT

- COMPONENT HANDLING
- STRUCTURAL JOINING & ALIGNING
- CONSTRUCTION EQUIPMENT SERVICING
- FIXTURE INSTALLATION
- SUBSYSTEM INSTALLATION
  - ELECTRONIC
  - POWER
  - FLUID
- SAFETY PROCEDURES

CATEGORY 4 – MAN/MACHINE INTERFACE FACTORS

- WORKSITE ACCESSIBILITY
- WORK TASK ACCESSIBILITY
- WORK SITE CONFIGURATION
- LIGHTING & VISIBILITY
- FATIGUE/TASK DURATION
- LEARNING CURVE

CATEGORY 5 – CONSTRUCTION BASE INTERFACE FACTORS

- DESIGN LOADS
- RESCUE OPERATION
- CREW TRANSFER
- COMM/POWER/SIGNAL INTERFACE

1034-19

Fig. 19 Simulation objectives: categories.

A closely related simulator objective is to determine the manipulator design factors (category 2) that result in maximum MRWS utility and crew productivity. Many candidate manipulator approaches exist that emphasize different control techniques and mechanization. The ideal laboratory manipulator should have the ability to investigate the various options to be considered. The preliminary requirements list for a laboratory device that can be used to investigate options, such as type of control (BFR versus resolved rate), extent of controller indexing, and the force sensitivity in master and slave that reduces operator fatigue, is shown in Fig. 20. The laboratory device should be designed as a BFR master/slave manipulator with the ability to interchange the master controller with a six-degrees-of-freedom resolved rate hand controller. The slave should be 2 m long and be mechanized with a counter-balance system to achieve high fidelity zero-g simulation.

Safety and rescue procedures for both the open and closed cherry picker can be developed during simulation. Methods for transfer of a crewman from one disabled MRWS to a second is one such issue. Simulation in general will aid the designer in identifying hazardous aspects of the configuration itself, pointing out the location and form of subsystems displays and controls for maximum safety between man and machine.

The simulator design must consider the fidelity of zero-g, not only in the design of the equipments but also in the mechanization of work sites for addressing the category 3 objectives (Fig. 19). One approach that shows considerable promise for achieving six-degrees-of-freedom simulation of typical large structure elements is to use helium-filled balloons. To minimize inertia mismatch, the balloon should be located in the internal volume of the beam. As shown in Fig. 21, the weight per meter of three typical construction beams is plotted. Near-term space construction activities will use 1-m-deep beams and the SPS will use 7.5-m beams. The lifting capacity is plotted for each beam, assuming the entire internal volume of the beam is filled with helium. As can be seen, the internal volume of a 1-m beam is not large enough to hold enough helium to support its weight; a 1-m beam would have to be supported with external helium-filled balloons, or use structural material lighter than the 15-mil aluminum elements used in the calculations for the Fig. 21 relationships. The external helium filled balloons would result in an inertia mismatch of 200% for a 10.5-m long beam. The larger beams can accommodate an internal balloon that results in a mismatch in inertia of only 30%.

- NO. OF DEXTEROUS ARMS – 2
- TYPE OF CONTROL – REPLICA MASTER SLAVE
- FORCE REFLECTION (BFR) – YES
- BFR SENSITIVITY – 1.4N (5 oz) FROM S TO M
- ARM LENGTH SLAVE – 2 m; MASTER – 0.6 m
- VELOCITY RATIO: S/M – 3.3
- DOF MASTER – 6 + OPEN LOOP YAW; SLAVE + TEND EFFECTOR – 7 + 1
- DEXTERITY MASTER – S(R,P)E(P)W(R,P,Y); SLAVE – S(R,P,Y)E(P)W(R,P,Y)
- TWO-ARM WORK VOLUME – 2200 ft^3
- MAX TIP SPEED: SLAVE – 0.75 m/sec (30 in./sec)
- MAX TIP FORCE SLAVE – 67N (15 lb); MASTER – 22-31N (5-7 lb)
- ACCURACY/RESOLUTION – ±5 mm ± 1mm
- DEADBAND @ MASTER WRIST TRANSLATION – 1.5 mm; ROTATION – 0.5°
- MAX FRICTION @ MASTER TRANSLATION – 1.1N (4 oz); ROTATION – 0.28 N·m (40 in.oz)
- POWER – ELECTRO MECHANICAL
- TRANSMISSION – TENDON DRIVE
- COUNTERWEIGHTS – DRIVE MOTORS
- MANUAL LOCKING OF SLAVE DEXTEROUS ARMS AND/OR END EFFECTOR – YES
- M/S SYNCHRONIZATION AT STARTUP – YES
- FAIL SAFE LOCKS AT PWR LOSS – YES
- OTHER FEATURES
  - – PROVIDE ABILITY TO EVALUATE INDEXING
  - – PROVIDE SEVERAL MASTER & SLAVE SHOULDER LOCATIONS

1034-20

Fig. 20 Recommended laboratory manipulator system.

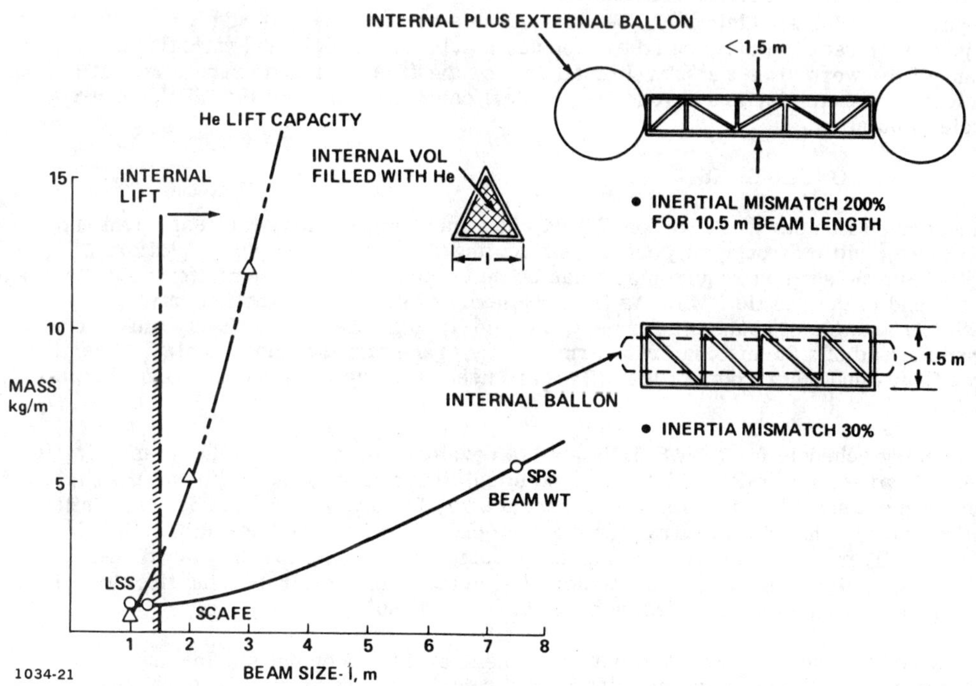

1034-21

Fig. 21 Helium-filled beams.

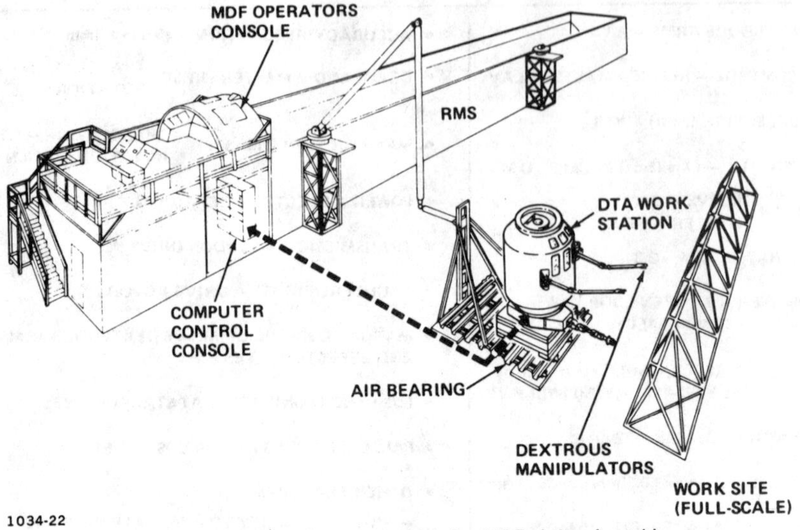

Fig. 22 Simulator approach using MDF air-bearing table.

The initial simulator should emphasize full-scale development test articles and work sites to determine man-machine interfaces while performing the detailed assembly and servicing tasks expected in near-term and mid-term (Fig. 1) missions. This requirement can be readily met using the JSC Manipulator Development Facilities (MDF), the RMS, and air-bearing floor in an arrangement similar to that shown in Fig. 22. Most of the immediate requirements for simulation can be met with this arrangement, though some simulation objectives must eventually be addressed using subscale models and virtual image displays. Many of the work sites and travel distances for the free flyer and crane turret MRWS applications, for example, are beyond the physical capability (size) of the MDF, requiring subscale simulation.

CONCLUSIONS

The Manned Remote Work Station (MRWS) has many applications for near-term support of the Space Shuttle through support of constructing and maintaining space platforms. Initial MRWS Shuttle support requirements can be met with a simple open platform that is mounted to the end of the Shuttle RMS. As the complexity of space construction activity increases with introduction of space platforms in the mid-1980's, the closed cherry picker operating from the end of a 35-m-long crane arm and the crane turret come into play. Eventually, free flyers and other roles of the MRWS are needed as the construction bases for the SPS evolve.

A planning schedule for MRWS flight article development that meets the mission requirements shown is presented in Fig. 23. A ground simulation program that supports the flight article development should be initiated to address the key factors outlined in Fig. 19. Initial simulation activity should concentrate on the open cherry picker, and be followed by closed-cabin and free-flyer investigations. Manipulator design and testing for the closed cherry picker should be initiated in advance of the actual simulation on the air-bearing floor to expedite the manipulator control system design prior to actual simulation.

Safety and rescue have been highlighted in these early conceptual studies and will become an integral part of the on-going simulation program. Procedures will be developed, as will refinement of the configuration through these simulations, to ensure overall crew safety.

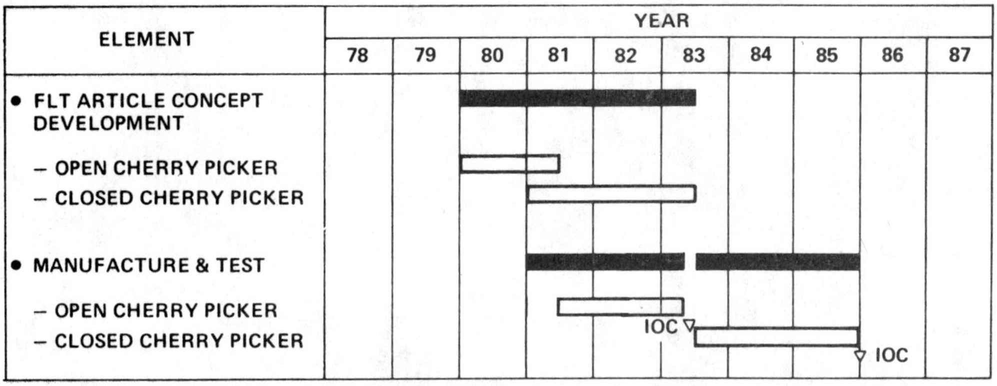

RELATED MRWS ROLES

ROLE	IOC
• CRANE TURRET	1986
• FREE FLYER	1990s
• POTV AIRLOCK	LATE 1980s

1034-23

Fig. 23 MRWS cherry picker program.

Introduction of the MRWS flight vehicle will enhance safety and rescue operations while constructing the large space structures envisioned in the future.

REFERENCES

1. MCR-77-234, NASA-JSC Contract NAS9-15120, Orbital Construction Support Equipment, June 1977 (Martin-Marietta).

2. NASA, Space Transportation System User Handbook, June 1977.

3. CASD-ASP77-018, NASA-JSC Contract NA59-15310, Space Construction Automated Fabrication Experiment Definition Study (SCAFEDS), May 1978 (General Dynamics).

4. NSS-LS-RP008, NASA-MSFC Contract NA88-32390, Systems Definition Study for Shuttle Demonstration Flights of Large Space Structures, January 1978 (Grumman).

5. NSS-OC-RP012, NASA-JSC Contract NAS9-14916, Orbital Construction Demonstration Study, June 1977 (Grumman).

6. MDC G69022, NASA-JSC Contract NAS9-14958, Space Station Systems Analysis Study, July 1977 (McDonnell Douglas).

7. D180-24071, NASA-JSC Contract NAS9-15196, Solar Power Satellite System Definition Study, March 1978 (Boeing).

8. S.H. Nassiff, Orbital Construction Support Equipment - Manned Remote Work Station, AIAA/IES/ASTM 10th Space Simulation Conference, October 1978.

9. Aviation Week & Space Technology, p. 59, April 5, 1976.

10. NASA-JSC Contract NAS9-15779, Manned Geosynchronous Mission Requirements & System Analysis Study, Final Review, Phase 1, May 22, 1979 (Grumman).

IAA 79-A22
(AAS 79-329)

CONSIDERATIONS ASSOCIATED WITH THE INTRODUCTION OF FEMALE CREWMEMBERS IN SPACECRAFT AND SPACE STATIONS*

Jeri W. Brown+

The United States has logged over 22,500 man-hours experience in space flight. Missions have ranged from the 1-man, few-hour flights of Mercury to the 3-man, 3-month habitation of Skylab. The Space Shuttle opens the next era of journeys into Earth orbit: the vehicle is different in that the Orbiter is reusable; the crew will be different in that females will most certainly be onboard. Past experience has shown spacecraft habitability and man/systems integration to be extremely important to the crew as well as sometimes critical to mission success. Hence, viewing those considerations which arise with the inclusion of female crewmembers is important and appropriate.

Two main areas addressing these considerations are emphasized in this paper. The first deals with the anthropometric and physiological differences between males and females, as related to spaceflight (e.g., size, strength, personal hygiene, etc.). The second discusses impacts to hardware design or crew performance and operations resulting from the considerations. Examples of effects on Orbiter systems and provisions are included. Data on females and their performance under conditions analogous to spaceflight are also examined. No major problems to date have been identified or are expected which would preclude females from space; indeed, the impacts are quite minor. In future, longer-duration space missions with larger crews, such as in a space station or colony, crew needs will be much the same, as important as ever, but the novelty of mixed crews will have significantly decreased.

Keywords: Female; astronauts; human factors; human engineering; crew performance; spaceflight; weightlessness; sex.

INTRODUCTION

The United States has logged over 22,500 man-hours experience in spaceflight. Missions have ranged from only a few hours with one Mercury crewman to almost 3 months with three Skylab crewmen. Spacecraft have increased in size from Mercury's 1.53 m^3 (54 ft^3) to the 102 m^3 (3600 ft) of the Skylab. Mission objectives have become increasingly complex and sophisticated. But in the past all crewmembers have been male.

---

* *This paper also appears in Volume 39, AAS Microfiche Series.*

+ *Lyndon B. Johnson Space Center, NASA, Houston, Texas 77058*

A 1972 study by the Space Science Board of the National Academy of Science stated the following:

> *Admission of women to the space crew at this time would undoubtedly create serious problems in the masculine world of space operations. However, broad trends in the 1960's toward increased penetration of masculine occupations and roles by women may accelerate in the 1970's to a degree that research oriented toward operations in the 1980's might well include objective review of the merits of female participation.*[1]

The "penetration" has occurred, and "female participation" is now a reality: women are now included in the most-recently-selected corps of astronaut candidates for the Space Shuttle Program and beyond. Hence, it is appropriate to view those considerations which are unique to females and must be dealt with. Indeed, as past experience has shown spacecraft habitability and man/systems integration to be extremely important to crewmembers as well as oftentimes critical to mission success, these female-unique considerations must be identified, assessed, and acted upon to ensure optimum crew station design, habitability, and operations.

This paper identifies these considerations. In some cases, suggestions for design solutions or special provisions are presented while in other cases where past experience cannot be extrapolated, further study is recommended.

The mixing of males and females in space crews is no longer the delicate issue it once may have been.[2] Indeed, sexual mores have changed significantly in recent years. It is not the author's intent to specify "right" or "wrong" concerning the handling or behavior of mixed crews in confined areas as in the Shuttle. Furthermore, some of the considerations discussed herein, especially those involving individual crewmember privacy and modesty, can be seen from several viewpoints. The objective of this paper is to identify these areas and discuss possible ways of resolving those which are potential "problem areas" to mixed crews. Mission length does affect some of the proposed recommendations.

CREWMEMBER DIFFERENCES

Ensuring spacecraft/astronaut interfaces are optimized contributes toward mission success. Effective and efficient crew performance, which can be critical for mission success, relies on designers, engineers, and planners taking into account the human factors: capabilities, characteristics, behavior, etc. From a human factors engineering viewpoint--aside from design considerations for spacecraft and spacesuit waste collection systems--integrating females into the space vehicle and mission operations is not much more than expanding the range of certain parameters which must already be considered for man/systems integration and supplying certain female-unique provisions. The prime parameters which are affected are anthropometric: body size, posture, strength, and reach. Special provisions must also be made for clothing and personal hygiene equipment. Female physiological differences may affect adaptation to the stresses of spaceflight. Each of these parameters is discussed in the following paragraphs.

---

[1]Space Science Board of the National Academy of Sciences National Research Council (1972). Human Factors in Long-Duration Spaceflight, 91.

[2]Berry, C. A. (1973). View of human problems to be addressed for long-duration spaceflights. Aerospace Medicine, 44(10), 1136-1146.

Anthropometric Differences

Anthropometric changes which occur in the weightless state of spaceflight are shown in Table 1.[3] These changes have been documented during past U. S. space missions by the male astronauts. These same changes would be expected to occur in female crewmembers, but the magnitude of the changes could very well be different due to the physiological differences between the sexes.

TABLE 1 Anthropometric Changes of Males In Weightlessness

CHANGE	TIME REQUIRED FOR CHANGE TO OCCUR
MAY BE PROGRESSIVE:	
Weight loss	Small initial loss first 1 to 2 days; final course depends on diet, exercise, and other factors.
Trunk and limb girth	Immediate in some areas; slow in others; depends on diet, exercise, and other factors.
Loss of muscle mass and strength	Days to weeks; depends on diet, exercise, and other factors.
Body composition and density	Days to weeks; depends on diet, exercise, and other factors.
CONSTANT AND PERSISTS THROUGHOUT FLIGHT:	
Height increase	Two phases: immediate step; then, hours to days for slower component.
Posture	Immediate.
Fluid shifts	Hours to 1 or 2 days.
Center-of-mass shifts	Days.

For instance, posture changes in space result when the effective gravity force acting on the muscles has been nulled on-orbit. As indicated in Fig. 1, the neutral body position is a significant deviation from the normal standing posture in a one-g (i.e., Earth gravity) environment. Because of differences in musculature strength between men and women (addressed later in this paper), one could anticipate a slightly different neutral posture for females. The female center of mass, normally displaced from that for males, can also be expected to shift. The change can be critical in properly designing, for instance, a free-flying astronaut maneuvering unit.

In another example of male/female differences, crewman height has been shown to increase as a function of time on-orbit.[4] But will women vary likewise? Data collected during the Skylab 4 and Apollo Soyuz Test Project (ASTP) missions indicate a two-phase "growth" over normal Earth height with average maximum height increases of 5 cm (2 in). Percent height change (Earth-height baseline) as a function of crewman days on-orbit is shown in Table 2.

In a one-g environment (i.e., Earth-based), height has been shown to increase during sleep overnight. In a limited data base collected from a small population including males and females, height increased from 0.68 to 1.95 percent. The magnitude and time response of change appear larger for females and for younger persons. If these data are generally true, then female astronauts might "grow" more than their male colleagues. From a design standpoint, the consideration of height change is critical in terms of close fitting garments, such as the spacesuit. Adjustments in the suit, especially in the torso length, must allow for at least a 5-cm (2-in) increase in height. Whether females will require more adjustment is still an open question.

---

[3]Webb Associates (Ed.) (1978). Anthropometric Source Book, Volume I: Anthropometry for Designers. NASA Reference Publication 1024.

[4]Brown, J. W. (1976). Zero-g effects on crewman height. JSC-11184.

Height change must also be considered in work station design where the design eye point is critical, such as in cockpit seating where the pilot must be able to view specific objects at certain times (e.g., the runway during Shuttle Orbiter landing). The Orbiter seat is adjustable to accommodate changes in height as resulting from exposure to weightlessness to ensure the eye is properly positioned for the various sized crewmembers. In another example, the crewman operating the Orbiter's Remote Manipulator System must have visual access out the window of the cabin. The shorter crewmember must rely on adjustments in the Orbiter's foot restraints to achieve this eye point. Fig. 2 illustrates the adjustment required for eye positioning.[5] Variations between male and female populations in height which are used for spaceflight crew station design are presented in Table 3.

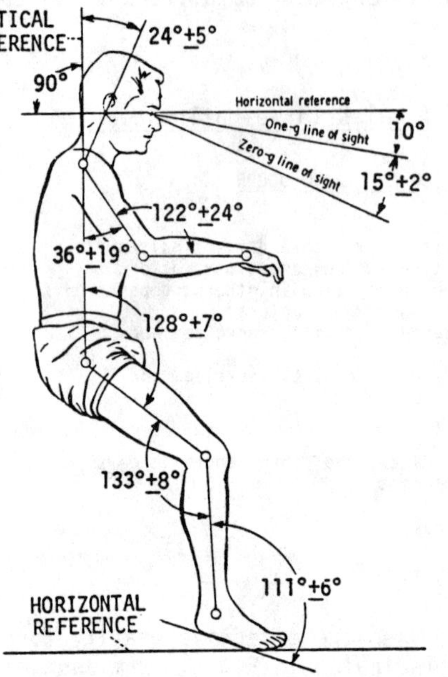

Fig. 1. Male neutral body posture.

Female body measurements are especially important to consider in the design of the Shuttle spacesuit, which has a hard upper torso (HUT) and will be provided in standard sizes to fit both sexes. As indicated in Table 4, upper body measurements vary significantly between men and women for biacromial breadth, bideltoid breadth, interscye, chest breadth and chest circumference. In most of these cases, the 95th percentile female is approximately equal to the male mean or less. However, the range of female measurements in bust depth exceeds that for male chest depth. The results could produce a HUT which fits a female across the shoulders and back but is too snug in the chest area. A HUT which fits the bust would be too large across the back, placing the shoulder joints incorrectly, thereby restricting arm motion. The bust can be comfortably compressed to an unknown extent for an unknown period of time,

TABLE 2 Percent Height Change vs. Days On-Orbit

DAYS ON-ORBIT	PERCENT HEIGHT INCREASE
6	0.57
8	1.79
9	2.58
21	2.80
35	2.93
58	2.53
82	3.50

---

[5]Brown, J. W. (1976). Shuttle remote manipulator system workstation--man-machine engineering. Proceedings of the Sixth Congress of the International Ergonomics Association, Maryland. pp. 149-153.

but the interior surfaces of the HUT must be smooth to preclude developing pressure points on the body during an EVA. The Orbiter suit is currently under development.

Similar consideration must be given to crewmember reach, where differences are "marked and significant."[6] Women's average values for static forward arm reach are 7.6 cm (3 in) less than for those of men. Differences in functional (dynamic) reaches are also evident [81.3 cm (32.0 in) for males and 71.9 cm (28.3 in) for females as measured by Kennedy, 1964 and 1976]. The Orbiter mission specialist's seat is adjustable to accommodate the crewmember's reaching certain switches during the mission launch phase.

Laubach's summary of human body strength in the referenced NASA

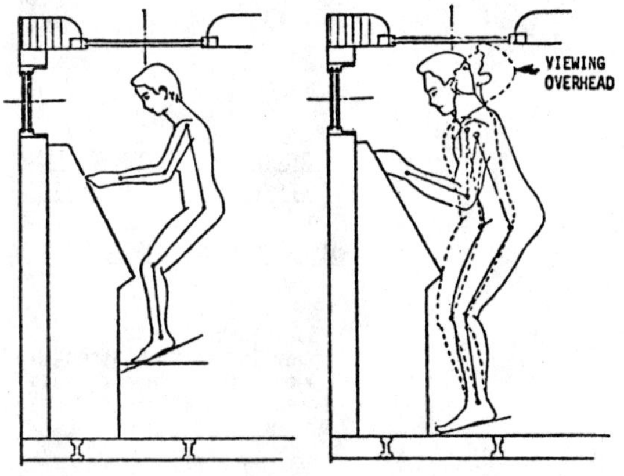

Fig. 2. Height adjustment.

TABLE 3 Comparative Heights for Projected 1985 Female and Male Astronauts

	PERCENTILE					
	5th		50th		95th	
	cm	in	cm	in	cm	in
FEMALE	152.4	60.0	162.8	64.1	172.7	68.0
MALE	168.1	66.2	178.3	70.2	188.7	74.3

TABLE 4 Upper Body Measurements

MEASUREMENT		PERCENTILE					
		5th		50th		95th	
		male	female	male	female	male	female
Biacromial breadth	cm -	37.6	33.4	40.9	36.1	44.0	38.8
	in -	14.8	13.1	16.1	14.2	17.3	15.3
Bideltoid breadth	cm -	44.4	38.6	48.6	42.4	62.9	46.8
	in -	17.5	15.2	19.1	16.7	20.8	18.4
Chest breadth	cm -	29.7	25.3	33.0	28.5	36.7	32.3
	in -	11.7	10.0	13.0	11.2	14.4	12.7
Interscye	cm -	32.6	31.4	38.9	35.6	45.2	39.9
	in -	12.8	12.4	15.3	14.0	17.8	15.7
Chest circumference	cm -	89.1	82.2	99.1	91.6	109.8	103.6
	in -	35.1	32.4	39.0	36.1	43.2	40.8
Chest depth	cm -	21.5		24.6		27.8	
	in -	8.5		9.7		10.9	
Bust depth	cm -		21.1		24.2		28.2
	in -		8.3		9.5		11.1

[6] Webb Associates, op. cit., p. V-11.

Anthropometric Source Book states that while research seems to substantiate that the mean percentage of women's total body strength is about 63.5 percent of men's total body strength, a more appropriate description would be stating that the percentages vary over a broad range (from 35 to 86%) depending on the specific muscle strength characteristic (see Table 5).

TABLE 5 Range and Average Mean Percentage Differences in Muscle Strength Characteristics Between Men and Women

MUSCLE STRENGTH CHARACTERISTIC	PERCENT DIFFERENCE Low	Mean	High
Total Body Strength	35	63.5	86
Upper Extremities Strength	35	55.8	79
Lower Extremities Strength	57	71.9	86
Trunk Strength	37	63.8	70
Dynamic Strength	59	68.6	84

In the population of 207 mission specialists and pilots from which the most-recent group of astronaut candidates was selected, marked differences in absolute strength were noted from ergometer measurements.[7] In comparing males and females, the females measured 35% less strength in the lower extremities and the back, and 50% less strength in the upper extremities. These differences are reduced to a few percent less in the lower extremities, zero in the back, and 25 percent less in the upper extremities if absolute values are normalized by dividing by the individual's lean body mass. Further reductions are conceivable if physical conditioning programs are implemented.

The reader is reminded at this point that these considerations--body size, posture, reach, and strength--are not female-unique but must be addressed for men, too. The impact of adding women onboard is merely an expanding of the parameters' ranges. The importance of listing them herein is for emphasis to the designers and planners. For example, ensuring that a female mission specialist performing extravehicular activities (EVA) in a spacesuited condition will be able to actuate mechanisms or otherwise carry out the specified tasks can only be ensured by addressing these strength differences.

Training of the newly-selected female astronaut candidates has been no different from that which the males have received, including various classroom courses and briefings, jet familiarization, survival training, parasail training, and zero-g parabolic flight sessions. Fig. 3 shows one of the female candidates, attired in a pressurized spacesuit, preparing to submerge herself during a training session at the Johnson Space Center Water Immersion Facility (WIF). The WIF affords a very good simulation of the weightlessness of space, and various tasks are performed in the WIF to evaluate crewmember capabilities, performance, and equipment designs. In Figs. 4 and 5, two female candidates participate in parasail training and water survival sessions to prepare for emergency ejection from an aircraft over land and water, respectively. A third (Fig. 6) familiarizes herself with weightlessness and the Shuttle handwasher aboard the KC-135 aircraft which provides approximately 30 seconds of "zero-g" while following a parabolic flight path.

---

[7]Berry, M. A. and W. E. Thornton (1978). Physical conditioning in men and women Shuttle astronaut candidates. Preprints of 1978 Annual Scientific Meeting of the Aerospace Medical Association, p. 123.

Fig. 3. WIF training session.

Fig. 4. Parasail training.

Fig. 5. Water survival training.

Fig. 6. Simulated weightlessness.

Physiological Differences

During 1973, a group of 12 females participated in a 36-day study--including 14 days of control testing, 17 days of bed rest, and 5 days of recovery testing--to obtain data on physiological responses of women to weightlessness as simulated by bed-rest. Measurements were recorded in the following areas: centrifuge tolerance; fluid electrolytes and hematology; responses to lower body negative pressure; physical working capacity; biochemistries; blood fibrinolytic activity; female hormonal and metabolic responses; biorhythms; and gynecology. The results, compared with responses obtained from similarly bed-rested males, indicate that women can cope with weightlessness. Additionally, females may be more sensitive subjects for evaluating countermeasures to weightlessness and developing criteria for assessing applicants for spaceflight. Final verification of these findings, however, must await Space Shuttle Orbiter flights of the 1980's when inflight measurements can be obtained for comparison.[8]

In terms of gynecological considerations, possible problem areas are related to hormone production disturbances, conception, pregnancy, and delivery. Experiments with female animals proposed for Spacelab will lend insight into effects of weightlessness in these areas. But considering reproduction in space under artificial gravity conditions of large space stations presents additional questions.

For zero-g, near-term flights, Mutke and Burchard (1976) state the following:

> *There are also speculations about the effect of weightlessness in respect to muscle tone within the abdominal wall, the change of fluid distribution within the uterus and connected tissue, the possible change in amount and direction of flow within the intratubular, intrauterine or intravaginal cavity, etc. The possible effects of changes in pH acidity, the composition and rate of flow of the different secretions is still unknown. We also might foresee changes in the intrauterine muscle-tone and changes in its blood and fluid flow with possible effect on the physiological performance, the irritability by and reactivity to various hormone levels which might be affected, or there might even be a change in hormone production by the female endocrinal system involved.*[9]

---

[8]Sandler, H. and D. L. Winter (1978). Physiological Responses of Women to Simulated Weightlessness. NASA SP-430.

[9]Mutke, H. G. and E. C. Burchard (1976). Gynecological considerations on the participation of females in future space flights. Raumfahrtforschung, 20, 292-294.

In addition to physiological concerns of menstruation, effects of the menstrual cycle on astronaut performance must also be considered. Petrofsky and colleagues (1976) examined isometric strength and endurance during the menstrual cycle.[10] Seven female subjects performed a series of 5 fatiguing isometric contractions at a tension of 40% MVC (maximum voluntary contraction) under two thermal conditions. In one case, the forearm was exposed to ambient temperature; in the second case, the forearm was immersed in water at 37°C (98.6°F) to stabilize the muscle temperature. Three of the subjects were taking oral contraceptives (Group 1); the others (Group 2) were not.

The following results were produced: (1) Isometric strength did not vary throughout the menstrual cycle; (2) Endurance varied only in the Group 2 subjects under ambient air conditions [Thermal causes for this finding cannot completely account for the changes noted because of muscle temperature variations recorded throughout the menstrual cycle and because of change profiles of endurance and temperature under ambient and thermally controlled conditions. The authors suggest some heretofore unknown hormone as possibly dominating over muscle temperature to affect endurance during the menstrual cycle.]; and (3) The magnitude of blood pressure and heart rate responses to isometric exercise were unaffected by endurance time or the menstrual cycle. Hence, the menstrual cycle will apparently not influence the performance of female astronauts taking oral contraceptives, but may (depending on the thermal conditions on-orbit) have some effect on endurance of those not taking birth control pills.

Fig. 7. Treadmill test.

During 1977, cardiopulmonary responses of 21 potential female astronaut candidates, as measured during pulmonary function screening and a maximum effort treadmill test (Fig. 7), were compared with similarly tested males. In summary, all static lung volumes and all dynamic flow-rates were higher in the males. Residual volume/total lung capacity, closing volume/dead volume, and closing capacity/total lung capacity were comparable for both sexes. Data indicated no alveolar-capillary differences male to female. Resting heart rates prior to and following exercise were comparable between men and women, but men at rest had greater blood pressure. Although heart rates during the second minute recovery period were comparable, men had significantly greater blood pressure than women.

During exercise, maximum heart rates

[10]Petrofsky, J. S., D. M. LeDonne, J. S. Rinehart, and A. R. Lind (1976). Isometric strength and endurance during the menstrual cycle. Europ. J. appl. Physiol., 35, 1-10.

(189-190 beats/minute) were comparable, as were diastolic blood pressure, respiratory exchange ratio, and ventilatory efficiency. $\dot{V}_{O_2}$ max, $\dot{V}_E$ max, and O_2 pulse were 68-70% greater in males, while $\dot{V}_{O_2}$ normalized for body weight (ml/kg/min) and systolic blood pressure for males were approximately 30% greater.[11]

CREW PROVISIONS

A number of specific topics can be identified under the general areas of personal hygiene and garments which form female-unique considerations. The mixing of men and women onboard spacecraft gives rise to additional considerations for privacy. Several questions must be asked--and answered--for future flights with female crewmembers. The following paragraphs summarize these considerations; recommended "answers" to these points are derived from individual interviews by the author of the 21 females examined for mission specialist and one female examined for payload specialist during 1977 at the Johnson Space Center. General demographic data on this population are included in Table 6.

TABLE 6 Demographic Data on Female Candidates

PARAMETER		22 TOTAL CANDIDATES	6 SELECTED CANDIDATES
Height:	range -	157.8-177.5 cm (62.1-69.9 in)	159.3-177.3 cm (62.7-69.8 in)
	mean -	167.1 cm (65.8 in)	166.1 cm (65.4 in)
	S. D. -	5.6 cm (2.2 in)	6.1 cm (2.4 in)
Weight:	range -	45.4-74.4 kg (100.0-164.0 lbs)	47.5-74.4 kg (104.8-164.0 lbs)
	mean -	57.0 kg (125.6 lbs)	59.3 kg (130.8 lbs)
	S. D. -	7.8 kg (16.7 lbs)	10.2 kg (22.4 lbs)
Age:	range -	24-39 years	26-34 years
	mean -	29.5 years	28.5 years
	S. D. -	3.4 years	3.0 years
Marital Status:		10 married	2 married
		12 single	4 single
Number of Children:			
	None -	17	5
	One -	1	0
	Two -	0	0
	Three -	3	1
	Four -	1	0

Garments

Previous spaceflight experience has shown that inflight garment provisions should be similar to those worn on Earth.[12] Hence, the female wardrobe should include basic outer- and under-wear. Flight suits or pants and shirt can be worn by both sexes, as planned for the Shuttle Program. However, overall habitability would be improved by tailoring the items for each sex. Female preferences, based on the interviews,

[11]Sawin, C. F. and J. A. Rummel (1978). Cardiopulmonary responses of the 200 Space Shuttle astronaut candidates. Preprints of 1978 Annual Meeting of the Aerospace Medical Association, pp. 121-122.

[12]Brown, J. W. (1974). Space garments for IVA wear. JSC-09540.

favored almost 3 to 1 a two-piece arrangement with pants and shirt rather than a one-piece flight-suit type garment. Although there are advantages of the one-piece suit in zero-g (e.g., no shirt "ride-up"), the distinct disadvantage is in terms of required donning/doffing for waste management operations. In any case, comfort and support of inflight activities are major factors in garment design.

For the women, panties rather than jockey shorts are more appropriate and will be included in their wardrobes. Preferences in this area favored a cotton garment over synthetics. In the zero-g environment, a brassiere will generally become more of a personal preference item than in a one-g state where support is sometimes mandatory for good health. A bra may be needed for comfort during launch and landing, as well as during extravehicular activities (EVA). The 21 female candidates included two who stated they never wore bras; bust measurements ranged from 81 cm (32 in) to 91 cm (36 in) with cup sizes from A to D. While one-third indicated they never go braless, one-quarter said they would not want to go braless in flight, which indicates a mandatory availability of bras as an optional flight item. A "leisure bra" which provides light control with stretch could be ideal to accommodate the zero-g effects of fluid shift and girth changes previously noted in spaceflight. As with men's underwear, standard-size, off-the-shelf garments are satisfactory. One change of underwear per day is recommended and will be provided in the Orbiter.

Although none of the females interviewed normally wore one, a girdle is not usually necessary for other than cosmetic reasons. Furthermore, as girdles are contraindicated because they restrict blood flow which would aggravate a zero-g effect of altered blood flow patterns, this garment will not be flown.

Eighty-one percent of the group felt pajamas or some form of nightwear should be provided. Some of these indicated that a T-shirt and panties were adequate, especially if the crew was small and had trained together. For payload specialists, more modest persons, and with larger crews, however, a quick-donning cover may become more important to the individual. Possible solutions would be a long T-shirt with a snap crotch or a pair of shorts. Again depending on the individuals involved, the male crewman might find having a pair of pajama bottoms or (outer) shorts for himself to wear would be desirable.

An anti-gravity suit (AGS) may be required to prevent body fluid pooling and to aid blood circulation for both men and women, especially for high G-loads after 3 days of zero-g activity.[13]

Hair

Not only does hair float free in zero-g as it does in a swimming pool (ref. Fig. 6), but it can adversely affect crew performance and become an irritant if long enough to get into the eyes, mouth or nose. Additionally, long hair can produce hazardous conditions if caught in scientific equipment, airlocks, or the spacesuit neck ring.

Long hair can probably be pinned satisfactorily with a barrette or large clip or restrained using a hair net, but short hair appears to be more convenient to deal with, will be easier to integrate with the communications cap (i.e., "Snoopy-cap") worn in the suit for EVA, and will facilitate cleaning the hair and scalp. The only available mode to date for scalp and hair cleansing is scrubbing with a washcloth. For longer-duration missions, then, having and keeping short hair appears very desirable.

---

[13] Shuttle Flight Operations Manual, Volume 12: Crew Systems. JSC-13059, 1978, p. 3.15-8.

Only 3 of the 21 interviewed--all 3 had hair longer than shoulder length--said they would not cut their hair if selected to fly. One other said she would be very reluctant to cut her hair and then only if it were nicely styled. The group was almost evenly divided on using a hair brush versus a comb. Both should be available as provisions for crewmember optional selection.

Cosmetics

Personal preference is the predominating factor in the use of cosmetics. Six of the 21 potential crewmembers used none at all, while several applied one or some combination of powder, rouge/blusher, lipstick, eye shadow, eye liner, eyebrow pencil, and mascara. Those in the user category felt that having cosmetics aboard the spacecraft was not a requirement but would be used if available. Lip balm and a skin moisturizer should be provided. Obviously, as with other spacecraft equipment and provisions, certain materials requirements would have to be met.

Personal Hygiene

Personal hygiene for females has several related considerations: management of menses, shaving leg and underarm hair, and waste management facilities design, including urine collection in the spacesuit.

The management of menses is a new unknown to spaceflight. The menstrual flow relies to a large extent on gravity for collection internally in the vagina by a tampon or externally by sanitary napkin or pad. A pad clearly is not adequate for the gravity-free conditions on-orbit. Whether the body can deliver the menstrual flow to the cervix by contractions alone and whether the wicking action of a tampon is adequate to absorb the flow remain to be seen. Tampon design, material, and wicking characteristics vary, and one may be better than others. No investigations of this potential engineering problem are known to the author. Medication to preclude on-orbit menstruation is possible but neither likely nor desirable for future extended flights or habitation in space. Flight experience must be relied upon to provide the answers. The recent bed-rest simulation at Ames Research Center involving females addresses the physiological effects of menstruation as follows:

> *From the findings of the present study, it does not appear that women sent into space will have additional problems because of menstruation. However, since the data in this study were limited and since this is the first all-female bed-rest study reported, additional research would be useful in determining the gynecological responses of women in the weightless environment. It is known that women experience shifts in blood and fluid balance as a result of menstruation. This factor should be investigated further to determine whether women will be able to cope with all aspects of weightlessness.*[14]

Twenty of the 21 females interviewed used tampons for collection of the flow, the other used a mini-pad. The amount and time course of the flow varied with the individual: most required more tampons per day for the first day or so of the period; total number of tampons used per period ranged from an estimated 3 to 30 with a mean number of 11. Most of the women had a brand name preference for tampons but would use whatever was provided if they flew. Some specifically stated that

---

[14]Sandler, op. cit., p. 47.

they did not like the deodorant type. Most were not averse to using an applicator-free tampon, which is smaller and requires manual insertion. Disposal of used tampons should be via the waste collection system rather than the trash bag for faster, cleaner operations.

Removal of female leg and underarm hair is basically a cultural factor. Only one of the 21 did not remove this hair; the others varied in their practices. All 20 used razors in lieu of depilatories or an electric shaver. Frequency of shaving varied widely: from daily to once every 1-2 months; only when exposed (e.g., when wearing shorts, skirt, sleeveless shirt); and during specific seasons. Hair growth rate and texture were also factors in the shaving routines. Only one female indicated that a 7-day hair growth could not be tolerated. However, half of the current female astronaut candidates would find shaving provisions desirable for a 30-day mission. Tweezers were listed as desirable to have for eyebrow shaping as well as for other applications.

Waste management facilities for the Shuttle Orbiter--which are comparatively similar to Earth-oriented waste collection systems, including seating and positioning--have been designed to accommodate both sexes (Fig. 8). The urinal can be used seated or removed for use by a "standing" crewmember, male or female. Air entrainment facilitates collection of the waste in the storage container for vacuum drying. The waste system was tested during 1973 and 1974 in simulated weightlessness aboard the KC-135 aircraft performing parabolic maneuvers by men and women.[15]

For spacesuit applications, urine collection for males relies on a roll-on device; for women, the answer is not so simple. The female urine collection device (UCD) is currently being developed. Potential concepts for the UCD include a small, strap-on cup of flexible, impermeable material filled with highly absorbent materials, a device held in place over the urethral opening by insertion into the vagina (similar to a tampon), and a waterproof garment with absorbent lining. Unconstrained urine collection diaper-style resulting in wetting a large skin area is highly undesirable and would be especially irritating for an EVA of several hours. Eight of the 21 interviewed favored a strap-on device, six favored an insertion device, and 7 were uncommitted. Most agreed that whichever functioned best and was comfortable should be provided and used. Antiseptic wipes for cleaning the perineal area will also be provided.

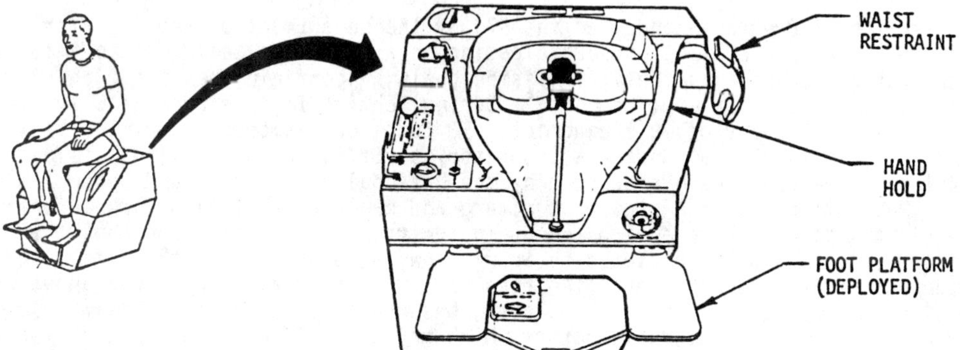

Fig. 8. Shuttle waste collection system.

[15]Burt, R. A., M. G. Koesterer, and S. R. Hunt, Jr. (1974). Solid metabolic waste transport and stowage investigation. General Electric Report No. 74SD4221.

Privacy

In the confined areas of small spacecraft, personal privacy can become more important than on Earth, where more opportunities to satisfy that need are readily available. Men and women both require privacy. Responses during the interviews indicated a need to find a place for being alone during a flight. Most felt that privacy was not too critical for a 7-day mission, but for a 30-day stay on-orbit, some time and a place to be alone were mandatory. The Orbiter sleep compartment was named most often as the place to be used for privacy.

The females also selected the sleep compartment and the waste management compartment as the two prime areas for getting dressed or changing clothes. Again, the crew size and individuals involved are major factors in how comfortable the female (or male) crewmember(s) would be.

During preparation for EVA [the mission specialist is the prime EVA crewmember; all female astronaut candidates are mission specialists and, hence, are potential EVA crewmembers], privacy for donning the urine collection system is mandatory. The women interviewed agreed that bra and panties--which are not much different from a 2-piece swimsuit--are adequate for EVA preparation activites requiring assistance. For EKG (electrocardiogram) electrode application, especially as required by a payload specialist during an intravehicular scientific experiment, a special shirt with appropriately-located access openings for electrode placement and leads would possibly be desirable.

PREVIOUS TEST EXPERIENCE

Earth-based simulations analogous to spaceflight have previously been conducted in an effort to learn more about man's responses to isolation, confinement, sensory deprivation, and other stresses. The Gulf Stream Drift Mission, the Sealab and Tektite I Programs, the Skylab Medical Experiments Altitude Test (SMEAT), and the NASA-sponsored 90-day manned test of a regenerative life support system are examples. Typically, the simulation crewmembers have been male. To date, no analog to spaceflight has been conducted with a crew of men and women. Only two major Earth-bound "missions" have been performed with all-female crews: the Tektite II Mission 6-50 and the NASA Marshall Space Flight Center (MSFC) Concept Verification Test (CVT) Mission IV.

The Tektite II Program, conducted in 1970 by the Department of the Interior, provided some data on females participating in a situation analogous to that facing astronauts: technical/scientific mission goals, a confining habitat (Fig. 9), and a stressful, hazardous environment.[16] During Tektite II, one crew of 5 female aquanauts and 9 crews of male aquanauts (53 total crewmembers) completed 14-, 20-, or 30-day missions in an undersea laboratory at St. John, U. S. Virgin Islands. Studies of the crews were conducted by the National Aeronautics and Space Administration (NASA) to assess the social structure and man's capabilities for performing useful work in a dynamic environment. In addition to demonstrating that the female crew very successfully completed their mission, a comparison of the female and male aquanauts produced some interesting results. Drawing firm conclusions about the two populations may not be possible, however, because only one female crew was selected and because of other external factors which may have affected the data collected. The results below should not be construed as significant differences in the relative capabilities of the two sexes, but probably are related more to social and environmental factors.

---

[16] Miller, J. W., J. G. VanDerwalker, and R. A. Waller (Ed.) (1971). Tektite II: Scientists in the Sea.

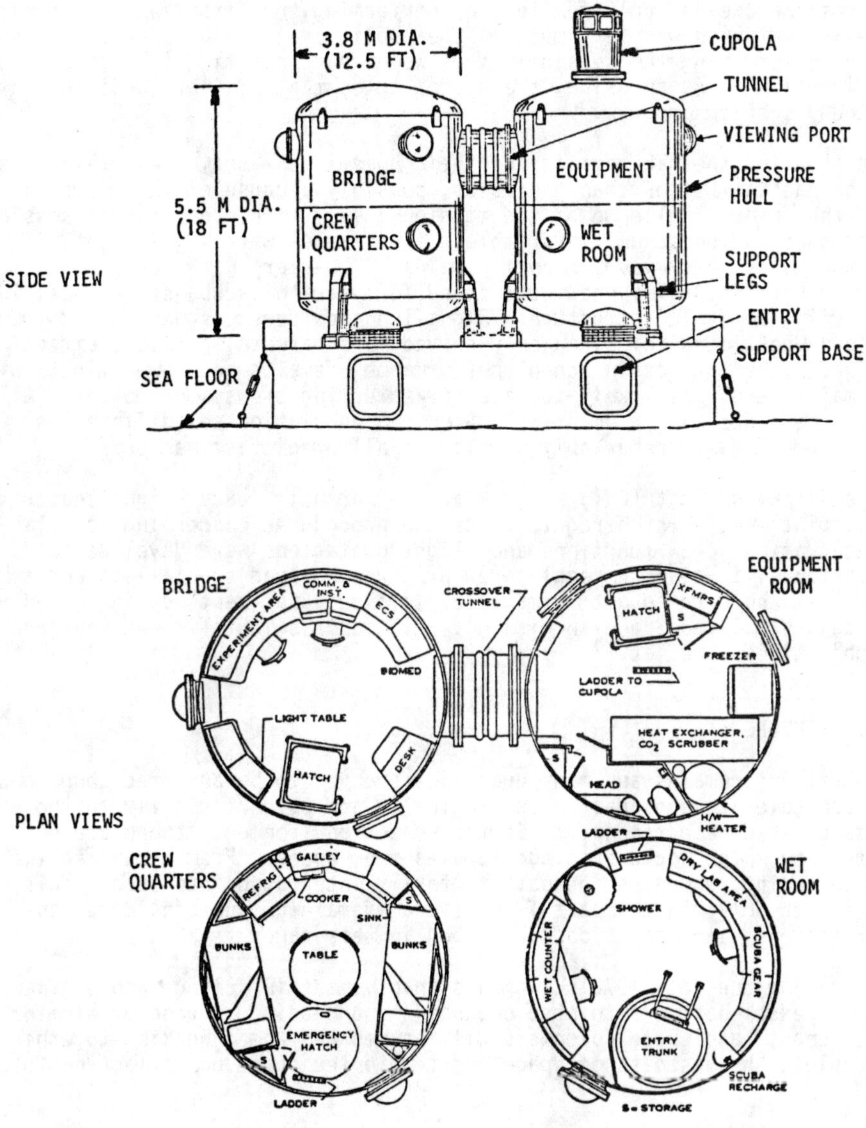

Fig. 9. The Tektite II habitat.

In the comparison, 5 females were compared to 43 males by an unweighted means analysis of variance. In terms of activity, the females worked significantly more than the combined male population ($\underline{F}$ = 12.49, $\underline{p}$ < .002), spent more time on marine science, their technical mission ($\underline{F}$ = 5.56, $\underline{p}$ < .025), and spent less time on leisure ($\underline{F}$ = 12.08, $\underline{p}$ < .002). No significant differences were noted in total sleep time, time spent in habitat maintenance, and time spent on self-maintenance. These results indicate that females are capable of maintaining a high work pace at least equal to that of males in an underwater habitat. That the females worked harder than the males could indicate either a natural superiority of females or a variation in motivation. While causal forces cannot be extracted from the data, the reader is reminded that increased motivation could very well have resulted because this was the one-and-only female crew performing the first female saturation dive, because this mission was very much in the public eye, and because this mission was depicted as determining the feasibility of women participating in both undersea and space exploration. [As an aside, the author notes that similar conditions are existent in today's space program.]

In terms of evaluating the habitability of their home-in-the-sea, the female aquanauts liked the habitat better than the males, possibly because of their smaller body size and the unique challenge of the mission itself. In other comparisons of general demographic and personality variables, the females were significantly younger than the males (32.7-years mean for males, 27.0-years mean for females; $\underline{F} = 5.43$, $\underline{p} < .025$). No differences were found for years of scuba experience, IQ, religious activity, and value structure. Significant differences were noted from the Life History Questionnaire: females received more parental physical affection; females received higher grades in school performance; females had fewer fights with peers; and females were less likely to have worked during the summer months. All of these differences are in agreement with American population sex differences. The male and female groups were remarkably similar on all predictive measures.

The Concept Verification Test (CVT) Program at the Marshall Space Flight Center was a series of simulations in which requirements and procedures concerning Spacelab experiment integration, crew functions and flight operations were developed and verified. Mission IV involved a female crew of three payload specialists and a mission specialist who carried out 10 material science experiments during a 5-day, 10-hour work day, simulated Spacelab mission. All major scientific and payload integration objectives were met.[17]

THE FUTURE

Although much testing remains and many questions are yet to be answered about how well females can tolerate and adapt physiologically and psychologically to the weightlessness of space and readapt to Earth's one-g environment, there are no indications to date which would preclude females from space. From a man-machine engineering standpoint, women can be satisfactorily integrated into space vehicles and operations with only minor impact from several female-unique considerations on design, provisions, and operations described in this paper.

As is evident during the late 1970's, women's involvement in science and engineering is increasing. Sexual barriers to most occupations are being reduced or eliminated. The future astronauts and space colonists will be men and women working together to explore and exploit the vastness of space and to pioneer space bases for the future of mankind.

REFERENCES

Berry, C. A. (1973). View of human problems to be addressed for long-duration spaceflights. Aerospace Medicine, 44(10), 1136-1146.
Berry, M. A. and W. E. Thornton (1978). Physical conditioning in men and women Shuttle astronaut candidates. Preprints of 1978 Annual Scientific Meeting of the Aerospace Medical Association, p. 123.
Brown, J. W. (1974). Space garments for IVA wear. JSC-09540.
Brown, J. W. (1976). Shuttle remote manipulator system workstation--man-machine engineering. Proceedings of the Sixth Congress of the International Ergonomics Association, Maryland. pp. 149-153.

[17] CVT/GPL Mission IV Test Report--Material Science Payload. MSFC 10M33232, 1975.

Brown, J. W. (1976). Zero-g effects on crewman height. JSC-11184.
Burt, R. A., M. G. Koesterer, and S. R. Hunt, Jr. (1974). Solid metabolic waste transport and stowage investigation. General Electric Report No. 74SD4221.
CVT/GPL Mission IV Test Report--Material Science Payload. MSFC 10M33232, 1975.
Miller, J. W., J. G. VanDerwalker, and R. A. Waller (Ed.) (1971). Tektite II: Scientists in the Sea.
Mutke, H. G. and E. C. Burchard (1976). Gynecological considerations on the participation of females in future space flights. Raumfahrtforschung, 20, 292-294.
Petrofsky, J. S., D. M. LeDonne, J. S. Rinehart, and A. R. Lind (1976). Isometric strength and endurance during the menstrual cycle. Europ. J. appl. Physiol., 35, 1-10.
Sandler, H. and D. L. Winter (1978). Physiological Responses of Women to Simulated Weightlessness. NASA SP-430.
Sawin, C. F. and J. A. Rummel (1978). Cardiopulmonary responses of the 200 Space Shuttle astronaut candidates. Preprints of 1978 Annual Meeting of the Aerospace Medical Association, pp. 121-122.
Shuttle Flight Operations Manual, Volume 12: Crew Systems. JSC-13059, 1978, p. 3.15-8.
Space Science Board of the National Academy of Sciences National Research Council (1972). Human Factors in Long-Duration Spaceflight, 91.
Webb Associates (Ed.) (1978). Anthropometric Source Book, Volume I: Anthropometry for Designers. NASA Reference Publication 1024.

IAA 79-A18
(AAS 79-325)

ANNUAL SURVEY OF SPACEFLIGHT SAFETY SYSTEMS: 9th SUPPLEMENT (SURVEY PERIOD: JULY 1978 -- JUNE 1979)*

Nelson E. Brown+

Total manned spaceflight safety encompasses not only the safety of crewmembers and supporting ground-based personnel, but also the space vehicles, payloads, support equipment and the general public and property within the flight envelopes. Extensive safety programs, equipment and training are required to effect overall safety for manned spaceflight programs. NASA and ESA are nearing completion of two major spaceflight systems, the Space Shuttle and Spacelab, initiating the beginning of an extensive range of future space undertakings with unlimited possibilities. Safety requirements and provisions to satisfy the requirements are continually evolving as a mandatory part of every space program. This survey identifies and documents salient safety studies, analyses, systems, and developments aimed toward providing spaceflight safety for the Shuttle era and beyond.

The first Space Shuttle Orbiter will have completed the orbital flight test phase and will be in routine earth-to-orbit service for over a year before an on-orbit rescue capability becomes available. Extravehicular systems available for enhancing crew and vehicle safety during this period are discussed and several major hazards affecting Shuttle Systems and equipment identified. Soviet spacesuit design and parachute systems for crew recovery are reported, and European Spacelab materials and atmospheric contamination measurement activities are presented. Water immersion testing, spacecraft lighting and color, and manned remote work station studies as related to safety for future long-duration spaceflights and large space construction are discussed. The survey concludes with brief comments concerning safety in space colonies and bases of the 21st century.

Keywords: Safety; rescue; hazards; extravehicular activity; spaceflight; space suits; life support systems; weightlessness, extravehicular mobility unit; safety surveys.

INTRODUCTION

The Space Shuttle Program is actively involved in crew safety and damage prevention to the Orbiter vehicle and payload systems. The identification and elimination of potential hazards, incorporation of safety systems, rescue provisions, and extensive crew training have been primary factors in ensuring crew and vehicle safety during previous U.S. space programs. Additional crew safety systems

---

* *This paper also appears in Volume 39, AAS Microfiche Series.*

\+ *McDonnell Douglas Corporation, Houston, Texas 77058; now at NASA Johnson Space Center, Houston, Texas 77058*

and on-orbit repair capability will be available for the initial Orbiter flight. NASA is developing systems, tools and training programs to perform contingency extravehicular tasks for closing and locking the payload bay doors for reentry and inspecting and repairing the Orbiter thermal protection system. This survey emphasizes reporting updated and additional equipment being developed to enhance the Shuttle era EVA capability.

The European Space Agency (ESA) and payload developers are implementing jettison capabilities, most with EVA backup, on all equipment extended beyond the Orbiter payload bay door closure envelope. ESA is also conducting tests on the possible buildup of toxic contaminants in the Spacelab. The Soviet Union is using extravehicular crewmen to inspect spacecraft exteriors and to retrieve scientific data. In addition improvements in the Soyuz reentry parachute system have recently been documented.

This paper is the 9th in a series of annual surveys which address various safety aspects of man in space. Literature published by government aerospace agencies, the space industry, universities and individuals provides safety material applicable to the numerous safety aspects of manned spaceflight. This safety material is reviewed yearly and synthesized to provide a sourcebook of information on applications, activities, status and advancements in spaceflight safety systems. Since 1971, editions of the series have reported on the work associated with the evolution of safety systems, safety policies and management, space rescue, working in space, spacecraft safety systems and crew provisions, hazard prevention, and safety training techniques and equipment. Through June 1979, a bibliography of 266 spaceflight safety articles has been compiled and the information summarized in annual reports.

SHUTTLE ERA SAFETY SYSTEMS AND CONSIDERATIONS

Shuttle Extravehicular Systems

The Crew Training and Procedures Division of the Flight Operations Directorate at NASA's Johnson Space Center is currently developing up to twenty Shuttle Flight Operations Manuals. These manuals describe in detail major Orbiter systems and their operation in both normal and emergency situations. Since an on-orbit crew rescue capability from a disabled Shuttle Orbiter will not be available until the sixteenth Shuttle flight, *Shuttle Flight Operations Manual--Vol. 15: EVA Systems* (Reference 254) presents extravehicular activity (EVA) systems data and crew operational procedures for responding to possible contingencies on the early Shuttle flights. Extravehicular systems and equipment described in the manual include: (1) Portable Oxygen System; (2) Extravehicular Mobility Unit; (3) EVA Airlock; (4) Manned Maneuvering Unit[1]; (5) Personnel Rescue System[2]; (6) Orbiter EVA Support Provisions; and (7) EVA Support Equipment.

Operational procedures are provided in the Manual for use by ground-based flight controllers and support personnel and the on-board flight crewmembers. The procedures provide a step-by-step description of each operation required for equipment checkout, operation, donning/doffing, and servicing functions. Contingency procedures are provided for all identifiable off-nominal situations that may endanger the crew or vehicle. The contingency procedures are, of course, within the performance capability of the crewmembers and EVA equipment. The contingency procedures

[1]The Manned Maneuvering Unit is a self-contained, free-flying propulsive backpack which provides the extravehicular crewmembers the capability to reach areas outside the Orbiter payload bay during earth-orbital flights.

[2]The Personnel Rescue System is a spherical pressure enclosure used in space rescue operations to transfer personnel from a disabled Orbiter to a rescue vehicle while in orbit.

for the early Shuttle flights emphasize a safe reentry vehicle; two major concerns are full closing and locking of the payload bay doors and inspection and repair of the Orbiter thermal protection system (TPS).

The following paragraphs provide an overview of the major safety related EVA systems and equipment discussed in Volume 15 of the Shuttle Flight Operations Manual. However, information concerning the Manned Maneuvering Unit (MMU) and Personnel Rescue System (PRS) was reported in the 1978 Survey of Space Flight Safety Systems paper and will not be repeated. Later paragraphs will discuss the EVA contingency procedures contained in the Manual for the early Shuttle flights.

Portable Oxygen System (POS)

The Portable Oxygen System (Fig. 1) is a rebreather-type oxygen system worn by the crewmember to provide breathing oxygen during normal and emergency on-orbit operations. The POS can be used to provide the following:

- Emergency oxygen and full-face protection from a contaminated or oxygen-deficient cabin
- Pure oxygen for prebreathing (denitrogenation) prior to extravehicular activity. Prebreathing is required when going from a 1.03 kg/cm^2 (14.7 psi) cabin to a 0.3 kg/cm^2 (4.0 psi) spacesuit environment to prevent dysbarism.
- Oxygen and full-face protection during post-landing egress
- Oxygen for certain medical contingencies
- Life support in later flights for on-orbit rescue situations. Used in conjunction with the Personnel Rescue System (spherical rescue enclosure).

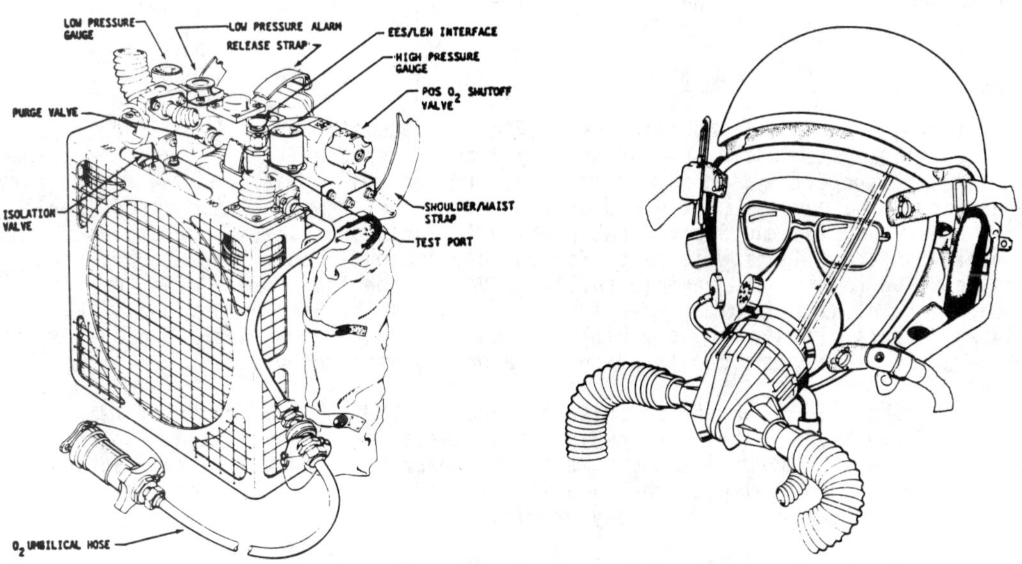

REBREATHER UNIT

FACEMASK AND HELMET

Fig. 1. Shuttle Portable Oxygen System.

The POS consists of a full face mask with visor, carbon dioxide control, water collection, internal oxygen supply, and associated controls. Major POS subsystems are: (1) rebreather assembly; (2) oxygen supply circuit; (3) recharge kits; and (4) extension hose/mouthpiece assembly. The POS can operate from Orbiter-supplied oxygen in a dependent mode or from an internal bottle in an independent mode. The internal bottle is initially charged to 221 kg/cm^2 (3150 psig) and may be recharged on-orbit to 60 kg/cm^2 (850 psig). Additionally, two operational modes are available in both the dependent and independent modes: closed-loop and open-discharge. In the closed-loop mode, the exhalant is passed through the rebreather circuit which contains a lithium hydroxide (LiOH) bed for CO_2 removal, felt wicks for water absorption, and a particulate filter to prevent particle migration. Makeup oxygen is provided and the breathing gases recycled.

Rebreather circuit recharge kits are available which provide water absorption and CO_2 and particulate removal for a period of three hours each. In the open-discharge mode, the exhalant is exhausted into the environment. In the dependent mode, the operational time is limited only by the oxygen supply carried by the Orbiter and the 3-hour recharge kit life. However, in the independent mode (POS internal oxygen bottle), time limitations are important. In the independent closed-loop mode with an initial charge of 221 kg/cm^2 (3150 psig), the breathing time is 41 to 62 minutes depending on the metabolic rate and 10 to 14 minutes on a 60 kg/cm^2 (850 psig) recharge. In the independent open-discharge mode, 7 to 11 minutes are available on a 221 kg/cm^2 (3150 psig) initial charge and 1.8 to 2.9 minutes on a 60 kg/cm^2 recharge. The open-discharge mode is used only when providing oxygen to the Ejection Escape Suit or the Launch/Entry Helmet and when using the extension hose/mouthpiece assembly. The extension hose/mouthpiece is used only momentarily during spacesuit donning to avoid interrupting the denitrogenation period.

The POS can be worn like a backpack, on the chest, or on either the right or left hip. One POS is carried on each flight for each crewmember plus spare units and recharge kits. There are 10 outlets in the Orbiter, including the airlock, in which the POS units can be operated on Orbiter oxygen.

Extravehicular Mobility Unit (EMU)

The Shuttle Extravehicular Mobility Unit (Fig. 2) consists primarily of four major systems: (1) an anthropomorphic pressure garment or spacesuit; (2) a self-contained portable life support system; (3) a displays and control module; and (4) a secondary oxygen pack. The purpose of the EMU is to provide the crewmember a means of performing tasks outside the Shuttle cabin while in Earth orbit. EVA operations during the early Shuttle flights will be performed only in contingency or emergency situations relative to crew and vehicle safety. EVA will be the primary means of crew rescue from a disabled Orbiter when the second Shuttle becomes operational. EVA in later flights will encompass vehicle and payload inspection, servicing, refurbishment and, ultimately, the construction of large space structures.

Many subsystems and components comprise the Shuttle EMU. The major subsystems are listed in Table 1 under their respective major system or as associated support equipment. The subsystems are too numerous to describe in detail in this survey paper; however, an overview of the capabilities and characteristics of the EMU is provided with emphasis on crew safety provisions.

The Space Suit Assembly (SSA), Fig. 3, is essentially made up of five interchangeable components: the hard upper torso, lower torso assembly, arms, gloves and helmet. The spacesuit is nominally operated at a pressure of 0.28 ± 0.01 kg/cm^2 (4.0 ± 0.2 psid) with a maximum pressure of 0.37 kg/cm^2 (5.3 psid). Suit sizes range from 5th percentile female to 95th percentile male, extrapolated to 1985.

The Life Support System (LSS), Fig. 4, is attached to the hard upper torso of the SSA and mainly consists of the primary life support system, secondary oxygen pack, displays and control module, and communications system. The LSS performs the following functions:

- Provides breathing oxygen and pressure control within the spacecraft
- Provides thermal control by recirculating and reconditioning the oxygen in the suit and water in the liquid cooling and vent garment
- Removes humidity, carbon dioxide, odors, and contaminants from the recirculating oxygen.

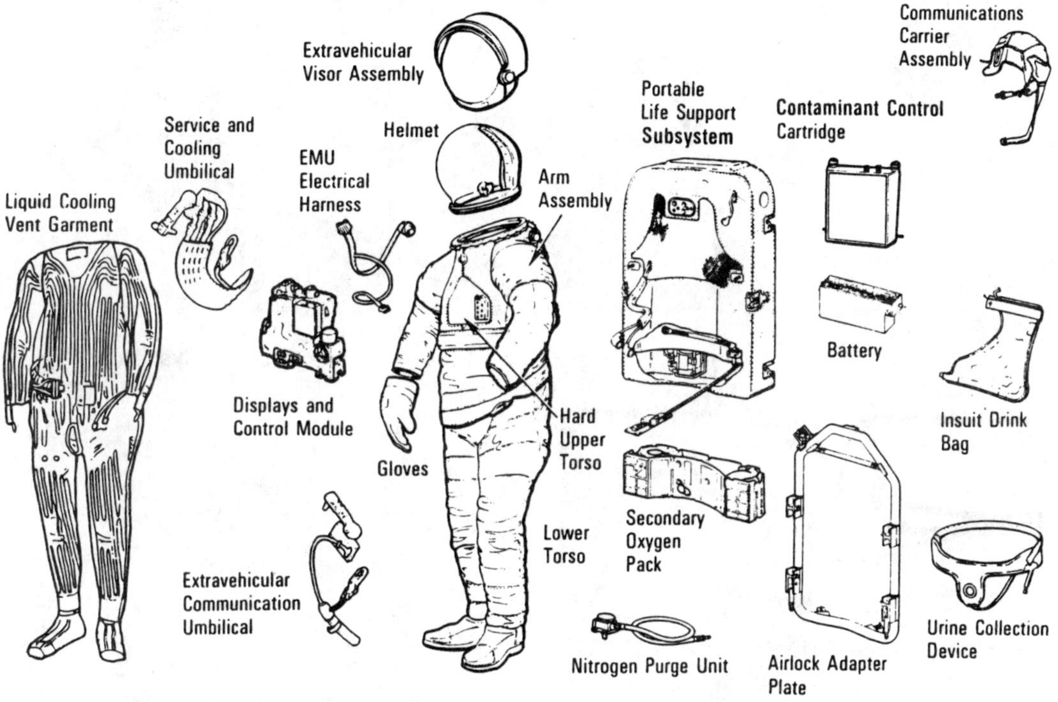

Fig. 2. Shuttle Extravehicular Mobility Unit (EMU) major components.

TABLE 1 Extravehicular Mobility Unit Subsystems

SPACE SUIT ASSEMBLY

- Hard Upper Torso
- Lower Torso Assembly
- Arms -- right and left
- Gloves -- right and left
- N_2 Purge Cuff
- Helmet
- Extravehicular Visor Assembly
- Communications Carrier Assembly
- Liquid Cooling and Vent Garment
- Insuit Drink Bag
- Urine Collection Device
- Operational Bioinstrumentation System

LIFE SUPPORT SYSTEM

- Primary Life Support Subsystem
- Displays and Controls Module
- EMU Electrical Harness
- Contamination Control Cartridge
- Battery
- Secondary Oxygen Pack
- Extravehicular Communications System

SUPPORT EQUIPMENT

- Service and Cooling Umbilical
- Extravehicular Communications Umbilical
- Airlock Adapter Plate
- EMU Maintenance

Fig. 3. Space Suit Assembly (SSA).

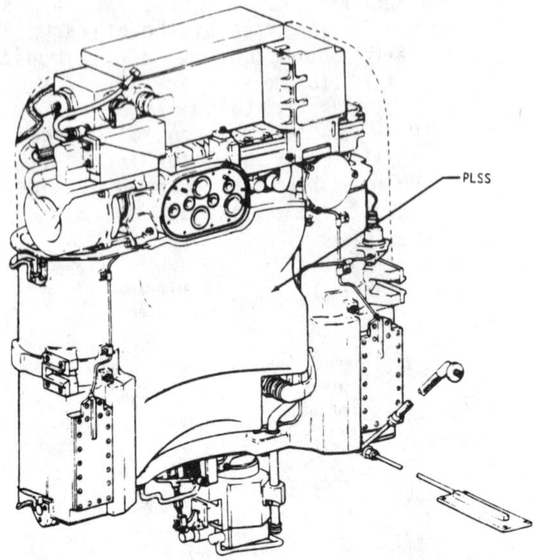

Fig. 4. Life Support System (LSS)

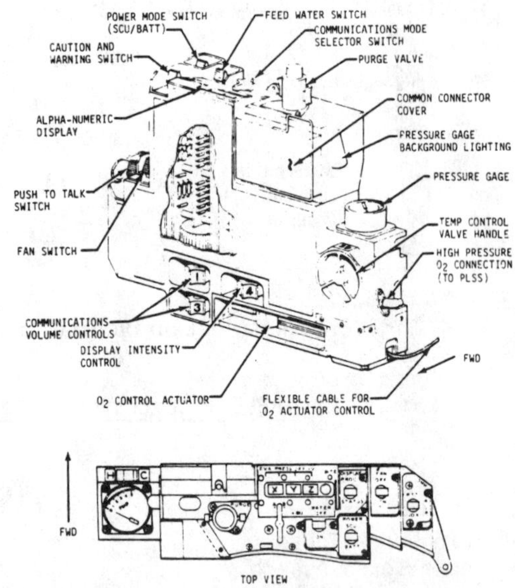

Fig. 5. EMU Displays and Control Module

The LSS external structure interfaces with the Manned Maneuvering Unit when free-flying EVA space maneuvers are required.

The displays and control module (DCM), Fig. 5, provides the visual displays and electrical and mechanical controls required for operation of the EMU. The DCM contains cooling and purge valves, suit pressure gauge, oxygen regulator module, cooling umbilical connectors, and most of the EMU electrical controls and switches. A caution and warning system alphanumeric display is located on the DCM within the crewmember field of view. The caution and warning system, physically located in the LSS, consists of instrumentation and a microprocessor to obtain, process, and visually display information concerning the operation and management of the EMU and also the Manned Maneuvering Unit. The caution and warning system stores and displays EMU operating and checkout procedures, monitors and displays EMU and MMU consumables status, monitors the operational integrity of the EMU, and alerts the EVA crewmember of EMU and MMU system anomalies. The caution and warning system provides corrective action messages for the EVA crewmember in the event of an EMU system malfunction and also provides EMU and MMU status on request. The system contains built-in test equipment consisting of software routines which verify proper system operation. The routines are multiplexed to provide continuous monitoring. Programmed checkout and operating sequences provide system procedures for EMU checkout, startup, and shutdown. The sequences are automatically interrupted for display of a fault message. The fault message information identifies failures, configuration and performance anomalies, and significant status parameters. A corrective action statement is included in the fault message for events requiring EVA crewmember action.

The secondary oxygen pack (SOP), Fig. 6, is a functionally independent life support system which provides the EVA crewmember an emergency backup for a minimum of 30 minutes. The SOP provides oxygen to the spacesuit and an open loop purge for removal of heat, CO_2, and humidity in the event of a primary LSS malfunction. The SOP consists of two 422 ± 14 kg/cm^2 (6000 ± 200 psi) oxygen tanks, pressure regulators, filters, pressure gauge, valves, flow limiting orifice, and appropriate interface connectors. The SOP oxygen tanks contain 1.18 kg (2.60 lbs) of oxygen which is provided to the suit at 0.24 ± 0.001 kg/cm^2 (3.45 ± 0.1 psid) in sufficient quantity to support metabolic work rates of 1000 BTU/hr for 30 minutes. Activation of the SOP is automatic if the primary oxygen circuit fails. The SOP is located within the primary life support system (PLSS) enclosure.

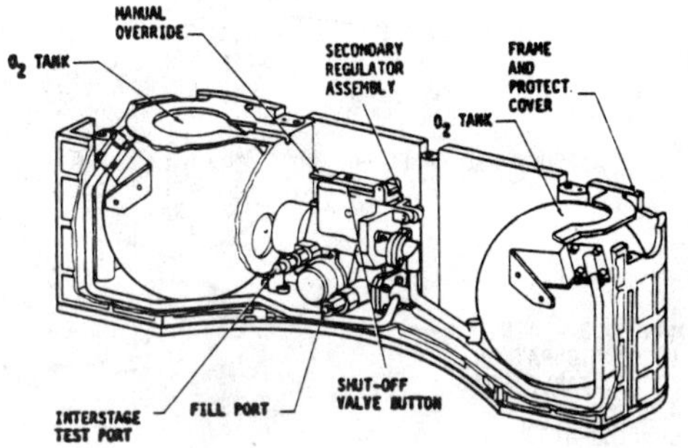

Fig. 6. Secondary Oxygen Pack (SOP).

An extravehicular communicator (EVC) package is installed in the extravehicular PLSS and provides the following:

- Duplex voice communications between Earth and the extravehicular crewmembers
- Uninterrupted voice communications between crewmembers
- Backup voice communications
- EVA rescue communications
- Continuous monitoring of EVA crewmember ECG
- Audible caution and warning tones to alert EVA crewmembers of abnormal or unsafe condition
- Audible MMU thruster cue signals of thruster activity.

The extravehicular communicator system, part of the Shuttle Orbiter EVA/ATC (air traffic control) communications system, is capable of maintaining simplex RF links between four EVA crewmembers and two Orbiter vehicles (a disabled Orbiter and a rescue Orbiter) in an on-orbit crew rescue situation.

EVA Airlock

The airlock (Fig. 7) is provided to accommodate crewmember extravehicular operations without depressurizing either the Orbiter cabin or an attached manned payload. The airlock is designed to support planned extravehicular activities, intravehicular transfer between the Orbiter and attached pressurized payloads, and crew EVA transfer between Orbiters in a rescue situation.

The airlock is a modular, cylindrical structure, 160 cm (63 in) inside diameter by 210.8 cm (83 in) long with two 1.0-m (39.4-in) diameter D-shaped openings, two pressure sealing hatches, and a complement of support subsystems. The airlock is removable and can be installed in one of three Orbiter locations (Fig. 8) depending

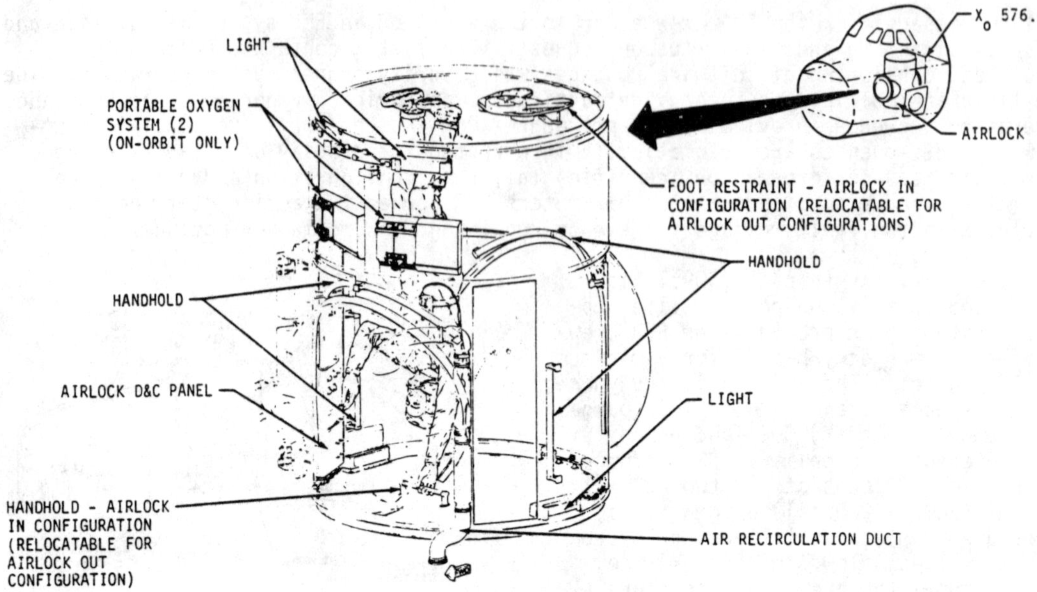

Fig. 7. Space Shuttle EVA Airlock.

on the payload being carried. The baseline location is inside the mid-deck compartment to allow maximum use of the cargo bay. It can be rotated 180° and positioned in the cargo bay. For habitable payload missions, such as Spacelab, the airlock may be positioned atop a pressurized tunnel adapter which connects the cabin with the pressurized payload.

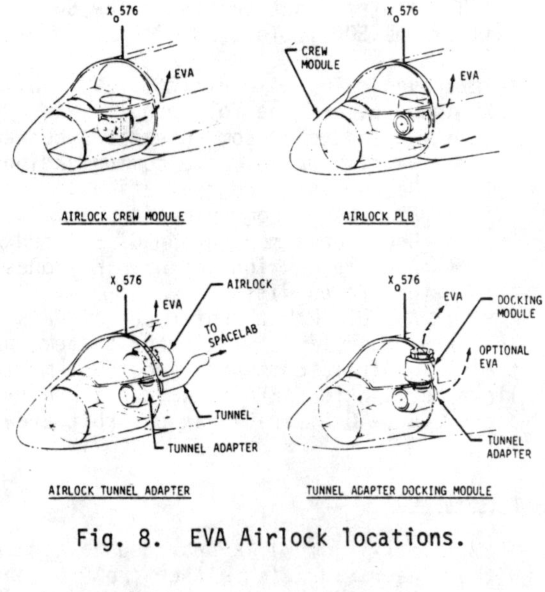

Fig. 8. EVA Airlock locations.

The airlock is also used to store the extravehicular mobility unit (spacesuit and life support system) during launch and inflight, and provides interfaces for various EVA support equipment during EVA preparation and post phases. Foot restraints and handrails are located in the airlock to assist spacesuit donning and airlock egress and ingress. Two display and control panels are provided for airlock environmental control system and avionics operations.

Airlock depressurization is accomplished by discharging airlock air into the Orbiter overboard vent line. The depressurization system discharges an airlock volume of 3.77 m^3 (133 ft^3) in 6 minutes for EVA egress. The airlock is repressurized by using equalization valves between the airlock and the pressurized Orbiter cabin. One normal mode and three emergency modes are available for airlock repressurization. Based on an Orbiter cabin pressure of 1.03 kg/cm^2 (14.7 psia), the airlock can be repressurized to approximately 0.99 kg/cm^2 (14.0 psia) in 16.5 seconds using the fastest emergency repressurizing mode.

Orbiter EVA Support Provisions

Orbiter EVA provisions consist of system and equipment items built into the vehicle specifically for supporting EVA. These provisions include: crew restraints and mobility aids, such as handholds, handrails, and tether attach points; crew safety tethers; safety slide-wires; and a modular equipment stowage assembly (MESA). Handrails are installed at various locations in the Orbiter payload bay to provide an extravehicular translation path. Handrails are provided on the fore and aft bulkheads, airlock exterior, and along the entire length of the payload bay near the door hinge lines. The handrails are configured to interface with the spacesuit glove and are equipped with tether attach points.

A retractable 7.6 m (25 ft) crew safety tether device (Fig. 9) is attached to the EVA crewmember's spacesuit and the tether line clipped to one of the Orbiter safety slidewires at all times when outside the airlock. The safety slidewires are located near each payload bay door hinge line and extend almost the length of the payload bay. The crew safety tether and slidewires allow the EVA crewmember to access all areas of the payload bay including the outer edges of the payload bay doors while in the full open position. Adjustable wrist and waist tether straps are also provided for retaining tools and equipment and to restrain the crewmember at worksites.

A modular equipment stowage assembly (MESA), Fig. 10, installed near the forward end of the payload bay provides stowage for tools and equipment used by the EVA crewmembers. Stowage bags are sized to retain the necessary tools and equipment for performing contingency tasks on the early Shuttle flights. The bags feature zippers that extend the full length of the bag for easy tool removal and replacement. Crew foot restraints and handrails are provided on the MESA to restrain the EVA crewmember during tool and equipment changeout.

EVA Support Equipment

EVA support equipment consists of "carry-on" or easily installed equipment categorized as tools, EVA workstations, portable lights, special tethers, and

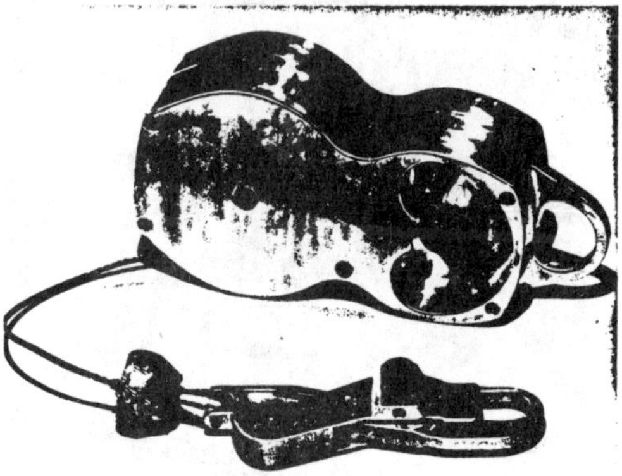

Fig. 9. Retractable crew safety tether.

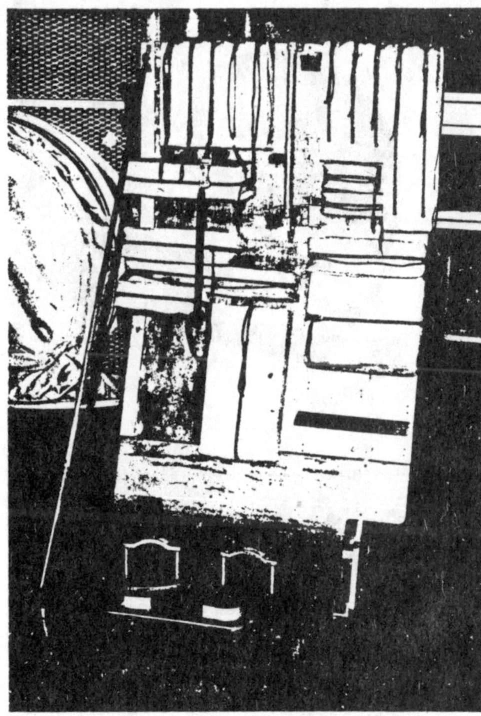

Fig. 10. Modular Equipment Stowage Assy.

dedicated special purpose locking/retaining devices. Of particular interest on the early Shuttle flights is the complement of EVA tools and equipment to ensure a backup capability for closing and locking the Orbiter payload bay doors for reentry. In an EVA backup mode, closing and/or locking the payload bay doors may involve one or more of the following tasks:

- Remove material jamming the payload bay door drive systems, radiators, bulkhead latches, or door centerline latches
- Disconnect payload bay door drive system, radiator actuators, door centerline latches, or bulkhead latch linkages
- Cut the payload bay door drive (torque tubes) linkages
- Install bulkhead 3-point latch tools
- Manually close the radiators, payload bay doors, and lock the door centerline latches.

Twenty off-the-shelf mechanic's hand tools have been modified and four special devices developed for accomplishing the above tasks. The modified tools include vise grip pliers, ratchet wrenches, pry bars, spanner wrenches, diagonal cutters, forceps, hammers, and adjustable wrenches. The four special devices are bolt extractor, tube cutter, 3-point latching tool, and door winch. A payload bay door winch (Fig. 11) is mounted on the forward and aft bulkheads and may be used by the EVA crewmember to manually winch the doors closed. The 3-point latching tools (Fig. 12) may be used to manually secure the payload bay doors at each bulkhead should the primary closing mechanism malfunction.

A mini-work system (MWS), Fig. 13, attaches to the front of the spacesuit and provides a work tether to restrain the crewmember at the worksite. Four tool caddies containing two tools each (Fig. 14) can be attached to the mini-work system. The tool caddies are lined with Velcro strips and a mating Velcro patch on the tool provides temporary stowage. Each tool is tethered via a small retractable tether at all times when outside the Orbiter cabin to avoid loss.

Fig. 11. Payload bay door winch.

Fig. 12. Bay door 3-pt. latching tool.

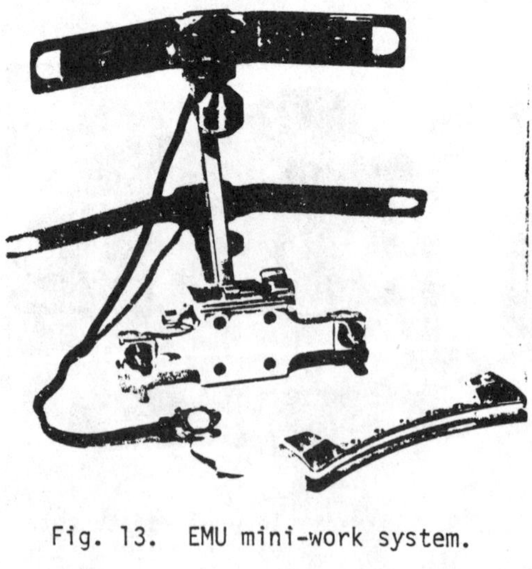

Fig. 13. EMU mini-work system.

Contingency Operating Procedures

Fig. 14. EVA tool caddies.

The Flight Operations Manual--Volume 15 also provides a complete set of step-by-step EVA procedures to safely accomplish all identified contingency operations outside the Orbiter cabin. Some 20 separate, detailed operational procedures are available which direct the crewmember in performing each of the payload bay door closing and locking tasks listed above. The contingency procedures are readily available to the EVA crewmember in the form of a spacesuit attached cuff checklist worn on the left arm. Crewmembers inside the vehicle have access to the contingency procedures in the form of a master EVA checklist and can assist the EVA crewmember via the extravehicular communications system.

Although a NASA decision had not been made at the publication time for this paper, a planned extravehicular activity may be required on the early flights (perhaps the first orbital Shuttle flight) to inspect, and, if necessary, repair the Shuttle thermal protection system. Such a task could be accomplished using either extendible translation booms in conjunction with the Shuttle Remote Manipulator System (RMS)[3] or free-flying Manned Maneuvering Units. Should the decision be made to use EVA in such a role, additional special tools and EVA operating procedures will be developed as necessary to ensure crew safety.

The NASA document, *Major Hazards--Space Shuttle Program* (Reference 255), contains a summary of selected major hazards affecting the Space Shuttle systems and equipment for the orbital flight tests and associated ground support equipment. Orbital flight tests consist of the first 4 to 6 launches to demonstrate flight worthiness. The document provides a means for periodic management overview and assessment of the cumulative residual risk of the Space Transportation System (STS) Program. Technical information and status of open hazards, rationale for major hazard closure, and accepted risks are provided.

A Shuttle major hazard is defined as a condition or potential condition associated with a design or operation that can result in injury to personnel and/or damage to hardware. The major hazards are identified at the Shuttle element and system levels as a result of safety analyses, trade studies, hazard analyses, failure modes and effects analyses, and numerous special studies.

The possible hazards summarized in the document represent only a small number of the hazards identified and resolved. From all of the STS hazards identified, a Senior Safety Review Board selects the major hazards to be highlighted to upper NASA management. The Senior Safety Review Board is composed of safety representatives from NASA Headquarters, Johnson Space Center, Kennedy Space Center, Marshall Space Flight Center, and Rockwell Corporation (the STS integration contractor).

[3]The Remote Manipulator System is an electromechanical, jointed arm mounted on the Orbiter longeron and used to deploy and retrieve payloads on-orbit. The arm is 15.2-m (50-ft) long, has six degrees of freedom, and is operated remotely by a crewmember from the Orbiter cabin. The RMS can also service payloads and assist in crew rescue.

The July 1978 issue of the document reported 81 major safety concerns: 35 open major hazards, 31 closed major hazards, and 15 accepted risks. An open major hazard is a possible hazard that has not been resolved to the complete satisfaction of the Senior Safety Review Board. Several examples of current open major hazards are listed below:

- Ice from the external tank impacts Orbiter thermal protection system
- Postseparation impact of Orbiter by external tank
- Solid rocket booster and Space Shuttle main engine effects on ejected crewmen and the escape system
- Lack of capability for suited access to a depressurized Spacelab
- Jamming of payload bay doors latching mechanisms
- Structural overheating caused by failure of the star tracker door in the open position
- Excessive on-orbit thermal distortion may prevent closure of the payload bay doors.

Summaries of each of the open major hazards are provided in the document. These hazards are being tracked by the Johnson Space Center Safety Division, NASA project officer, and Rockwell/Space Systems Group to assure satisfactory resolution. Safety assessments of the 31 closed major hazards have been performed resulting in closure by the Board. Rationale for the 15 accepted risks has been assessed and considered satisfactory. A major hazard is classified as an accepted risk if the decision, supported by technical rationale, has been made and documented by NASA program management and concurred by the Johnson Space Center Safety, Reliability and Quality Assurance Director. The Major Hazards document is updated approximately every 6 months to reflect changes in hazard status and add newly identified hazards.

The European Space Agency document, *Guidelines for Space Materials Selection* (Reference 256), provides a review of the different classes of materials used in habitable spacecraft manufacture. The purpose of the document is to present basic considerations and examples in order to orient spacecraft and payload designers in their preliminary materials selection. Some specific characteristics of selected materials of interest to designers include flammability, offgassing, odor, toxicity, and the effects of radiation and thermal stress. The document addresses the following materials:

- Adhesives
- Adhesive tape
- Coatings and varnishes
- Glass
- Lubricants
- Metals
- Paints
- Plastic films
- Potting compounds
- Reinforced and thermosetting resins
- Rubber
- Thermoplastics.

The following information is presented for each material:

- Use in spacecraft--Brief discussion of typical spacecraft applications
- Major categories--Chemical nature and aspects of various products which may be useful in spacecraft manufacture
- Processing--Primary fabrication methods involved
- Precautions--Paints to be considered when designing with the material
- To be avoided--Obvious meaning
- Effects of space environment--Summary of possible damage to and/or by the materials under spaceflight stresses
- Some representative products--Short list of approved and qualified products.

In any manned space program, particular attention must be paid to ensuring that the habitable environment within the spacecraft is safe. This requires that the materials used in the construction of the spacecraft and experiments do not release potentially toxic contaminants into the spacecraft environment. Reference 257, *Atmospheric Trace Contaminants--Their Measurement and Identification for Spacelab*, discusses the methods currently used at ESTEC in the analysis and identification of atmospheric trace contaminants for the Spacelab Program. Trace contaminants are defined as minute amounts of organic or inorganic species present in the surrounding medium.

Persons and the materials around them are continual sources of atmospheric trace contaminants. As part of the respiratory cycle, man removes oxygen from the atmosphere and replaces it with carbon dioxide. The metabolic processes lead to trace amounts of inorganic and organic compounds being generated. Nonmetallic materials, especially paint, coatings, and highly plasticized polymers gradually release their volatile components into the environment. Investigations by numerous health authorities have specified "safe" levels to be defined for large numbers of trace contaminants. These levels, often referred to as Maximum Allowed Concentrations (MACs) or Threshold Limit Values (TLVs), are considered the maximum concentrations that can be tolerated for a defined working period (usually eight hours) without exhibiting any detrimental effects. Many compounds that were in widespread use years ago are now subjected to intensive controls as the hazards associated with their use have become fully realized. In some cases, not only are the contaminants themselves hazards but an induced secondary reaction may well form a highly toxic compound.

Several methods of atmospheric sampling exist. One of the most simple methods employs calibrated colorimetric tubes containing adsorbents that change color as a known volume of air is passed through. The extent of color change compared to a standard indicates the amount of contaminant. Standards are available for a wide range of organic and inorganic contaminants. Where very low levels of organic contaminants are required, a concentration technique using adsorbent traps is used and requires special laboratory analysis. For inorganic volatiles or low-boiling point organics, gas samples must be taken into evacuated chambers for laboratory evaluation.

The European Space Agency is currently conducting "unit-level tests" to measure the trace contaminant generation rate of assembled Spacelab units. The units are placed in a chamber with a recirculating atmosphere arrangement, and on-line analysis of carbon monoxide and methane together with the total organics is obtained. Concentrated samples using adsorbent traps were analyzed using a gas-chromatograph/mass-spectrometer system. One commercial electric motor tested showed traces of 43 compounds but with concentration levels several magnitudes below the NASA maximum allowable concentrations. To roughly determine if the units are acceptable, the proportion of the total maximum allowable concentration is based on the ratio of the unit weight to the total payload weight. The Spacelab will be subjected to a "total offgassing test" prior to delivery to NASA. The tests will be conducted over a 24-hour period using a special portable sampling system developed by ESTEC.

Georgi Grechko egressed the Salyut 6 transfer module on December 20, 1978, to inspect the forward docking port. This was the USSR's first extravehicular activity in almost nine years and lasted about 88 minutes. The article, *Salute to Salyut* (Reference 258), provides summary information on the new Soviet spacesuit. The extravehicular suit, Fig. 15, as described in the article, features a hard upper torso with soft arm and leg units. The cosmonaut dons the suit through a hatch in the back of the torso. The swing-away hatch contains the integrated life support system, a cooling pump subsystem which circulates liquid through a mesh of plastic tubes, and a communications system. Prior to donning the suit, adjustments to special straps are made which allows a crewmember to ingress the suit unaided in five minutes. The adjustment straps allow one suit to fit a wide range of cosmonauts. The gloves provide sufficient dexterity to allow writing with a ballpoint pen, and the arms provide movement for clasping the hands behind the helmet. Published reports indicate the spacesuits will be used for extravehicular space assembly tests on future Soviet space vehicles. Extravehicular mobility aids and work platforms have been installed on the exterior of the Salyut space station.

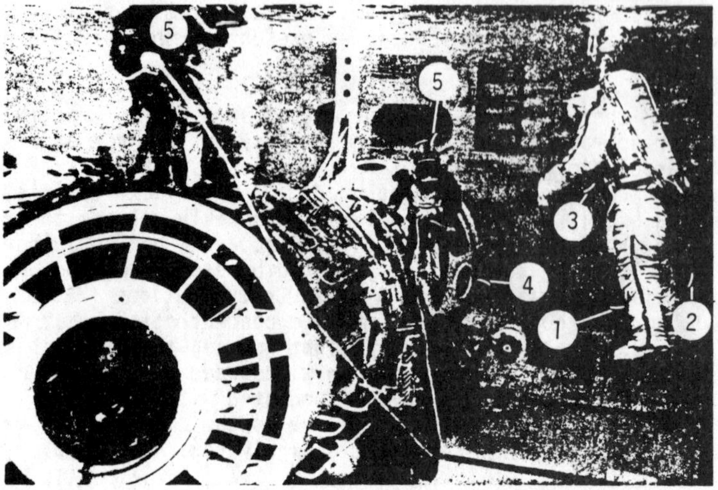

1 Spacesuit
2 Umbilical
3 Suit control panel
4 Soyuz mockup
5 Safety divers

Fig. 15. Soviet spacesuit in water immersion simulation.

The parachute landing system for the Soyuz spacecraft has been improved to ensure a safe and comfortable landing on either earth terrain or water. *From Space on a Parachute* (Reference 259) discusses the Soyuz parachute system and its operation under possible emergency situations. The parachute system consists of a 1000 m^2 (10,760 ft^2) main chute and an independently stored 574 m^2 (6,176 ft^2) reserve chute. Both parachute systems contain ripcord chutes and 14 m^2 (150 ft^2) braking chutes. Under normal Soyuz descent, the braking chute is deployed at 11 to 9 km (6.8 to 5.6 mi) which decreases the velocity of the spacecraft from 240 m/sec (787 ft/sec) to 90 m/sec (295 ft/sec). The main chute is then deployed and within 4 seconds slows the descent rate to 35 m/sec (114 ft/sec) and as descent progresses to 6 m/sec (20 ft/sec). After the main chute is fully open, the Soyuz heat protection shield is ejected from the soft landing engines. At approximately 1.5 m (5 ft) from Earth, the soft landing engines are automatically switched on for a soft landing. After touchdown the cosmonauts release half of the parachute shroudlines to prevent dragging the spacecraft should wind conditions prevail.

Should the main parachute not eject successfully, the braking chute from the main parachute is separated within 22 seconds to allow operation of the reserve chute system. Since the reserve chute is only one half the size of the main chute, the velocity at the time the soft landing engines are switched on may reach 8 to 11 m/sec (26-36 ft/sec). The landing impact is harder but not unsafe or damaging. Shock absorbing devices are mounted in the cosmonauts' armchairs to prevent injury. If the soft landing engines fail to function, the landing could cause cosmonaut discomfort. Landing will be fairly hard at a velocity of from 4 to 9 m/sec (13-30 ft/sec) but with no injuries or spacecraft damage.

The parachute system is also a part of the launch safety equipment for crew emergencies. If the Soyuz launch vehicle deviates from the vertical at a certain angle, the spacecraft automatically separates and is carried away from the launch vehicle at an altitude adequate for operation of a parachute system.

NEW DEVELOPMENTS

A final report entitled *Extravehicular Activity Translation Arm Study* (Reference 260) documents the preliminary design of an extendible device which would allow an extravehicular crewmember to access most of the underside of the Shuttle Orbiter.

Inspection and servicing operations could be performed on-orbit relative to the thermal protection system, external tank umbilical doors, external side hatch, landing gear area, etc. The translation arm would be stowed in the aft 1.2 m (48 in) of the payload bay in a deployable package with dimensions of 1.0 m (40 in) wide, 1.3 m (50 in) deep, and 1.1 m (43 in) long. The arm would be deployed on-orbit as a dual element articulated system--an upper boom and lower boom in a triangular truss configuration. The upper boom would be fully deployed and locked by an EVA crewmember while in the payload bay. The crewmember would translate along the deployed boom to the translation arm elbow, deploy and lock the elbow, then deploy the lower boom. The deployment rate would be a function of spacecraft and boom dynamics and also crewmember fatigue. The arm would be retracted, stowed and returned to earth for reuse.

The extendible arm system has been used on space systems. A similar arm, although not retractable, formed the main structure of the Synthetic Aperture Radar (SAR) antenna system on the SEASAT-A spacecraft. For the extravehicular arm, weight and deflection requirements dictate that the truss material be graphite/epoxy and the fittings aluminum alloy.

From the very beginning of Soviet manned spaceflight, a curious regularity has appeared: the longer the flight, the more light the cosmonauts wished to have aboard the spacecraft. The article *Light and Color on a Manned Spacecraft* (Reference 261) addresses illumination and color climate on long duration missions to enhance crew comfort, well-being, and safe, efficient spacecraft operation. In simulated spaceflight experiments of orbital flights, separate operators worked under conditions of sonic monotony and stressed visual work. During the first few hours, the operators were satisfied with a minimum illumination level. As fatigue increased, the operators increased the illumination level. Another experiment placed subjects in a nonstandard situation of continuous work for three days without sleep and breaks only for eating. The results were the same: with increased fatigue the illumination was increased by the subjects. Obviously the conclusion drawn was the illumination on an orbiting spacecraft and long-duration flight vehicles should be dynamic, adjusted as circumstances warrant.

More serious illumination and color media problems may arise with the development of future interplanetary spaceships. Visualize a flight to Mars: total silence, blackness of space, infrequent communications with Earth, limited volume of rooms, and a journey of years--not months or days. In contrast, man lives on Earth in a rich and constantly changing light and color medium. This medium supplies him with the necessary impressions and, especially important, determines the biological rhythms. Hence, on interplanetary or space colonies, it is necessary to organize the light and color "climate." This does not imply simply the functionally organized quantitative-qualitative distribution of light and color within an area, but an entire complex of measures acting on the emotional sphere of man. Special rest zones for emotional-psychological release may be required, perhaps a common wardroom in which the crewmembers could commune together, converse and rest.

An idea has developed: visualize a quiet ample hall of an interplanetary ship. One of the walls is moved aside and before the eyes of the occupants unfolds a panorama of terrestrial landscapes with the illusion of three dimensionality, a sense of space. The panorama is alive with the brightness and coloring of the image changing in accordance with the hours of day and season of the year; brightness of the landscape varies; one day is different from the next. Such a created situation will enable man to orient himself by the terrestrial clock and assist him in sensing the passage of time--he will be psychologically bound to earth. Such scenes can be created using laser and holographic technology and supplemented by sound reproductions of forest noises, streams, tides, birds, etc. A session of 10-15 minutes could make possible a thorough rest, restoring strength and initiative.

Spaceships for long-duration missions should offer convenience and comfort like on Earth. This will assist in safely and efficiently conquering the new space element.

Numerous space applications over the past decade have envisioned construction of ultra-large structures in space. Fabrication and assembly of future earth orbiting systems such as satellite power systems, large communications antennas, and large space habitats will require new approaches, techniques, fabrication equipment and orbital construction support systems. Present methods of ground fabrication and subsequent delivery to orbit are inefficient and very costly. A NASA-funded study by the Grumman Aerospace Corporation entitled *Manned Remote Work Station Development Article* (Reference 262) develops concepts and prototype hardware for universal crew stations to be used as space cherry pickers, railed work station, space crane turrets, airlocks and free-flying manned work vehicles. In such operational modes, the manned remote work stations (MRWS) can provide support for various construction, maintenance, repair, servicing, and rescue operations.

Figure 16 depicts concepts of two of the early work stations--the open cherry picker and the closed-cabin unit. The open cherry picker will be mounted to and operated from the end of the Orbiter remote manipulator system by a spacesuited crewmember with a self-contained life support system. The open cherry picker will: (1) provide a stable platform to react forces; (2) support handling and assembly of large structures; (3) provide servicing and refurbishment of satellites; (4) enhance deployment and retrieval of experiment equipment; and (5) support rescue operations. The closed-cabin unit will provide the necessary requirements and interfaces for a safe working habitat to operate in a more hostile space environment, such as geosynchronous orbit. The unit is totally enclosed with a viable environment for shirt-sleeve operation. The early closed-cabin unit would be mounted on a large space crane and feature two dextrous manipulator arms. Later versions of the closed-cabin unit would be equipped with systems for free-flying: maneuvering capability, attitude control, rendezvous and docking equipment, controls and displays, and environmental control systems.

More advanced Manned Remote Work Station systems may include pressurized orbital transfer vehicles (POTV) and railed work stations to support extremely large orbiting structural systems. The present long range planning for MRWS programs is shown in Fig. 17. The open cherry picker is planned to be operational about mid-1982 and the attached closed-cabin unit by the mid-1980's, assuming a space crane is available. The free-flying MRWS is envisioned by the early 1990's and the

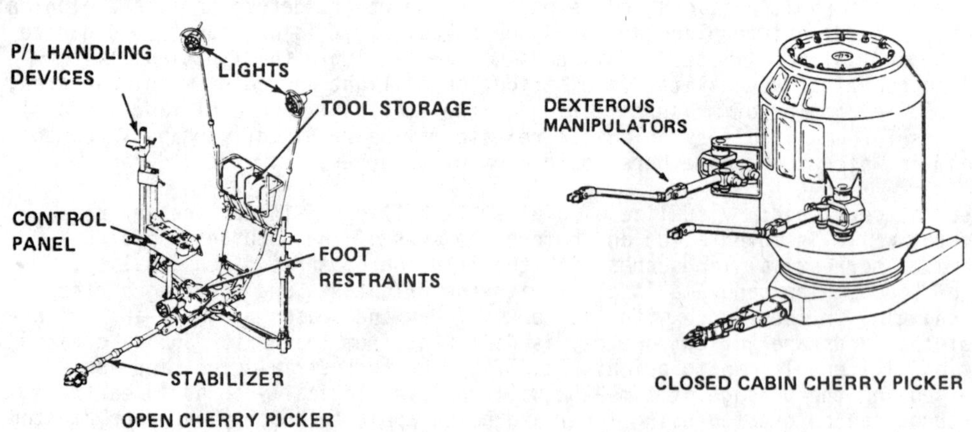

Fig. 16. Early Manned Remote Work Station concepts.

railed work station and pressurized transfer vehicle by the mid-1990's.

Neutral buoyancy simulations have been employed by NASA engineers and astronauts for many years as a means of preparing for the zero-gravity state of space. *Tank Tests Validate Structure Assembly* (Reference 263) describes manned simulations being conducted in the Marshall Space Flight Center (MSFC) neutral buoyancy facility. The simulations have included assessment of various aspects concerning large space structures: beam structure assembly, mass handling, and Manned Maneuvering Unit operation.

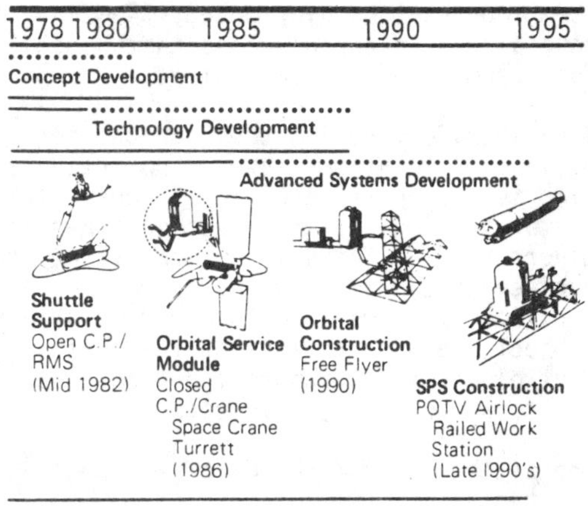

Fig. 17. MRWS long-range planning.

The MSFC neutral buoyancy facility is a 23-m (75-ft) diameter by 12-m (40-ft) deep tank containing 4.9 x 10^6 liter (1.3 million gallons) of water in which spacesuited subjects can be balanced to achieve "weightlessness," thereby providing a realistic zero-gravity simulation environment. Ensuring that future planned space activities are safe for the crewmembers involved is inherent in the conduct of these tests. For example, one simulation evaluated man's capability to manipulate an object that would weigh 8,165 kg (9 tons) in Earth's gravity environment (Fig. 18). Inflight crewman safety with reference to starting, stopping and maneuvering such high potential inertia

Fig. 18. Water immersion assessment of handling large space structures.

items must not be overlooked. Testing indicated he could easily maneuver these high-mass pieces of hardware and that visibility around such masses is the key to successful assembly of such equipment.

Crewman safety during these ground tests is also addressed in the article. Neutral buoyancy simulations involve potentially hazardous operations and, at MSFC, real-time test support involves 15 people for testing with one spacesuited-subject, 19 for testing with two spacesuited-subjects. Included in these groups are safety divers, utility divers, a water safety team, photo divers, a deck chief, a systems engineer, suit technicians, a recompression crew, and a control room crew. Responsibilities are well-defined: divers to monitor the well-being of each spacesuited subject; persons to monitor the status of the pressure suit and its environmental control and life support functions; and engineers, technicians, and medical personnel to support the activities via direct, television, and audio channels.

One of the major findings of the neutral buoyancy tests cited as important addressed safety:

> *No space suits were damaged in the tests even though some beams have relatively sharp edges. The tests indicate that safety in structures assembly will be achievable without undue conservatism in the activities undertaken.*

PLANNING FOR THE NEW MILLENNIUM

The U.S. Space Transportation System (STS) Shuttle, presently in the mainstream development phase, will provide the wherewithal for new space frontier transportation. Communications platforms, solar power satellites, meteorological and climatological platforms, and permanently manned space stations all will result from the advent of the Shuttle. STS capabilities will continue to open frontiers beyond our imagination, including serious considerations for climate control, in-space hospitals and farm lands, and permanent space and lunar habitats manned by industrial workers who will assemble satellite power stations, mass-drivers and space colonies. Hopes and aspirations in recent years have been somewhat dampened for such projects in the near future since action by the U.S. Congress has modified many of NASA's priority programs and cancelled others. Such systems will, nevertheless, eventually be developed by the U.S. or other nations--perhaps by a consortium.

As in all previous manned space programs, safety will encompass a significant part of these future programs including safety of the crew/astroworkers/habitants, space structures/vehicles, and planet/lunar populations. Since sufficient launch and space rescue vehicles will be available during the large space structures era to conduct space rescue and repair operations, safety studies have been primarily concerned with the hazards of meteoroids and ionizing radiation to large lunar and earth orbiting habitants. Independent space shelters to house large numbers of habitants in emergency situations have been addressed.

Several very significant studies and conferences conducted in 1975-76 concerning space manufacturing and colonization of space developed concepts for protection from meteoroids and cosmic radiation. The first conference was organized by the Princeton University and cosponsored by NASA and the American Institute of Aeronautics and Astronautics. The conference explored large space structure construction methods, productivity and payoff to the Earth of permanent industrial habitats near the L5 Earth-Moon Lagrange libration point. The second conference/study, sponsored jointly by NASA and the American Society for Engineering Education, was held at Stanford University and the NASA Ames Research Center. The goal was to design a system for the colonization of space. A third study, called the 1976 NASA-Ames Summer Study on Space Manufacturing from Non-Terrestrial Materials, provided formulae for optimizing the design of magnetic mass-drivers, data on space manufacturing plants, mass throughput of material, power requirements and workforce needs. Noted individuals in specific space fields from universities, government organizations, and private industry participated in the studies.

Probably the most frightening large space station or space colony disaster imaginable would be collision with a swarm of large meteoroids, followed by explosive decompression. However, from all available meteoroid data from earth, nearby space and the moon, the real meteoroid problem will be finding and repairing minor leaks made by meteoroids about the size of a grain of sand. The most common meteorite effort would be a slow sandblasting of the windows and hull produced by micro-meteorites. Such effects are described in T. A. Heppenheimer's book, *Colonies in Space* (Reference 264). Only once every few years would the colony be hit by something large enough to make the inhabitants take notice.

The data from all sources tend to agree and give the following picture of meteoroid hazards:

SIZE	MASS	RATE OF OCCURRENCE	WORST EFFECTS
0.64-cm (0.25-in) pebble	0.03 gm (0.001 oz)	Every 3 years	Loss of window panel. Atmosphere escapes at 1% per hour.
5.1-cm (2.0-in) rock	99 gm (3.5 oz)	Every 7 years	60% of atmosphere escapes in 10 hours.
Boulder	907 kg (1 ton)	Every 250,000,000 years	Severe damage.

Ionizing radiation is a major threat to space colonies and may be intensive during solar flares. Cosmic rays, electrically charged, break chemical bonds when they pass through body tissue. Several concepts have been studied to protect space inhabitants from radiation. For small construction shelters, the entire shelter might be enclosed in a magnet shield to deflect all but the very high-energy particles. However, to magnetically shield a large space colony would be weight prohibitive. Charging the colony electrically to 10 billion volts to repel all particles below the 10 billion volts level and plasma radiation sheilds have been considered.

The most attractive shielding method studied is that of covering the exterior of the colony with 1.8 m (6 ft) of lunar soil delivered to the colony by lunar mass drivers. The soil would consist of slag or waste left over after metals had been removed for colony fabrication.

In Chapter 7 of his well-known book *The High Frontier* (Reference 265), O'Neill addresses risks and dangers of space habitats, including those resulting from physical and social causes: meteoroids, radiation, habitat environment, wars, and sabotage. His conclusion that the risks for space settlers are "rather less than those to which we are exposed every day here on Earth" is followed with details concerning each. O'Neill describes three sources for meteoroid data: recordings from spacecraft sensors, seismometers placed on the lunar surface during the Apollo Program, and photographs taken by an Earth-based array of sky-cameras called the Prairie Network. Averaging these data allows one to speculate as follows: (1) A one-ton meteoroid would certainly "produce a hole and cause local damage" to--but "should by no means destroy"--a large community. However, based on frequency data, the community would have to wait a million years for one as large; (2) More-frequently-found meteoroids would be much smaller--somewhere around the weight of a tennis ball--and would strike an estimated once in 3 years; and (3) Meteoroid strike dangers are reduced in outer space communities because meteoroid densities are reduced from those found near Earth, where gravity facilitates their congregation.

Windows present an obvious risk in large habitats. By subdividing window space, the risks can be reduced such that pane breakage would not be catastrophic, especially if repairs are performed expeditiously.

The space colonists must be protected from space radiation including cosmic rays and solar flares. Shielding required for protection would be substantial; however, by careful habitat design--including atmospheric depth, structure thickness above and below ground level, building materials, and overall geometry--radiation levels can be reduced to comparable Earth-levels.

An oxygen atmosphere would present another potential hazard: fire. Addition of hydrogen to the atmosphere, reduction of atmospheric pressure, large total volume, sufficient water supplies and selection of non-burning materials for construction will help reduce the fire danger.

Because of space settlement vulnerability and distance from Earth, O'Neill suggests that the probability of wars would be considerably smaller than that of wars between Earth nations. Sabotage from within will be hampered by "customs inspections" and large-scale structures which are less-easily damaged.

In summary, the author believes that development of space communities would be "relatively safe from catastrophe, in which an increasing fraction of the human race would be widely dispersed and consequently safe from simultaneous destruction...."

In *Outline For an Exploratory Study on the Design of a Lunar Base Habitat* (Reference 266), Trotti and Criswell describe the results of a habitability study of a 12-man lunar base designed for 6-month residence periods. The proposed base consists of pneumatic structures with Kevlar membrane to support internal atmospheric pressures, a rigid pallet to house the base life support systems, and foam-core walls for noise and thermal insulation. The structures would be designed to expand after deployment from the pallet. After the final configuration has been deployed, three meters of lunar soil would be laid on the modules to form radiation protection and thermal insulation. Additional design details and artist's renderings are presented for habitat volumes, airlocks, suit storage and cleaning, living modules and accessories, and dining and food preparation areas.

REFERENCES

254. OFT Shuttle Flight Operations Manual--Volume 15: EVA Systems. JSC-12770. 1978.

255. Major Hazards--Space Shuttle Program. JSC-09990G. 1978.

256. Product Assurance Division (1978). Guidelines for space materials selection. ESA PSS-07/QRM-01-Issue 4.

257. Debeir, M. and M. D. Judd (1979). Atmospheric trace contaminants--their measurement and identification for Spacelab. In ESA Bulletin No. 17. pp. 66-71.

258. Oberg, J. (1979). Salute to Salyut. In Spaceworld, Vol. P-4-184. pp. 4-14.

259. Lobanov, N. (1978). From space on a parachute. NASA Technical Memorandum 75577, Translation of "Iz kosmosa na parashyute," Aviatsiya i Kosmonavtika, No. 9, Sep. 1978, pp. 34-35.

260. Preiswerk, P. R. and J. C. Stammreich (1978). Extravehicular activity translation arm (EVATA) study. Final Report ARC-TN-1064, NAS9-15579.

261. Melnikov, L. and O. Andreyeva (1978). Light and color on a manned spacecraft. In Transl. on USSR Sci. and Technol., No. 54 (JPRS-72282). pp. 32-33.

262. Grumman Aerospace Corporation (1979). Manned Remote Work Station Development Article, Final Report--Executive Summary. Report NSS-MR-RP008. NAS9-15507.

263. Covault, C. (1978). Tank tests validate structure assembly. In Aviation Week & Space Technology, Vol. 108, No. 26. pp. 55-62.

264. Heppenheimer, T. A. Colonies in Space. New York: Warner Books, Inc. (1977).

265. O'Neill, G. K. The High Frontier. New York: Bantam Books, Inc. (1978).

266. Trotti, G. and D. Criswell (1978). Outline for an exploratory study on the design of lunar base habitats. American Astronautical Society Paper 78-143.

IAA 79-A20
(AAS 79-327)

MISSION ASSURANCE FOR SPACELAB UTILIZATION*

H. Schürmanns+

The mission assurance goals for a Spacelab mission are defined and their new or different aspects are discussed. It follows a description of some features for a cost-effective implementation approach and concludes with the status of work for the D1 payload.

Keywords: Mission assurance, safety, interface verification, performance assurance, Spacelab (SL), Space Transportation System (STS).

* *This paper appears in full in Volume 39, AAS Microfiche Series.*

+ *Deutsche Forschungs- und Versuchsanstalt für Luft- und Raumfahrt e.V. (DFVLR), Bereich für Projektträgerschaften (PT), Köln, Germany*

IAA 79-A21
(AAS 79-328)

SAFETY PROBLEMS - IDENTIFICATION AND MONITORING OF HAZARDS FOR MANNED REUSABLE SPACECRAFT*

J. Leupen[+]

With the introduction of Manned Reusable Spacecraft a need was created for a different approach to the safety and reliability problems in comparison with the usual approach in former space programs.

The significance of the maintainability impact on safety requirements was not sufficiently recognized.

Based on the latest design, development and production experience of the Spacelab Airlock, a more systematic approach is suggested to realize (from a safety point of view):

- Control of design and development
- Hardware control
- Instruction control

Special techniques have been developed and used for design control, such as fault isolation flow diagrams and for hardware/operation control. Critical characteristic control sheets and their function are explained more fully in the paper. Based on the experiences of the last few years recommendations are suggested to ensure better co-ordination of safety and maintainability requirements and methods are proposed to ensure improved visibility of safety verification.

---

* *This paper appears in full in Volume 39, AAS Microfiche Series.*

+ *Overtoom 32, 1054 HK Amsterdam, The Netherlands*

WORLDWIDE DISASTER RESPONSE, RESCUE AND SAFETY EMPLOYING SPACE-BORNE SYSTEMS 1979

The following papers also appear in Volume 39, AAS Microfiche Series, supplement to this volume:

IAA 79-A30 and -A34

The following papers appear in full in Volume 39, AAS Microfiche Series:

IAA 79-A28, -A31 and A33

IAA 79-A26

PERSPECTIVES ON WORLDWIDE DISASTER AND RESCUE RESPONSE EMPLOYING SPACE-BORNE SYSTEMS

A Continuing Survey Paper: 1979 Supplement

Gloria W. Heath*

An initial survey paper was presented at the IAF Congress in 1978 (Paper No. 78-A-69) as background for Session II of the Space Rescue and Safety Symposium of the International Academy of Astronautics. The session is entitled "Worldwide Disaster and Rescue Response Employing Space-borne Systems". An annual supplement is prepared for each succeeding session.

This survey paper considers the realities of response to disaster and rescue situations - realities which dictate the requirements that space-borne systems need to address if they are to be effectively incorporated by national and international cooperative agencies carrying out distress-response functions.

DISASTER AND RESCUE RESPONSE FUNCTIONS AMENABLE TO EMPLOYMENT OF SPACE-BORNE SYSTEMS

The principal functions carried out by distress-response agencies include

- establishing communications among those affected by a distress occurrence and those called upon to render assistance
- detecting an impending distress situation
- instituting preventive measures
- receiving reliable information on the nature and severity of the damage and injuries sustained
- dispatching appropriate assistance.

* *SAR-ASSIST, Inc., One Island Lane, Greenwich, Connecticut 06830*

VARIED NATURE OF DISTRESS-RESPONSE AGENCIES

The exercise of the safety and rescue functions listed above occurs within the particular jurisdictional and budgetary framework permitted each agency that bears distress-related responsibilities.

Agencies vary widely in personnel capabilities, resources, geographical and political situation. These determine its organizational format. A wide variance of organizational formats among agencies worldwide has evolved.

Organizational format is an underlying factor that determines distress-response agency decisions with respect to the incorporation of space-borne segments into the disaster and rescue response system.

CONSIDERATIONS GERMANE TO DECISION-MAKING WITH RESPECT TO INCORPORATION OF SPACE-BORNE SEGMENTS IN DISASTER AND RESCUE RESPONSE SYSTEMS

Considerations germane to decision-making are addressed in this paper in outline form under the topical headings

- cost-related influences
- distress-response and appropriateness or non-appropriateness of the space segment remedies offered
- institutional questions that have not yet yielded to effective resolution.

Future sessions will amplify the topics in individual papers and inter-disciplinary discussions.

Cost-related influences

- The question of assurance of operational space-borne segment accommodation
- GENERAL Can the "product" be put to effective use?

  SPECIFIC With respect to REMOTE SENSING:

  timeliness of receipt
  interpretability by decision-making agency.
- The question of dedicated vs. non-dedicated systems

  discrete functions vs. combined functions
- Concept of COST:
  - political/societal values: life
    property
    individual's responsibility
  - affordability of expenditure in relation to
    level of national development
    national aspirations
    expenditure by the individual

and adverse consequences of change that may be incurred by system introduction into national life

- the question of where is the cost to be borne

Appropriateness of the Space-borne Remedies Offered

- Treating the symptoms or the basic cause and the question of whether there are alternative means of accomplishing an equivalent result

Example 1) A large proportion of downed aircraft incidents occur in adverse weather conditions. Should major effort be directed to

- systems that require a pilot to fit his aircraft with an electronic transmitter which will be activated (automatically) in an emergency situation and that enables the rescue agency to receive timely notice both that a craft has gone down (alerting) and the geographical location of the emergency transmission site (locating)

OR

- increasing adverse weather information warning to pilots and to utilizing weather information/analysis to give accurate ground location of the site where the adverse weather would have forced the aircraft (search target) down

OR

- alternate means of gaining intelligence of aircraft movements and their progress along their track (such as flight plan filing, traffic control procedures)

Example 2) Costly national relief schemes to aid flash-flood victims who have settled in vulnerable canyon locations

OR

- the siting of dwelling in non-vulnerable locations.

- Indiscriminate coupling of the functions of ALERTING AND LOCATING

The functions of alerting and location, if both are combined in each system offered, may not fit all distress-response situations productively. The alerting function may unnecessarily burden the response system beyond its capacity to respond - a question of determining

- the proportion of alerts that require response to the site
- whether or not the response agency would have knowledge of the cases to which its response was needed (by other means)

- the adverse situations of non-vital alerts masking true distress transmissions.

- Indiscriminate coupling of the terms SEARCH and RESCUE

  SEARCH applies to determining a distress victim's geographical location and distress condition before assistance can be rendered. In not all distress situations will RESCUE be feasible or necessary - space segment involvement is ALERTING and LOCATING.

  RESCUE applies to the dispatch of trained personnel and equipment to a distress site to extricate a distress victim from a precarious situation - space segment involvement is EMERGENCY COMMUNICATIONS.

- With respect to disaster response, there are differing national perceptions of

  - what constitutes a "disaster"
  - the efficacy of issuing "warning"
  - permissible behavior of those to whom warning is issued.

  These factors influence system configuration/provisions.

Unresolved Distress-related issues - Institutional Arrangements

- With respect to selection of space-borne system radio frequencies and system provisions for distress-related functions, the questions of

  - who is left out of the system
  - interference with emergency communications
  - capability to respond to multiple/close sequence emergencies
  - keeping open the possibility of incorporating system features developed in future.

- With respect to the global nature of the causes and effects of disasters and the magnitude of measures required to alleviate the effects that go beyond the ability of single nations to manage

  Are the international instruments in place for

  - organizing cooperative response systems
  - conducting cooperative response (including training of personnel).

  Is concordant and efficacious response assured - a question of setting criteria and examining performance

IAA 79-A30
(AAS 79-332)

SATELLITE COMMUNICATIONS FOR DISASTER RELIEF OPERATIONS

Joseph N. Sivo*

Severe natural disasters that occur in developing and underdeveloped areas of the world require the assistance of the world community to minimize the impact on the affected populations. Reliable communications during the periods of relief activities are essential to maximizing the timeliness and efficiency of relief operations.

This paper will discuss the use of existing and planned communication satellite systems to provide assistance in the implementation of disaster relief operations on a global basis. Focus of the paper will be on satellite communications system implications and their potential impact on field operations in disaster situations. Consideration will be given to the utilization of both INTELSAT and MARISAT systems operating at frequencies ranging from 1.5 to 4 GHz and to the size and type of ground terminals necessary for satellite access. Estimates of communication requirements for a global system will be given. Some discussion of cost estimates for satellite services to support relief operations will also be included.

The information to be presented is a result of several years of study and experimentation within NASA relative to the use of satellite systems for communications to remote regions. These efforts have included studies of communication satellites for both pre and post disaster applications conducted for NOAA as well as recent experiments conducted in conjunction with the Office of Foreign Disaster Assistance of the Agency for International Development.

INTRODUCTION

This paper will address from an overall systems perspective the role that satellite communications systems can play in assisting in the timely and effective reaction of the world community to major natural disasters. In any systems approach it is essential to determine as clearly as possible a definition of the problem to be solved. The service requirements necessary to effect a problem solution must also be identified in order to establish the system alternatives available to the system analyst.

* *NASA - Lewis Research Center, 21000 Brookpark Rd, Cleveland, Ohio 44135*

Disaster related activities are generally separated into six stages or phases. The phases cover the events from disaster detection to rehabilitation. The time periods of greatest concern include the time interval just preceding a natural disaster and the time during and immediately following the occurrence. Three general classes of services are required, one deals with the collection of information dealing with the possible onset of a disaster, the second deals with the communications services required to disseminate alerting information prior to an occurrence and the third with the services dealing with responding to the needs of the effected area following an occurrence. This paper will deal with the requirements associated with the second and third classes of service.

On a global basis the world population is less than adequately educated or informed on what to do in an impending disaster situation and even less able to cope with the consequences of a major natural disaster despite much good work by international organizations such as UNDRO and Red Cross. The lack of adequate communication services to many areas of the world underlie this condition. In the world today where satellite technology permits communications between any two points on the globe, this situation need not persist.

In order to deal with both pre and post disaster situations a highly responsive and geographically flexible communication system is needed.

Since disaster events can and do occur in all parts of the world, a communication system that utilizes satellites as an integral part of a globally interconnected system offers the potential needed to deal with the geographically diverse requirements. The community concerned with disaster related activities fortunately has now at its disposal two global communications satellite networks, INTELSAT and MARISAT upon which to build a disaster communications network. This paper will present a possible system scenario for effective disaster communication operations using existing and planned systems and review the status of hardware technology available for use at this time.

PROBLEM STATEMENT AND REQUIREMENTS

In order to address the question of what communication systems is best, it is necessary to establish a clear statement of the problem to be solved by the systems under consideration. Too often a less than clear understanding of the problem is held by all parties involved and the resulting confusion prevents an acceptable system solution. Even when the problem can be clearly stated and agreed to by all, it is necessary to define the range of system requirements permitted in order to facilitate comparison of alternate solutions.

With regard to the Problem Statement, it is important to establish the major aspects of a disaster situation that are to be considered in the system assessment. One such scenario segments a typical disaster event into six major phases as shown in Table 1.

TABLE 1 General Phases of Disaster Problem

Phase	Event
1	Detection
2	Preparation
3	Warning
4	Assessment
5	Relief
6	Rehabilitation

The communication capability necessary to provide services required in the six phases range from remote data collection to interactive voice to broadcast voice and in some instances video. The general set of services for each phase is shown in Table 2.

TABLE 2 General Requirements During Disaster Phases

Phase	Type of Service	Purpose
Detection	Remote Data Collection and Imaging	Monitor Disaster Parameters
Preparation	Voice and Video Program Distribution	Educate Population and Disaster Teams
Warning/Alerting	Voice Broadcast	Initiate Safety Procedures
Assessment	Two-Way Voice, Data, Slow Scan Video	On-Site Assessment of Damage and Relief Requirements
Relief	Two-Way Voice and Data	Coordinate Near Term Relief Operations
Rehabilitation	Two-Way Voice and Data	Long Term Coordination of Area Rehabilitation Activities

In each of the phases there is a potential role for satellite services as part of a global disaster assistance system. Remote sensing satellites can and do contribute to information required to assist in prediction and detection of possible disaster situations. However without a reliable global communications network to convey the information to the effected parties none of the six phases can be accomplished effectively. It is this required communication network that will be discussed in some detail herein with attention given to utilization of existing satellite systems and current technology capability.

A schematic of a general communications network using satellite system capability is shown in Fig. 1 for the predisaster situation. Services pertaining to detection, preparation, and warning are provided and the network permits interconnection among the capitals of disaster prone areas with the major disaster assistance agencies, UNDRO and AID. Service penetration within a given country for education purposes is possible from the local capital via satellite broadcast to receive-only terminals distributed throughout the country as required.

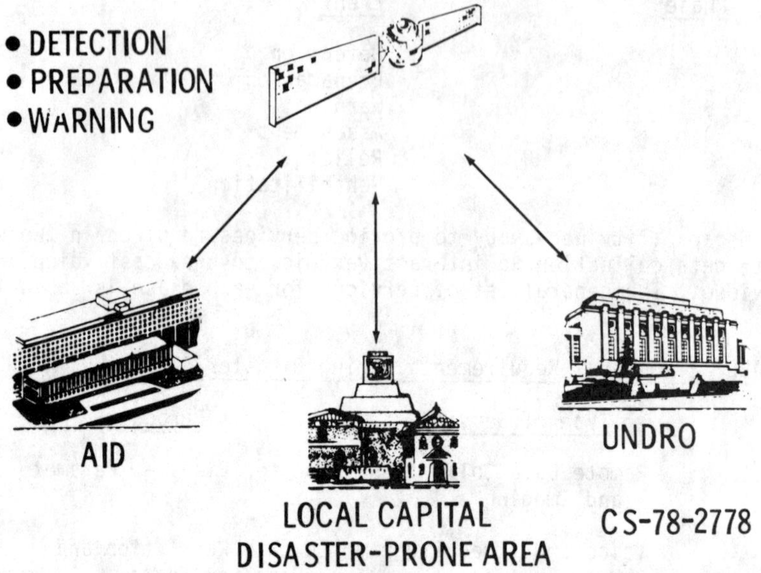

Fig. 1. Satellite communication system - pre-disaster.

In the post disaster situation, the network is depicted in Fig. 2 where the disaster area is linked with the local capital using transportable ground terminals. Communications to the outside world is then linked via the capital. This communication service is by and large interactive voice, data, facsimile and slow scan video.

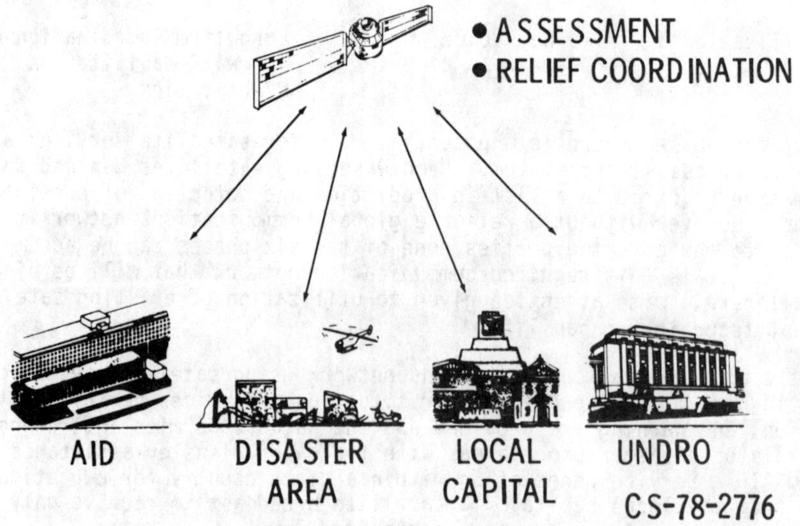

Fig. 2. Satellite communication system - post-disaster.

The broadcast type of service is a capability inherent in any satellite system; however, it is necessary to consider very carefully both the technical and political implications of rendering such a service. The participation of the local capital is a way of assuring local control in the system network.

A GLOBAL DISASTER NETWORK

Current international and planned regional satellite systems permit interactive as well as distributed (broadcast) services to a variety of ground terminal configurations. In the INTELSAT system global service is available in a receive-only video mode to relatively small, 5 to 7 meter, earth terminals. Extensive demonstrations of this type of service have been conducted in Nigeria and Sudan. Experiments have also been conducted to demonstrate interactive voice and data services with similar size terminals. Photos of typical small INTELSAT terminals are shown in Fig. 3, Fig. 4, and Fig. 5. The operating frequency band for the INTELSAT system is C-Band (4/6 GHz).

Fig. 3. Transportable terminal

Fig. 4. 6 meter truck mounted terminal - deployed

Fig. 5. 6 meter truck mounted terminal - stowed

Maritime communications services are currently being provided with the global coverage afforded by the MARISAT system. Interactive voice, data and teletype service is provided to ships at sea with terminals of 1.2 meter aperature. Typical MARISAT terminals are shown in Fig. 6 and Fig. 7. The operating frequency band to the small terminals for MARISAT is L-Band (1.5 GHz).

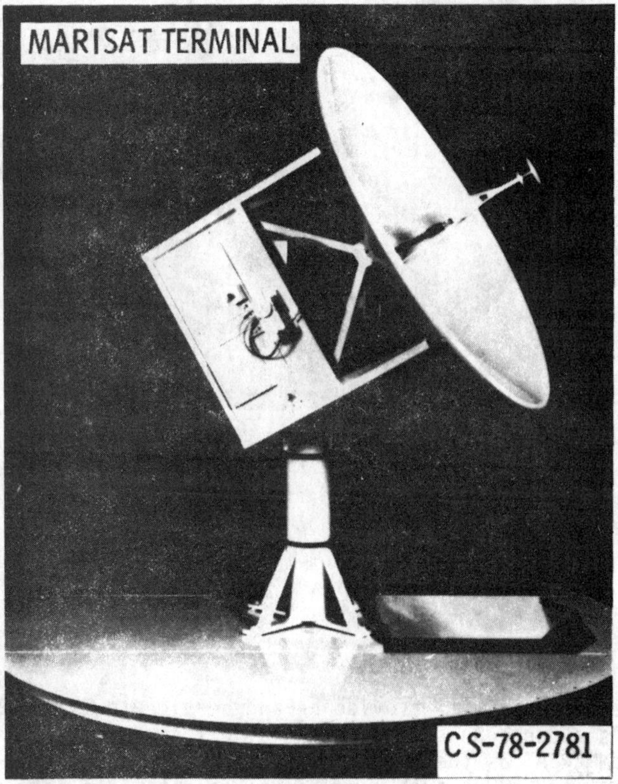

Fig. 6. MARISAT ship terminal

Fig. 7. MARISAT fixed site terminal

In the future INTELSAT will provide L-Band service to INMARSAT on the INTELSAT V satellite. As a consequence the potential exists for utilization of either INTELSAT or INMARSAT as the backbone system upon which to build a disaster communications network. One such communications satellite network is shown schematically in Fig. 8. This system would utilize the L-Band capability of INTELSAT V to provide one interactive voice channel using a highly transportable small (1.2 meter antenna) ground terminal.

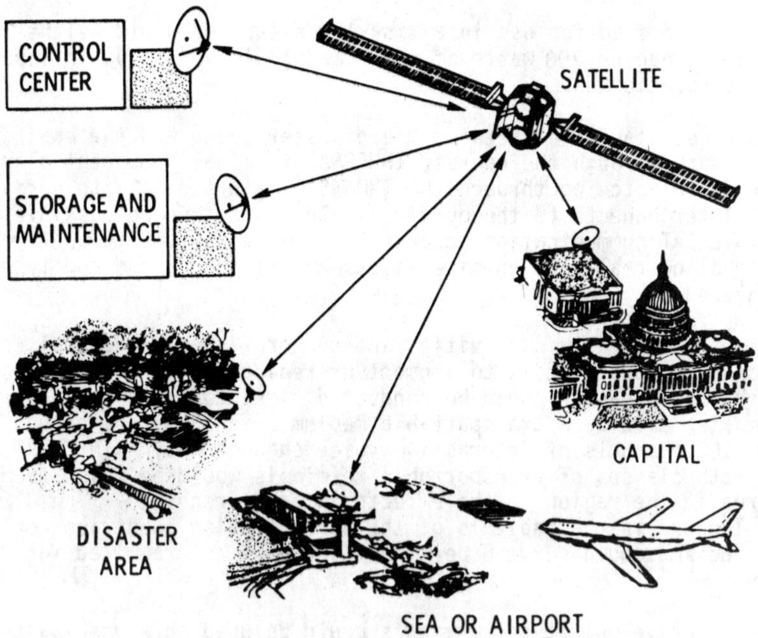

Fig. 8. Disaster communication network schematic

A schematic of a modified MARISAT terminal is shown in Fig. 9.

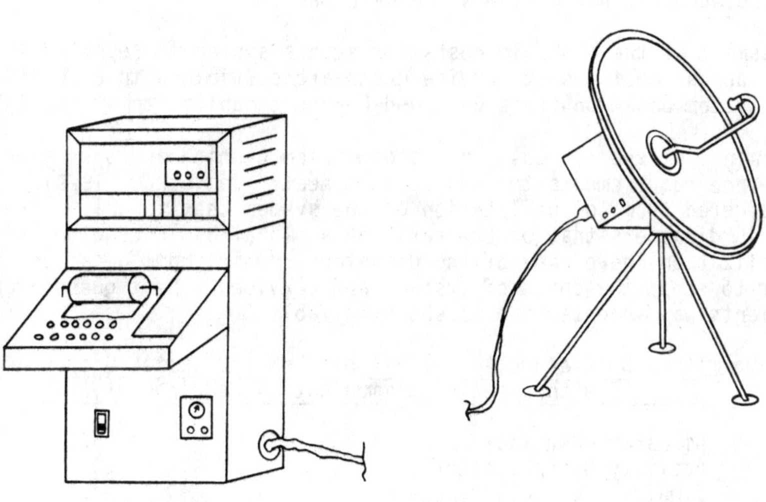

Fig. 9. Lightweight, portable MARISAT terminal

This terminal modified for use in a disaster situation would weight about 150 kilograms, require 700 watts of power and could be set up in about 2 to 4 hours by two field technicians.

A link would be established between the disaster scene and the capital of the effected country through the nearest INMARSAT terminal. Current plans for INMARSAT require patching through the INMARSAT main terminal to normal communication channels. Interconnects to the outside world would then be possible through the standard INTELSAT communications channels. Communications to and from the disaster scene would allow real time on-site assessment of the damage and type of relief supplies needed.

Coordination of the relief activities and the provision of general communications services on a temporary basis to augment or replace existing communications service disrupted by the disaster could be conducted close to but somewhat remote from the actual disaster site by a transportable medium size (5 meter) ground terminal which permits 10-100 channels of interactive voice channels using INTELSAT C-Band transponders. Both classes of transportable terminals would be moved from a general storage area in the region to the effected area as required. The Class B type terminals located in the capitals of the participating countries could be permanent installations and could serve other functions when not required for disaster communications.

The 5 meter receive only type terminals could be used in a predisaster situation to permit the distribution of video programming at the community level of educational material pertaining to disasters.

A satellite communications network similar to that just discussed could be implemented now. Since the satellites are already a part of existing systems, it remains only to establish the character of the ground systems and arrange for leasing of satellite transponder space. The system is evolutionary in nature and allows growth from demonstration projects to operational use depending on the political, institutional and economic arrangements that are made.

An assessment of the probable costs for such a system is reported in (LeRoy, 1979). It would appear that modest service costs are possible with a relatively flexible network system based on the L-Band and C-Band capabilities of the INTELSAT V system.

To determine an expected cost of system implementation and operation it is necessary to scope the requirements the system must meet. In (LeRoy, 1979) a range of capability was considered from low utilization of the system that is only a few of the approximately 45 disasters that occur yearly on a global basis require system support and high utilization where many of the disasters require communications system support. In order to scope the costs of systems and services a best guess average set of requirements was selected and is shown in Table 3.

TABLE 3 Average Annual Set of Global Disaster Occurrences with Expected Communication Requirements

Disasters Per Year	45
Activity Days/Disaster	60
Simultaneous Occurrences	7
Disasters Requiring Terminals (@ 25%)	11
Terminal Utilization, Days	5
Terminals/Disaster	2
Voice Channels/Terminal	3
Channel Usage/Day, Hours	4

The expected costs for this average case implementing an all L-Band system using MARISAT current costs as a basis results in a yearly cost for the average utilization case of about one million dollars and includes space segment costs $800K, ground terminal costs at $150K and transportation at $50K. If the INTELSAT system is used at C-Band on an "as needed" basis, lower yearly costs are possible; however, larger field terminals are required. It should be noted that a combination of L-Band and C-Band utilization have been postulated herein with expected costs in the same range as indicated in (LeRoy, 1979).

NASA DISASTER RELATED ACTIVITIES

As part of an extended evaluation of how future satellite technology might be used for the warning as well as interactive (two-way) communications functions, NASA conducted a series of systems studies for the National Oceanic and Atmospheric Administration, NOAA, (Spar, 1974; LeRoy, 1977; Hein, 1972) dealing with direct-to-home disaster warning. Consideration was given to potential disaster situations occurring within the 50 states. Disaster events ranged from tornadoes (see Table 4) with very short response times required to winter storms and general floods that had moderate response times required.

TABLE 4 Disaster Response Times

Disaster Type	Smallest Area Warned	Message On-Line Upper-Bound
Tornado or Severe Storm	Part of County	1-5 Minutes
Hurricane	Part of Coast	1-15 Minutes
River Flood	Part of State	15 Minutes - 1 Hour
Small Craft Warning	Part of Coast (Lake)	15 Minutes - 1 Hour
Winter Storm	Part of State	15 Minutes - 1 Hour
Others	Part of County	1 Minute - 1 Hour

The general set of communications services to be provided by the alternative systems are shown in Table 5.

TABLE 5 Disaster Warning System General Service Requirements

Delivery of Warnings to Public via Home Receivers
Spotter Alerting and Reporting
Data Collection
Coordination among National Weather Service Facilities
Broadcasting of Routine Weather Information

Alternative system solutions were evaluated and compared including dedicated satellite systems, hybrid satellite/terrestrial systems and all terrestrial systems. For the direct-to-home warning requirement, it was necessary to operate the broadcast channel at low frequency, approximately 1 GHz, and at relatively high transmit power coupled with a large spacecraft antenna to allow area selectivity. A sketch of the satellite is shown in Fig. 10.

DISASTER WARNING SATELLITE

Fig. 10. Disaster warning broadcast satellite

The technology demonstrated in ATS-6 experimental satellite was applied in this case study. In this mode, this satellite system permitted broadcasting of a voice warning to selected areas of the country which could be received by a very small receive-only voice terminal shown in Fig. 11. The alternative all-terrestrial system utilized a large number of transmitting stations distributed around the country operating at VHF (about 150 MHz) frequency capable of broadcasting voice warnings to similar small receive-only voice terminals. About 750 VHF broadcasting transmitters would be required to provide complete coverage of the United States.

The conclusion of these studies indicated that for the United States all three system types could effectively provide warnings as well as interactive communications services to meet the overall requirements. However, the highly developed terrestrial networks already in place coupled with the U.S. domestic satellite services available precluded the option of a dedicated satellite system for disaster communications from being cost competitive. Although these studies did deal with the range of communication services needed for disaster communications on a global scale, the conclusion reached may not be applicable. The existence of two global satellite systems INTELSAT and INMARSAT do, however, offer the potential to provide needed services without requiring a dedicated satellite system. It should be noted that because of the very large geographical area to be served in any global system, a careful trade-off must be made of the minimum terminal size as a function of the overall cost of service taking into account the communication capacity needed, the flexibility required, the capability of existing systems and the timing of system operation.

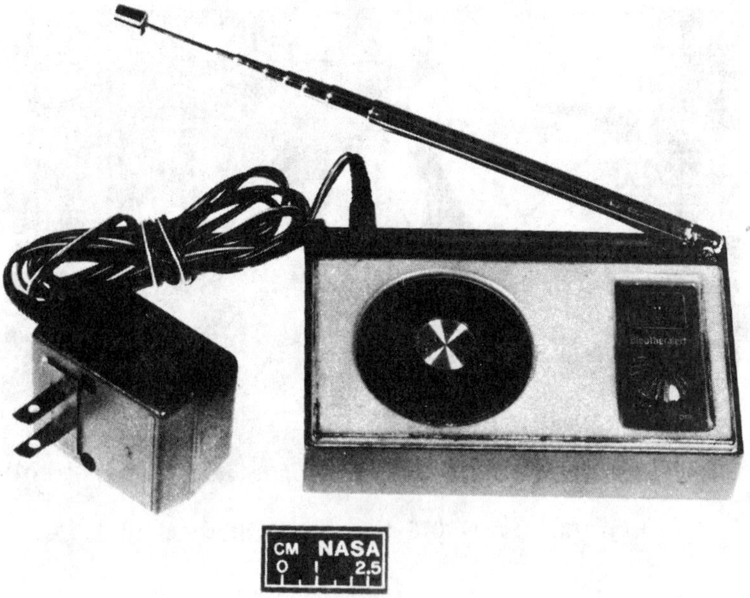

Fig. 11. Disaster warning receiver

As part of the overall study activity concurrent experiments related to disaster communications have been conducted to establish the feasibility of the advanced technology inherent in the NASA experimental satellites CTS and ATS-6.

Using the CTS satellite, experiments were conducted under both real and simulated disaster conditions using multichannel voice terminals with antennas of 1.2 meter diameter and are reported in (Donoughe, 1978; Kaiser, 1977; Edelman, 1977). Photographs of the terminals used are shown in Fig. 12, Fig. 13 and Fig. 14.

Fig. 12. Small CTS transportable terminal

Fig. 13. Small CTS transportable terminal

Fig. 14. Trailer mounted CTS terminal

Similar experiments have been conducted with ATS-6 as reported in (Anderson, 1976; Maruschak, 1978; Laurence, 1975; Baker, 1977). Because of the range of downlink frequencies possible with ATS-6, a variety of terminals as shown in Fig. 15, Fig. 16, Fig. 17 and Fig. 18 were used ranging from receive only video at UHF and S-Band frequencies to two-way voice to suitcase terminals at L-Band.

Fig. 15. Video receive only - SITE experiment

Fig. 16. Transportable two-way video terminal antenna

Fig. 17. Suitcase portable terminal

Fig. 18. Suitcase portable terminal and helix antenna

FUTURE SYSTEM POTENTIAL

The discussion to this point has centered on applying capability that either currently exists or is planned for the near future. No major advanced technology has been considered. Technology advances, however, will occur during the next decade that may permit application of complex spacecraft/ground system combinations that will offer increased flexibility in network organization. Steerable spacecraft spot beam antenna systems employing high frequency may allow higher effective power to be radiated to the disaster region. Satellite-to-satellite interconnection and on-board satellite switching will permit global links to be established without requiring multiple satellite hops. This will allow consideration of small terminals than described with existing satellite systems even to the utilization of man-pack terminals. The technical feasibility of such small terminals was demonstrated with ATS-6 where suitcase size terminals as shown in Fig. 17 and Fig. 18 were used to implement two-way voice links.

Advanced satellite systems employing large spacecraft antennas and high frequencies as shown in Fig. 19 in an artist's concept, may generate high intensity spot beams for use in very localized regions to facilitate a wide range of communication services to support pre and post-disaster operations.

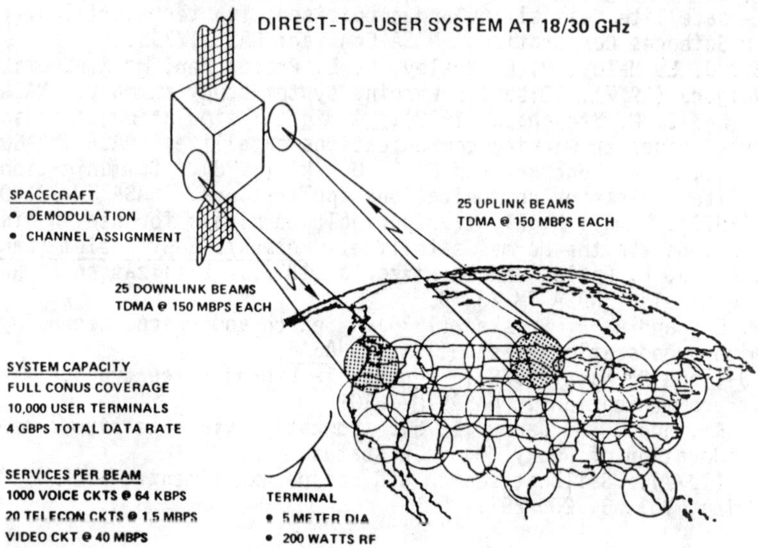

Fig. 19. Advanced K-band system

It should be noted that the cost of providing a communication service to a node in the network system is to first order dependent on the throughput of the node. That is to say high capacity nodes (multi-channel) costs-per-channel are much less than for low or single channel capacity nodes.

Providing very portable ground terminals may reduce the cost of individual terminals but the costs associated with establishing the channels through the satellite will most likely be commensurately higher. Even with the advanced technology systems, to determine the service rendered would require an overall systems analysis to establish the best fit between desired capability and service cost.

CONCLUDING REMARKS

Responding effectively to natural disaster situations depends to a large degree on the existence of a reliable and flexible communication network devoted to disaster related communication services. At the present time existing international satellite communication systems offer real potential for the establishment of an effective network in the near term. Experimentation with these systems is possible in order to define the real requirements of such a network.

Advances in technology in satellite systems may offer additional network flexibility in the future but costs of advanced systems if higher than current systems must be weighed against the added benefits that are derived. It will be necessary to periodically assess future options as technology advances in order to maintain a good balance between service rendered and costs incurred.

REFERENCES

LeRoy, B. E. (1979). Global disaster satellite communications system for disaster assessment and relief coordination. NASA TM-79105.

Spar, J. H., W. H. Hodge, M. J. Funk, and T. F. Bamford (1974). Disaster warning system - satellite feasibility and comparison with terrestrial systems. Computer Sciences Corporation. NASA Contract NAS3-17795.

LeRoy, B. E., J. E. Maloy, R. C. Braley, C. E. Provencher, H. A. Shumaker, and M. E. Valgora (1977). Disaster warning system study summary. NASA TM-73797.

Hein, G. F., and S. M. Stevenson (1972). A digital simulation of message traffic for natural disaster warning communications satellite. NASA TMX-68135.

Donoughe, P. L., H. R. Hunczak, and G. S. Gurski (1978). Communications technology satellite - disaster communications applications. NASA TM-79109.

Kaiser, J. (1977). Small highly transportable terminals for use in disaster relief communications via the Hermes satellite. Hermes/CTS Symposium (Canada).

Edelman, E. A., J. L. Fiala, and L. Rizzolla (1977). Utilization of NASA Lewis mobile terminals. NASA TM-73859.

Anderson, P. E., and A. F. Brisken (1976). Search and rescue technology experiments with a geostationary satellite. NASAR.

Maruschak, J., and D. Nace (1978). A portable L-band voice transceiver for satellite communications. NASA TM-78108.

Laurence, D. R., and J. M. Janky (1975). Education via satellite: how it was done by the Federation of Rocky Mountain States.

Baker, J. L. (1977). Sailboat search and rescue experiment. IEEE J. Oceanic Engineering, Vol. 3, 285-291.

IAA 79-A34
(AAS 79-335)

EMPLOYMENT OF LARGE STRUCTURE COMMUNICATIONS SATELLITES FOR EMERGENCY CALLS*

Selection of Radio Frequencies in Respect to Worldwide Distress Usage

G. Landauer, E. Messerschmid+

Future emergency and distress call systems will make use of modern satellite technology for wide-range transmissions. It is demonstrated how the application of large antenna structures in space reduces the transmitter equipment on the earth to rather lightweight devices. By an investigation of the frequency-dependent link margins it can be shown that for emergency transmissions around L-band frequencies the transmitter weight is minimum. For simple emergency messages (bitrate 10^2 bps) and speech transmissions (10^4 bps) the transmitters are easily portable for diameters of the satellite antenna larger than a few meters and a few tens of meters, respectively. Large space antennas together with new techniques such as electronically switched multi-beam antennas and efficient multiple access systems render possible the frequency reuse based on spot-beam concepts. For two potential distress and emergency frequencies in the UHF range the worldwide available bandwidth and the number of simultaneous transmissions due to frequency reuse is derived as a function of the size of the satellite antenna.

Keywords: Emergency and distress frequencies; link margins; attenuation losses; multipath fadings; frequency reuse; asynchronous access; spectrum efficiency.

INTRODUCTION

Improved disaster and rescue response time and extended information content relating the type of distress and the kind of rescue desired can only be provided through much improved search and rescue (SAR) technologies, particularly those employing space-borne systems (Heath, 1978). The disruption of normal communication facilities as a result of natural or man-made disasters focuses attention on the unique role that can be filled by transportable earth terminals linked to future large structure communications satellites.

The discussion of the general role and the technology of large multi-function communications satellites (OAF's: orbiting antenna farms) has already been opened some time ago (Cuccia, 1978). It has been felt that emergency call transmissions as part of a regular service of future OAF's deserve special attention. To a large extent

* *This paper also appears in full in Volume 39, AAS Microfiche Series.*

+ *Dr. G. Landauer and E. Messerschmid are associated with DFVLR - German Aerospace Research Establishment, Institute for Communications Technology, D-8031 Oberpfaffenhofen, West-Germany*

these transmissions are disadvantaged compared to commercial satellite services: They are both power-limited <u>and</u> bandwidth-limited, their asynchronous access in time enhances spectrum shortage. The omnidirectional characteristics of portable transmitter antennas demand more power margins to cope with multipath effects. In a non-linear multi-carrier transponder of the space segment weak signals degrade relative to strong signals. In addition, emergency call transmitters must be lightweight, handy, robust and very reliable. The response to all these decisive constraints can in part be an optimal selection of radio transmission frequencies for worldwide distress usage.

In the past the introduction of frequencies for distress, emergency, search and rescue operations essentially was triggered by the occurance of disasters and the development of technical means for handling distress calls at increasing frequencies. The Titanic catastrophe in 1912 initiated international regulations on the use and protection of the distress, safety and calling frequency for radiotelegraphy at 500 kHz. After the sinking of the fourmast bark Pamir in 1956 the claim for the introduction of EPIRB's (emergency position indicating radio beacons) was enforced. Numerous wrecked aircrafts were not found in time because of the lack of ELT's (emergency locater transmitters). To ease the situation, the distress frequencies 2182 kHz and 121.5/243 MHz have been introduced in 1959 for maritime and aeronautical application.

TABLE 1 International Distress, Emergency, Safety, Survival and SAR Operations Frequencies

Band	Frequency [MHz]	Frequency Range or Guard Band [MHz]	T	A	M	EPIRB ELT	S	Footnotes in RR concerning Emergency etc.	Other Use	Service
LF	0.500	0.490 - 0.510	(X)	(X)	X			187,32	-	Radiotelegr. Mobile
MF	2.182	2.170 - 2.194	(X)	X	X	X	X	201,201A,35	-	Mobile
HF	3.0235	2.850 - 3.025	(X)	X			X	201A	yes:213	Aeron. Mobile (R)
	(4.125)							1352B		
	5.680	5.680 - 5.730	(X)	X			X	201A		Aeron. Mobile (OR)
	(6.2155)							1353		
	8.364	8.195 - 8.815		X	X		X	201A		Maritime Mobile
	10.003	9.995 - 10.005					X	214 } 201A,203A	-	Standard Frequency
	14.993	14.990 - 15.010					X	219		and SAR concerning
	19.993 ± .003	19.990 - 20.010					X	220		Manned Space Vehic.
VHF	121.5	117.975 -132.0	(⊗)	⊗	(⊗)	⊗	X	201A,273,273A	yes	Aeronautical Mobile
	123.5		(⊗)	⊗				ICAO Conv/A10	yes	
	(156.3)	156 -174 Reg.I			X		X	201A,287,35	yes:285,288	Fixed and Mobile
	156.8	150.05 -174(170) Region II(III)	X	X	X		X		yes:233A	
	157.3125	157.3125-157.4125			⊗			287A,Spa 2-5	yes:290	Ships-to-Satellites
	161.9125	161.9125-162.0125			⊗					Satellites-to-Ships
	243.0	235.0 - 267.0	(⊗)	⊗	(⊗)	⊗	X	201A,308A,309	yes:305	Fixed and Mobile
	282.8	273.0 - 328.6	⊗	⊗				308A		
UHF	406	406.0 - 406.1	(⊗)	⊗	⊗	⊗		314,317A,317B	yes:314,317B	Mobile-Satellite (Earth-to-Space)
	1543	1542.5 -1543.5	(⊗)	⊗	⊗	(⊗)		not yet	yes:2521	Space-to-Earth
	1644	1644.0 -1645.0	(⊗)	⊗	⊗	(⊗)		not yet	yes:3521	Earth-to-Space

Due to the propagation characteristics of electromagnetic waves, the transmission of distress calls via satellites at UHF or even higher frequencies is favoured. For this reason the band 406.0-406.1 MHz was reserved for the use and development of low-power (< 5 W) EPIRB systems using space techniques (Radio Regulations, 1971). However, the small bandwidth associated with this frequency band and strong ionospheric scintillations which may seriously disturb the transmission in the lower UHF range led to the developments of EPIRB's at L-band frequencies (Goebel, 1975, 1976; Weinberg, Kaminsky and Keblawi, 1977). The frequency band 1.644-1.645 GHz (not yet internationally accepted as distress frequency) has the obvious advantage that it may commonly be used from maritime and aeronautical mobile terminals. Table 1 shows a compilation of the present international distress, emergency, safety, survival and SAR operations frequencies as documented in the Radio Regulations (1971). The first three columns contain band nomenclature, frequency, guard band or frequency range. The next five columns indicate by "X" or "(X)", whether the frequency can be used for distress calls or only for SAR operations from terrestrial (T), aeronautical (A), maritime (M), EPIRB/ELT or space-borne (S) transmitters. Circles notify that space techniques may be employed. The footnotes of the Radio Regulations concerning the use of the frequencies are denoted in the last columns.

It has been experimentally demonstrated that the transmission of low bitrate emergency calls and voice communication from omnidirectional antennas via satellite is technically feasible (Goebel, 1975, 1976; Anderson, 1978). The field trials were performed with ATS-6 having an antenna diameter of almost 10 m which diameter has not yet been exceeded by any other satellite antenna. However, the space shuttle transportation system (STS) will bring larger structures into geostationary orbit. While today the antenna diameter of most commercial communications satellites is still kept small in order to achieve global earth coverage, new techniques such as electronically switched large multi-beam antennas (MBA) and efficient multiple access systems render possible the frequency reuse based on spot-beam concepts. Figure 1 illustrates how these techniques might materialize in the next decade for the benefit of fast and reliable transmissions of distress calls. It is important to recognize that spot beams will be like thin radiation fibres coming from space (Cuccia, 1977). Thus any allocated frequency band can be reused many times.

Over the next decade we will see a complexity inversion compared to past practice, particularly in communication satellites (Bekey, 1979). The trend of development is going the direction to build satellites of increasing antenna size and power. This in turn permits the reduction of power and antenna size in the ground terminals. Future mobile ground equipment thus can be made technically simple and rather cheap so that a wide field of application is opened.

As mentioned before, a careful selection of radio frequencies is required to fulfill the stringent boundary conditions inherent to emergency transmitters. The choice of frequencies for future distress call systems essentially depends on frequency assignment due to the Radio Regulations, technical limitations and propagation characteristics of electromagnetic waves.

In Section 2 the basic parameters of a distress channel are defined. With these parameters, the minimum of transmitter rf-power is derived as a function of data rate and MBA diameter. In Section 3 a compilation of effects contributing to the link margin is contained. The question of how much rf-power might be produced by future portable terminals as a function of weight, transmission time, frequency and type of battery, is answered in Section 4. In Section 5 numbers for effective isotropically radiated power as a function of frequency, S/C antenna diameter and transmitter weight is given. Finally, the throughput capacity in connection with frequency reuse finishes the technical aspects of our discussion on distress and emergency call systems by employment of large structure communications satellites.

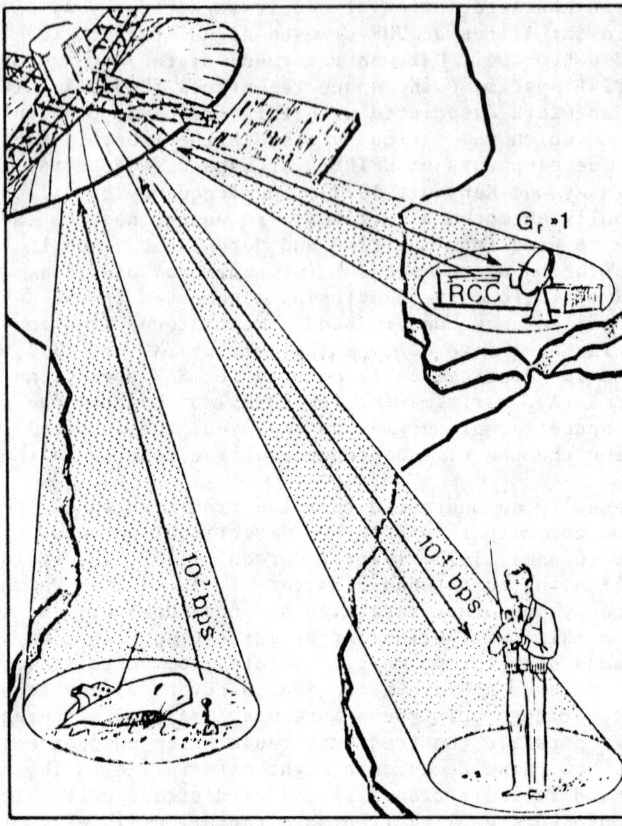

Fig. 1 Simplified view on future emergency calls and SAR transmissions using spot-beam technique in connection with portable transmitters.

2. SATELLITE LINK FOR EMERGENCY CALL SYSTEMS

The carrier power-to-noise power density ratio C/kT_r is given by the link equation (Spilker, 1977)

$$\frac{C}{kT_r} = R\,\frac{E_b}{N_o} = P_t G_t \cdot \left(\frac{G_r}{T_r}\right)\left(\frac{\lambda}{4\pi S}\right)^2 \cdot \frac{1}{kM} \qquad (1)$$

The product $P_t G_t$ is the transmitter EIRP if the up-link is considered. Portable devices for automatic distress emissions or distress calls will have omnidirectional transmitter antennas (see Fig. 1 and for definitions Table 2), hence EIRP = P_t. The ratio G_r/T_r is the figure of merit of the satellite receive system, the square of $\lambda/4\pi S$ corresponds to the free space signal attenuation, k is the Boltzmann constant and M the link margin.

The gain of the satellite antenna is a function of the antenna diameter D and the wave length λ:

$$G_r = \eta \left(\frac{\pi D}{\lambda} \right)^2 . \tag{2}$$

The antenna efficiency η depends on aperture illumination taper, spill-over, blockage etc. In the following $\eta = 0.5$ is assumed, corresponding to 30 dB sidelobes. The satellite receiving antenna views the earth temperature of around 300 ^oK. To this must be added the feed loss temperature and the receiver noise temperature. Assume a 1 dB feed loss (60 ^oK) and a receiver noise temperature of 640 ^oK; then T_r sums up to a total noise temperature of 1000 ^oK (Cuccia, 1977).

Future distress call systems certainly will employ frequency saving digital techniques. Because of strong fading typical for power limited transmissions with omnidirectional antennas, non-coherent or differentially coherent modulations such as FSK or DPSK are appropriate.

If the link margin is neglected, i.e. M=1, it readily can be seen from equations (1) and (2) that for a given C/kT_r the required power of a transmitter with omnidirectional antenna is independent of frequency. Under this ideal condition, and the assumptions listed in Table 2, the transmitter power is

$$P_t \approx 10^{-2} \frac{R}{D^2} \cdot Wsm^2 . \tag{3}$$

Considering narrow band distress emissions of $R=10^2$ bps, and human voice transmission of 10^4 bps which might be achieved using speech interpolation and delta modulation, the minimum power required for a satellite antenna of 1 m diameter is 1 W and 100 W, respectively.

TABLE 2 Basic Link Parameters

Transmitter	G_t = 0 db gain; P_t = transmitter rf-power } EIRP = P_t R = { 10^2 bps for automatic transmission; 10^4 bps for speech channel } data rate λ = c/f wave length of rf waves, f = rf frequency, c = vel. of light
Slant range	$S = (R^2 \sin^2\varepsilon + 2\,RH + H^2)^{1/2} - R \sin\varepsilon$; ε = elevation angle, earth radius R = 6378 km, satellite altitude H = 35 784 km
Receiver	gain G_r, total noise temperature T_r; bit error rate BER = 10^{-4} corresponding to a ratio of bit energy to noise power density $\frac{E_b}{No}$ = { 12.3 dB for non-coherent FSK; 9.3 dB for DPSK
Assumption	link is up-link limited (power limited).

In the following, equation (3) serves as reference with respect to link margins and S/C antenna diameter. For SAR systems the limitation to up-link considerations is justified because the down-link receiving system is not limited to small figures of merit.

3. ATTENUATION FACTORS DETERMINING THE LINK MARGINS

The transmitter power quoted in the previous section is unrealistic because ionospheric effects have to be taken into account at VHF frequencies as well as atmospheric and rain attenuation at frequencies beyond 4 GHz. Furthermore, due to surface tolerances, the antenna gain at high frequencies is smaller than is derived from equation (2).

The attenuation factors to a large extent increase with decreasing elevation angle. If not mentioned in the derivations of the peculiar margins, we implicitely assume a lower limit for the elevation angle of ten degrees. Details and references are given in Table 3.

Ionospheric Attenuation

For frequencies equal or higher than VHF frequencies ionospheric attenuation of an electron layer in equilibrium (Chapman Layer), auroral absorption and polar cap absorption can be neglected compared with absorption due to ionospheric scintilla-

TABLE 3 Propagation Effects in the Range 0.1 - 20 GHz

Ionospheric Scintillation	equator : 22 dB at 250 MHz exceeded in 5 % of time, $\sim f^{-1}$ mid-lat. : 25 dB at 100 MHz exceeded in .1 % of time, $\sim f^{-2}$ high-lat.: 20 dB at 100 MHz (max. fading depth), $\sim f^{-1}$ (Spilker, 1977; CCIR-Report 504-1).
Polarization Loss	1.25 - 3 dB at 136 MHz } take average and interpolate 0.2 - 1.6 dB at 1.5 GHz } 2.4 dB at 100 MHz, $\sim f^{0.2}$ (CCIR-Report IV/430, 1966-1969)
Atmospheric Absorption	Assume 100 % humidity, ε = 10 deg. (Crane, 1971)
Rain Attenuation	Crane's Model (Jain, 1979): attenuation $A = \alpha\gamma(D)R^{\beta(D)-\delta}\, H/\sin\varepsilon$ (dB) R = rain rate in mm/h, H = melting layer height in km D = surface projection of propagation path, ε = elevation angle. α, β are dependent on frequency } are parameters to fit theory γ, δ are dependent on path length } and measurement data Assume outage time of 10^{-3} (about 10 min/week), then

region	climate	R (mm/h)	D (km)
equator	tropical	55	27
mid-latitude	temperate	15	21
high-latitude	polar	5	11

tions. The latter arise from irregularities of electron density in the ionosphere; they depend on path geometry, location, season, solar activity and local time. The margins and the frequency scalings are listed in Table 3 for equatorial regions, mid- and high-latitude regions.

Polarization Losses

Polarization losses using circular polarization are due to the change of axial ratios of the rf components in inhomogenous ionospheric layers.

Atmospheric Absorption

The presence of water vapor and oxygen leads to a frequency dependent absorption of the signal energy. The attenuation margin taken is based on 100 % humidity.

Rain Attenuation

The latitude dependent rain attenuation can be computed with the help of Crane's model (Jain, 1979). In our analysis an outage time of 10^{-3} is assumed.

The sum of the margins quoted above is plotted into Fig. 2 as function of the frequency for the various regions under consideration: curve (a) for equatorial regions, curves (b) and (c) for mid-latitude and high-latitude regions. It can be seen that if no multipath effects and no gain degradation is taken into account, the propagation margin is smaller than 3 dB in the frequency range between 3 and 6 GHz.

Multipath Fading

Let us consider a signal transmitted from an omnidirectional ground antenna to a satellite antenna. It may arrive over a direct path and an earth- or sea-scattered path. The direct signal combines vectorially with the time-delayed reflected signal. The resulting signal is subjected to attenuation, delay and spread in time, shift and spread in frequency. Assuming that the most severe effect on transmissions arises from attenuation, the frequency dependence of the fading amplitude can be derived for various outage probabilities. These outage probabilities essentially are given by the statistics of the sea or land surface roughness (CCIR-Report 505-1, 1974).

If the direct signal dominates at the receiver input, curve (d) in Fig. 2 represents the fade margin as a function of frequency. This margin is exceeded only 1 % of the time; it is based on the wave statistics of the North Atlantic Ocean, circular polarization, elevation angles equal to or larger than 10 degress and omnidirectional transmitter antennas. The reversal of sense of rotation occuring on reflection from sea/earth surfaces can be exploited to reduce multipath interference, because the direct signal suffers no reversal while the indirect signal does. The amount of suppression of the indirect signal depends on the reflection coefficient, which for circular polarization is much smaller than for horizontal polarization. For the same surface roughness the fade margins for land are about the same as those quoted for sea surfaces. It is recognized, however, that the coherent reflection coefficients and therefore the fade depths should be somewhat smaller because of the non-smooth nature of most land surfaces (CCIR-Report 505-1, 1974).

If the satellite is not in a direct line-of-sight of the transmitter, the indirect signals can dominate and the signal at the receiver is subjected to larger fades. The envelope of the received signals, measured over a few hundred wavelengths, is Rayleigh-distributed. This has been confirmed by measurements at high sea states and between high buildings (Hagenauer and Goebel, 1979; Jakes, 1974). The latter measurements indicate that at 140 MHz a margin of 7.2 dB covers 99 % of urban areas. This margin is expected to scale as $f^{0.6}$; see curve (e) in Fig. 2 (Eckhardt and co-workers, 1979).

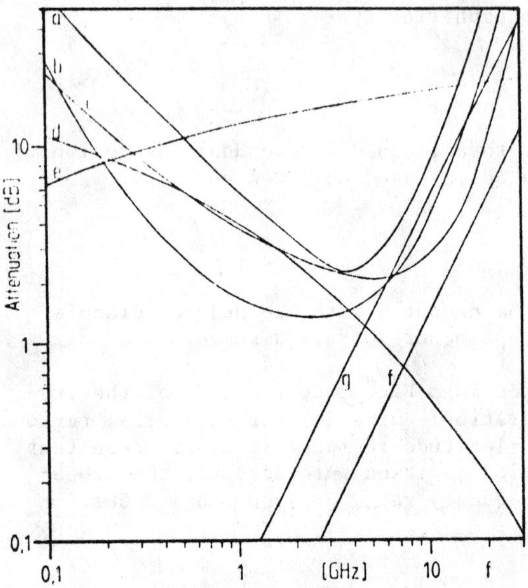

Fig. 2 Link margins for ionospheric and atmospheric effects in equatorial (a), mid-latitude (b), high-latitude (c) regions; for multipath at sea/smooth land surfaces (d) and urban areas (e); for antenna surface errors σ_{rms} = 1.3, 2.6 mm (f, g) as function of frequency.

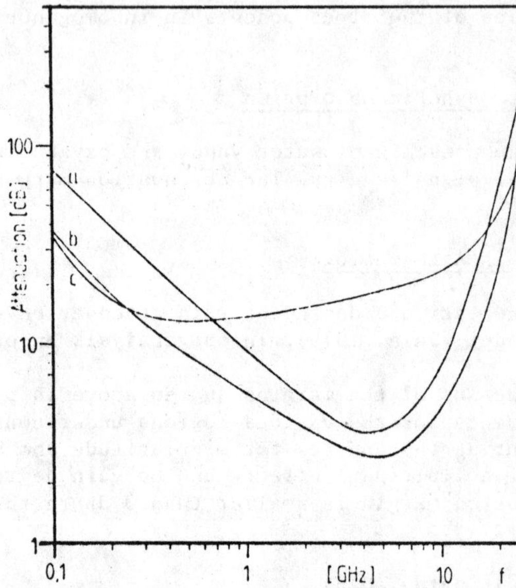

Fig. 3 Total link margin for multipath at sea/smooth land surfaces plus equatorial (a) and mid-latitude (b) regions; for multipath typical for urban areas and mid-latitude regions (c), as function of frequency.

Fig. 3 shows as a function of frequency the total link margin that has to be taken into account for three combinations of propagation attenuation: curves (a) and (b) represent the margins for equatorial and mid-latitude regions plus sea/smooth land multipath, curve (c) corresponds to mid-latitude regions with multipath typical for urban areas. The minimum of the functions (a) and (b) is around 4 GHz; at frequencies nearby a total margin of 3-4 dB is sufficient. If the received signal is dominated by Rayleigh-scattered multipath components, higher link margins are required. In this case the optimum frequency range extends down to VHF-frequencies.

Noise

Natural terrestrial noise produced by lightning discharge in thunderstorms and man-made noise generated by automotive ignition, electric machines etc. decreases rapidly with frequency and can be neglected for the frequency range under discussion.

Antenna Gain Degradation

The antenna gain for parabolic antennas of diameter D is (Jain, 1979)

$$G(\theta,\lambda,\sigma) = \eta \cdot \left(\frac{\pi D}{\lambda}\right)^2 e^{-\left(\frac{4\pi\sigma}{\lambda}\right)^2} e^{-2.76\left(\frac{\Delta\theta}{\theta_o}\right)^2} .$$

Obviously, the formula as given by equation (2) neglects the antenna mispointing factor $\exp[-2.76\,(\Delta\theta/\theta_o)^2]$, where $\Delta\theta$ is the antenna pointing accuracy and

$$\theta_o \approx 66\,\frac{\lambda}{D}\ [\deg]\quad \text{HPBW} . \tag{5}$$

It furthermore assumes the surface rms errors σ to be zero.

Structural design studies indicate that lightweight mesh deployable parabolic reflectors can hold tolerances within $\sigma/D = 1.3 \times 10^{-4}$ permitting the gain to be as high as 58 dB (Sadin, 1978). The resulting gain degradation is

$$A = 7.62 \times 10^{-3}\,[\sigma(\text{mm})\ f(\text{GHz})]^2\ (\text{dB}),$$

where σ, for example, is 1.3 mm for an antenna of 10 m diameter, resulting in 5.2 dB gain reduction at 20 GHz. The same number is obtained for an antenna diameter of 20 m at 10 GHz already. Fig. 2 shows for these two examples (curves (f) and (g)) the required additional link margin as a function of frequency.

Lightweight deployable phased arrays or lenses can hold tolerance almost twenty times smaller. The techniques involved, however, bear higher risks and practical experience may yield a somewhat smaller advantage. In our analysis we rely on the more conservative limit which is in agreement with NASA expectations (Naumann, 1979).

It is interesting to note that, using the conservative limit on surface tolerances, the gain degradation corresponds to an antenna mispointing of 0.07 degrees. In spite of the smallness of the tolerable mispointing, it is assumed that antenna technology and steering techniques will meet these requirements at least for the benefit of small terminal users.

4. TRANSMITTER WEIGHT AS FUNCTION OF RF-POWER

In a link optimization of distress call systems with respect to frequency and terminal weight, the conversion efficiency of dc-power to rf-power, the duration of transmission, the energy density of batteries as well as the degree of miniaturization play an important role. Market and trend analysis indicate that solid state rf amplifiers (FET's and bipolar transistors) have a transmitter efficiency about proportional to $\exp[-\sqrt{f(\text{GHz})}/2]$. The present upper limit of rf-power of simple, robust devices is approximately 100 W $\cdot$ $[f(\text{GHz})]^{-2}$.

The product $D_E \cdot \gamma$ of energy density of a battery and the discharging depth required to keep the voltage larger than the tolerable limit, is for modern rechargeable batteries (Ni-Cd) at most 16 Wh/kg. For non-rechargeable batteries of high energy density such as Lithium cells, the corresponding figure is about 270 Wh/kg. The total weight of the portable transmitter is estimated to be

$$W_t = \frac{3}{2} P_t \left[\frac{1.23\ \Delta t(h)}{D_E \gamma} + \frac{1}{\beta_T} \right] e^{\sqrt{f(GHz)}/2} \tag{6}$$

β_t is the ratio of dc-power consumption to transmitter weight and might be as high as 15 W/kg.

The total transmission time $\Delta t = m \cdot \Delta\tau$ as a product of the duration of a single message $\Delta\tau$ and the number of repetitions m roughly can be optimized as a function of acceptable terminal weight, outage time, anticipated alerting time and spectrum efficiency. In a first attempt a total transmission time of Δt = 2h was found to be appropriate.

5. EIRP AS A FUNCTION OF FREQUENCY WITH RESPECT TO S/C ANTENNA DIAMETER AND TRANSMITTER WEIGHT

Adding additional power to the minimum power as given by eq. (3) to compensate for the various attenuation contributions, we arrive at an estimate of required EIRP for portable low-weight transmitters. This EIRP is through eq. (3) and the surface tolerence a strong function of the diameter of the satellite antenna. Let us consider antenna diameters of 1 m providing approximately earth coverage at 1 GHz, of 4 m corresponding to antennas that fit without folding into the space shuttle cargo bay, of 10 m as had been placed already into stationary orbit (ATS 6) some time ago, of 20 m and 40 m, respectively. The latter antennas probably will become available in the mid-eighties for experiments (Naumann, 1979) and eventually at the end of the next decade they become operational (Cuccia, 1977).

The results as a function of frequency are shown in Fig. 4 for the low-rate data channel and in Fig. 5 for the speech channel, both for mid-latitude regions and sea/smooth land multipath. The figures also show the rf-power as given by eq. (6) with respect to terminal weight and battery type. It is obvious that as the MBA diameter increases, the interval of suited frequency becomes smaller. This comes about because of the surface tolerance becoming effective only for larger antennas. If these tolerances get smaller, the barrier at 2o GHz will shift to somewhat higher frequencies. However, the qualitative characteristics at high frequencies will be the same: small antennas yield rain limitation and large antennas yield a surface tolerance limitation. At VHF frequencies a high transmitter power is required to overcome ionospheric scintillations. Even at frequencies around 400 MHz about four times more power is needed than at L-band frequencies, resulting in a penalty of SAR transmitter weight by the same factor. If an additional safety margin of 2 dB is taken into account, Mini-EPIRB's of 100 g weight with Lithium-type batteries require a MBA of only 4 m diameter in the range between 800 MHz and 6 GHz. As indicated by Fig. 5 this best frequency interval shifts for speech transmissions to somewhat smaller frequencies, however, L-band frequencies still represent a good choice for distress emissions. If Rayleigh scattering is dominating the multipath effects (Fig. 2, curve e), more transmitter power is required but then also frequencies at VHF seem to be suited.

Portable terminals for speech transmission need less than 1 W EIRP for MBA's larger than 20 m. Our results agree with those given by Anderson (1978), who for diameters of 22 m (at 2.5 GHz) and 56 m (1 GHz) has derived EIRP's of 500 mW and 100 mW. In order to achieve an EIRP of 500 mW between 1 and 2 GHz, the expected weight of the terminals can be as low as 300 g (Ni-Cd) or even half of it (Lithium batteries).

6. REUSE OF DISTRESS FREQUENCIES WITH LARGE SPACE ANTENNAS

The spectrum utilization of a satellite system is in large measure determined by the cell size of the beam and the kind of service. The cell size can vary between global coverage and a few tens of kilometers. It is given by the half power beam width, eq. (5). From the ratio of the earth surface to the cell area the total bandwidth available for emergency calls is found to be

$$\Delta W_{tot} \approx \frac{0.1}{m\,k}\left(\frac{D}{\lambda}\right)^2 \cdot \Delta W_{cell} ,$$

where ΔW_{cell} is the frequency band of the cell. k is a correction factor taking into account the latitude dependent size of the cell area (k=1 at the equator and k=2 for 47 degr. latitude (Miller, 1979)). The reuse factor m guarantees that neighbouring cells are required to have different frequencies; for distress transmissions only one sense of polarization is practical, hence m=3. Figure 6 shows the total bandwidth worldwide available as function of the antenna diameter for the distress frequency at 406 MHz (ΔW_{cell}=0.1 MHz) and the potential distress freq. 1644 MHz (ΔW_{cell}=1 MHz).

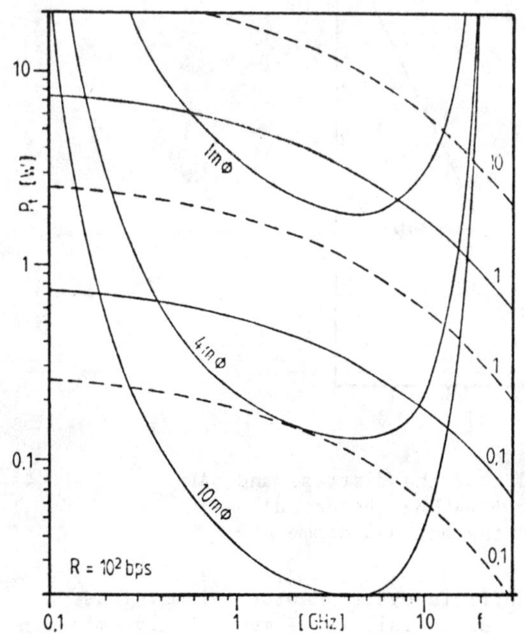

Fig. 4

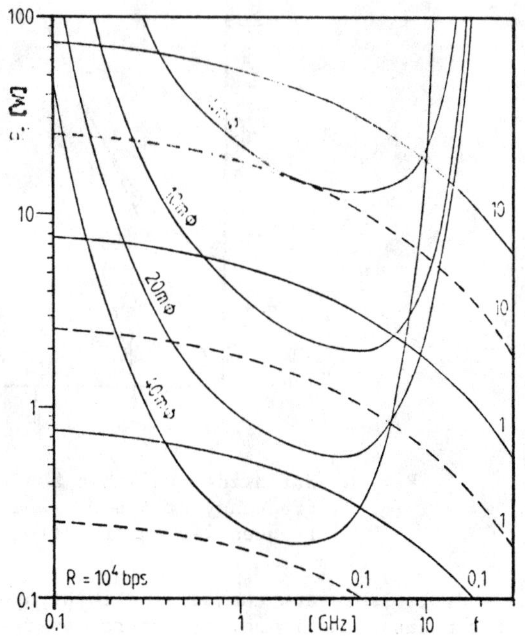

Fig. 5

Transmitter rf-power for data rates of 10^2 bps and 10^4 bps (Fig. 4 and Fig. 5, respectively) for various diameters of S/C antenna and terminal weights [kg] as function of frequency. The (un-)broken lines correspond to (Li-type) Ni-Cd batteries.

Knowing the spectrum efficiency for the asynchronous access of distress emissions, the total number of simultaneous users can be calculated from Fig. 6. For distress calls the spectrum efficiency is very low since the probability of interference of simultaneous distress calls must be kept at a low level. For automatically working transmitters and no back-channel, as is typical for slow-rate EPIRB transmitters, for a given spectrum utilization the probability of non-interference is increased considerably by repeated transmissions. It can be shown that for a probability of non-interference of 99.9 % the spectrum efficiency is typically as small as 1 %. Thus for MBA's of 4 m diameter the total number of simultaneous emergency transmissions at 100 bps and 1.644 GHz is 800, and for 40 m diameter it can be hundred times larger. For the frequency at 406 MHz the corresponding numbers are smaller by a factor of 155.

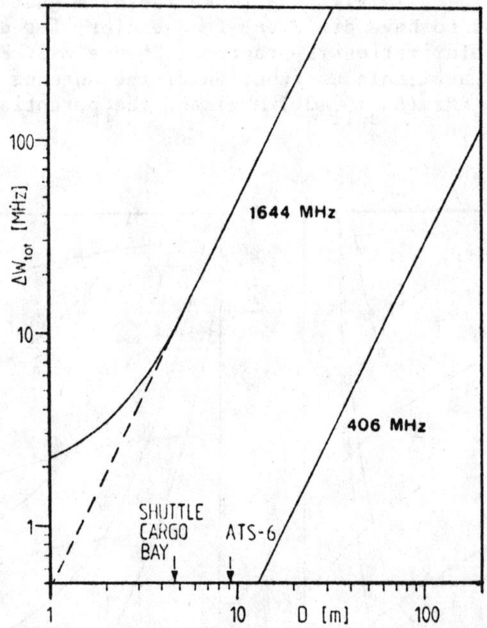

Fig. 6 Worldwide available bandwidth of the distress and SAR frequency at 406 MHz and at 1644 MHz (potential SAR frequencies) as function of the antenna diameter.

For a distress speech channel the throughput efficiency is an order of magnitude higher because the immediate control over success or failure of message transmission reduces the number of repetitions, hence increases spectrum utilization. Therefore, in this case and with 10^4 bps, the number of simultaneous emergency calls is a factor of about ten smaller than the numbers quoted above.

7. CONCLUSIONS

There is a worldwide need for improved technical and economical means for emergency and disaster aid. The employment of large structure communications satellites is particularly required in cases where an alarm using public communication nets (telephone) or sophisticated satellite earth stations is impossible: for mobile application and in geographical areas where access to public communication nets is not available.

Communications satellites of large antenna diameters permit the reduction of power and antenna size in the ground terminals. This is crucial to open the field for lightweight and technically simple and thus cheap equipment for emergency calls and SAR terminals. It has been demonstrated that for the constraints typical for emergency transmissions via future satellites, radio frequencies in the range 0.8-6 GHz are optimal. Outside this range additional transmitter power is needed to compensate the propagation attenuation due to ionospheric scintillations (lower frequencies), rain, surface ruggedness of large satellite antennas, and the low efficiency of dc-to-rf power conversion (higher frequencies). As has been shown, additional power margins can result in a considerable increase of terminal weight.

The introduction of an international frequency band for distress calls in the range mentioned above is recommended. L-Band frequencies should be favoured. The frequency band 1.644-1.645 GHz would be ideal because of the additional advantage that it may commonly be used from maritime and aeronautical mobile terminals. International regulations should aim at the highest efficiency of future satellite-aided distress call systems.

8. REFERENCES

Anderson, R. E. (1978). Satellite-aided land mobile radio experiments, applications, and problems. General Electric Company, (unnumbered).

Bekey, I. (1979). Big Comsats for big jobs at low user cost. Astronautics & Aeronautics, Febr. 1979.

CCIR-Report IV/430 (1966-1969).

CCIR-Report 504-1 (1974). Vol. VIII. Mobile Services (Study Group 8), Geneva.

CCIR-Report 505-1 (1974). Vol. VIII. Mobile Services (Study Group 8), Geneva.

Crane, R. K. (1971). Proc. of the IEEE, Vol. 59, No. 2.

Cuccia, C. L. (1977). Communication satellite technologies in the early twenty-first century - a projection into the post-Intelsat V Era. IAF Paper No. 77-34, XXVIIIth International Congress, Praha, Sept. 25 - Oct. 1.

Cuccia, C. L. (1978). The technology of large multi-function communication satellites in the post-Intelsat V Era. IAF Paper No. 78-262, XXIXth International Astronautical Congress, Dubrovnik, Oct. 1-8.

Eckhard, G. et al. (1979). Studie über die Möglichkeiten der Funkübertragung zu mobilen terrestrischen Stationen über Kommunikationssatelliten (TERSAT), AEG-Telefunken Schlußbericht, März.

Goebel, W. (1975). European communications experiments in L-band with ATS-6: Distress Buoy, System Design, Instrumentation and Experiment, Vol. 6, ESTEC, Noordwijk, Holland.

Goebel, W. (1977). DFVLR European communications experiments in L-band in 1976: Distress Keyboard Sender, System Design, Instrumentation and Experiment, Vol. 11, ESTEC, Noordwijk, Holland.

Hagenauer, J., and W. Goebel (1979). Distribution of multipath fadings measured during a distress buoy experiment in 1975, Report of the MARTAG working group on SAR, Annex IV/1, April 1979.

Heath, G. W. (1978). Perspectives on worldwide disaster and rescue response employing space-borne systems. IAF Paper No. 78-A-69, XXIXth International Astronautical Congress, Dubrovnik, Oct. 1-8.

Jain, P. C. (1979). Satellite Communication System Architecture, Lehrgang D11.08, 2.-6.7.1979, CCG, Zweig Oberpfaffenhofen.

Jakes, W. (1974). Microwave mobile communications, J. Wiley and sons.

Miller, J. E. (1979). Spectrum efficient technology for the land mobile-satellite service. Proc. Int. Telecom. Expos. INTELCOM 79, Dallas, Texas (USA),

Febr. 26 - March 2.
Naumann, G. (1979). LSST Study Group; Private Communication, Oberpfaffenhofen.
Radio Regulations (1971). Frequency Allocations 10 kHz to 275 GHz, Chapter II Article 5 of the Radio Regulations (Edition of 1968), modified by the WARC for Space Telecommunications (Geneva, 1971).
Sadin, S. R. (1978). Technology needs and opportunities for future NASA missions, Langley Research Center, Hampton, Va., Jan. 17.
Spilker, J. J. (1977). Digital communications by satellite, Prentice Hall Electrical Engineering Series, Englewood Cliffs, New Jersey.
Weinberg, A., Y. Kaminsky, and F. Keblawi (1977). SAMSAR, a novel concept for a satellite based maritime search and rescue system. <u>IEEE Journal of Oceanic Engineering</u>, Vol. OE-2, No. 3.

IAA 79-A32

SOME CONSIDERATION OF SATELLITE TECHNOLOGY APPLICATIONS FOR DISASTER MATTERS - LOOKING TO THE FUTURE

H.G.S. Murthy, D. Felske*

Natural disasters are complex and divergent phenomena. The role satellite technology can play in predicting, detecting, early warning and relief operations is as diverse as the phenomena of disaster occurrence itself.

Significant distinctions in satellite technology applications in coping with disaster situations can be made by dividing them into sudden and creeping disasters.

In sudden disaster situations where urgent relief operations prevail, the main aspect of space technology applications is currently based on satellite telecommunications. Some promising experiments with small, highly portable ground terminals have already been carried out. Proliferation of those terminals and setting up of appropriate ground telecommunication networks will depend on the availability of suitably equipped long-life telecommunication satellites and protected frequencies for up and down links.

For creeping disaster detection and risk assessment, contributions are mainly expected from remote-sensing methods. Despite some surprisingly successful applications, further development in satellite sensors and evaluation methods as well as modelling of natural processes leading to disasters is necessary.

Maximum effort must in future however be devoted to the development of methods for prediction, early detection and early warning of sudden disasters such as earthquakes and volcano eruptions. The technical means for remote continuous gathering of data are already available. Future research has to find reliable precursors and appropriate monitoring methods.

Keywords: Satellite remote sensing: satellite telecommunication; natural disaster warning; disaster relief assistance.

* *United Nations Outer Space Affairs Division, United Nations, New York*

INTRODUCTION

Satellite technology is providing new possibilities for gathering and relaying data which can be used for natural disaster prevention, warning and relief purposes. The importance of satellite systems applications lies in the ability to provide repetitive data on earth's surface features on a global scale.

Satellite telecommunication and remote sensing technology can be of tremendous help by providing information for early warning of potential disasters or on the area affected and the damage caused by disasters.

The role satellite technology can play in predicting, detecting early warning and relief operations is as diverse as the phenomena of disaster occurrence itself. A more distinct assessment of satellite technology applications in coping with disaster situations can be made by distinguishing between sudden and creeping disasters.

TELECOMMUNICATIONS IN SUDDEN DISASTER SITUATIONS

In sudden disaster situations such as earthquakes, volcano eruptions, tropical storms and floods, where urgent relief operations prevail, the main aspect of space-technology applications is currently based on satellite communications.

It is in this connexion encouraging to note that promising experiments with a small, highly transportable ground terminal for satellite communications in disaster situations have been carried out in India (SAC, 1978) a disaster-prone developing country. The Space Applications Centre of the Indian Space Research Organization has developed an Emergency Communication Terminal (ECT) which is primarily intended to provide a communication link on very short notice in case of emergencies such as natural disasters.

The ECT is designed for transportation on a jeep or by helicopter to the required location and can be commissioned into operation within six hours by a crew of three persons. The complete station operates from batteries, and a portable battery recharger is carried for recharging the batteries. With one set of fully charged batteries the terminal can operate for about six hours continuously.

The terminal consists of a 20 w transmitter, up-and-down link converters in the 4/6 GHz telecommunications band and a 3 m parabolic reflector antenna. The base band consists of a single channel per carrier FM unit. The complete electronics package is housed in an insulated weatherproof shelter. This shelter is fixed on a normal jeep for transportation by road. The complete antenna along with its mount is accommodated in a trailor.

The antenna and mount are designed so that they can easily and quickly be erected in any level place and that no critical adjustment for surface alignment is required. Telecommunications experiments were carried out using the SYMPHONIE satellite. It was rather ironic that the first field trial of the ECT had to take place under real emergency situations in the aftermath of a tropical cyclone in Andhra Pradesh in December 1977.

In this connexion it is important to note that the United Nations, within its satellite communications systems using the SYMPHONIE satellite system, has at its disposal a highly transportable Emergency Telecommunication Station (ETS) developed here in Munich by the DFVLR (United Nations, 1978). This station can be incorporated into the satellite telecommunication system to improve communications for peace-keeping missions, disaster relief operations and similar humanitarian activities of the United Nations. The station employs a 3.3 m antenna and is equipped for operations with 3 pairs of single channel per carrier links to transmit voice, telex, facsimile or data. The station can be operated unattended in an equipment shelter under adverse environmental conditions. Minimum time is required for erection and readiness for operation.

Remarkable experiments for disaster communications have been conducted with the Communications Technology Satellite (CTS), jointly operated by USA and Canada (Donoughe et al., 1978). The high-power transmitter of CTS (200 w) provides for the utilization of very small ground terminals. Voice links experiments from mobile ground stations employing 1.2 m and 2 m antennas is simulated and real emergency communication situations have demonstrated reliable high-quality point-to-point connexions. CTS operates in the 12/14 GHz band.

Worldwide use of those terminals and setting-up of appropriate ground telecommunications networks will however depend on the availability of suitably equipped long-life communications satellites and protected frequencies for up-and-down link. International effort will be necessary to provide protected frequency bands and appropriate locations for geosynchronous satellites for the implementation of a future operational emergency communications network.

SATELLITE REMOTE SENSING FOR POST-DISASTER ASSESSMENT

The most successful applications of satellite technology are found in post-disaster assessment of the area affected and the damage caused by the disaster. Remote sensing imagery has proven a valuable tool in determining relief requirements of particular disaster situations and further risks in the affected area.

Flooding

Remote sensing is best applicable to flood disaster detection and monitoring. Imagery from meteorological and earth resources satellites provides valuable information for mapping the extent and location of inundated areas caused by snow melt or heavy rainfall, as for instance that connected with tropical cyclones. This imagery gives valuable data on identifying areas where relief operations may be needed and where further damage may be expected. The current available satellite sensors in the visible and infrared spectral bands meet the requirements for flood monitoring. The spatial resolution provided by earth resources satellites is sufficiently high.

Land Slides

Satellite imagery can be used for mapping of land slides and land slide-prone areas. Imagery together with geological and geodetic data can provide valuable information for risk assessment and prediction of land slides. The spatial resolution of available satellite systems is, however, not sufficient in all cases. Therefore aerial photography is often necessary to supplement satellite imagery.

Fires

Satellite remote sensing imagery is able to provide significant information on the extent and spread of forest and grass fires. Because of the low reflectivity, burned areas can be easily recognized in satellite images and can be mapped with a minimum of effort. Infrared imagery can map the precise location and extent of a fire through the smoke. Remote sensing imagery is particularly useful for fire damage assessment.

Other applications of post disaster assessment using satellite remote sensing are for instance mapping of crop damage from hail and windstorms. This kind of satellite technology is particularly helpful in all aspects where area mapping in different spectral ranges provides valuable information for damage and risk assessment.

SATELLITE TECHNOLOGY APPLICATIONS IN CREEPING DISASTER DETECTION AND EARLY WARNING

Creeping disasters are those events whose eventual or total impact may be catastrophic but which proceed slowly toward their culmination and initially exhibit only subtle limits of their eventual impact (Robinove, 1975). A creeping disaster may not be identified as such for some time after its onset. The main task in tackling creeping disasters is therefore the early detection and reliable identification of changes in environmental features with potential harmful effects on man and environment. While the remote sensing technology seems most amenable to post disaster survey, detection and assessment of creeping disasters can best be implemented by combining various methods of remote sensing with ground data collection systems. Many promising remote sensing and ground data gathering applications in various creeping disaster occurrences have been reported in the past and a few of them are mentioned below. However, further development in sensor techniques, surveying and data processing is necessary to meet the divergent requirements of reliable recognition of early signs of creeping disasters.

Snow Cover Monitoring and Run-off Predictions

Monitoring and mapping of snow lines is a commonly used technique for snow cover estimates and consequently for run-off as a result of snow melt. Modern spacecraft equipped with sensors in the visible and infrared ranges even allow us to distinguish between snow and cloud cover. Problems yet to be solved are estimates of water content of snow and snow thickness measurements. Here developments in remotely operated sensor and data collection platforms may bring further success. Run-off evaluation and forecast techniques to certain watersheds have reached a fairly high level of accuracy and reliability, thus providing important information for flood control, disaster preparedness and prevention purposes as well as for hydroelectric power generation, irrigation and industrial and human consumption of water. The accurate estimate of liquid precipitation over large areas still remains a problem as there are only point measurements of rainfall available within the meteorological ground network. It is unlikely that remote sensing technology will provide advance information on rainfall with necessary accuracy for forecast and warning purposes. This problem can probably only be solved by applying an appropriate network of surface sensors and data collection techniques via satellite links.

Glacier Monitoring

Principally the same remote sensing methods are applicable for snow line monitoring over glacier areas. Repetitive imagery can indicate whether the glacier accumulation area ratios and their mass balance are changing. Surging glaciers, which may advance over large areas and cause devastating floods by blocking and suddenly releasing large quantities of melt water, can be identified, inventoried and monitored even in remote, inaccessible areas. There is, of course, a need for more precise input data for modelling of accumulation and melting processes which have to be gained using ground sensors. Further development particularly in sensor and monitoring techniques is necessary.

Stream Flow Monitoring

Mapping of major river systems to determine their spatial variation and the seasonality of stream flow is an urgent task in different parts of the world. Repetitive imagery combined with the ability of multispectral data to reliably identify water bodies provides the tool to tackle this problem. Results of these studies give valuable information on endangered embankments and necessary precautionary measures. The available means are adequate to solve the problem of river course mapping.

Monitoring of Drought-Prone Areas

Direct monitoring of soil moisture by remote sensing imagery is difficult. Recent developments in microwave sensors however show promising results (Shanda et al., 1979). Current soil moisture assessments are mainly undertaken indirectly by monitoring the vegetation distribution. Changes in vegetation coverage or in health conditions of vegetation can successfully be monitored using repetitive imagery and automatic data processing. Qualitative maps of temporal changes can easily be produced by this technique. Extensive use of ground truth data has however to be made for more reliable quantitative investigations. Future improvements will depend on sensor development and on adequate modelling of radiation characteristics of vegetation and soil.

Crop Disease and Destruction Monitoring

Agriculture crop disasters include a variety of applications because stress on crops may indicate lack of moisture, disease, insect infestation and destruction by hail and storm. Many efforts have been undertaken to detect stressed crops with satellite remote sensing methods before the stress became severe enough to damage the crops beyond redemption. Crops and forests that have been killed are identifiable and distinguishable from healthy crops, but premortality stress is in general not readily evident. The relative vigor of crops, their aerial extent and density can nevertheless be readily assessed by spectral recognition techniques. Improvements may be expected as new sensors in the infrared range and sophisticated image processing methods will become available.

FUTURE DEVELOPMENT NEEDS

In contrast to the above-mentioned instances where satellite technology can play a useful role, there are many other types of disaster situations where remote sensing or other applications of satellite technology cannot be applied at all. Yet there are others, such as earthquakes and volcanic eruptions, where satellite technology should, and I am convinced, can be of some assistance for early recognition and warning purposes. The technical means for remote gathering and relay of data for rapid digital image analysis and dissemination of information are already available. The problem we are facing is to find reliable precursors so that we can monitor them with appropriate temporal and spatial resolution and display the information to experienced teams of interpreters for rapid analysis. Improvements in detecting and predicting the time of occurrence of earthquakes for instance would be of tremendous value as they could save many lives. Even time scales of a few hours would be of great importance. Maximum attention should therefore be paid to prediction aspects of these types of disasters.

The excellent results in early detection, tracking and warning of tropical cyclones may serve as an example for what internationally co-ordinated activities can achieve. In all the deliberations concerned with disaster matters one should however carefully avoid discussion of cost efficiency questions as long as human lives are involved.

Another type which definitely possesses the characteristics of creeping disasters is environmental pollution by human activity. In this connexion, pollution has to be understood in its broadest sense as any strain on the environment - as any harmful change of the well-balanced natural ecological system. At present we do not know how stable our environment is and how it responds to the various kinds of pollution. It is therefore first and foremost necessary to measure the environmental baselines with sufficient sensitivity, accuracy and reliability before any assessment of changes can be undertaken.

Particular attention should be paid to the earth's atmosphere. We know that our atmosphere is unique in the planetary system and that it cannot be very stable. It seems that rather small modifications of the environment could make it impossible for the earth to retain an atmosphere which supports life as it presently exists. Any interference with the atmosphere therefore requires the greatest care and international co-operation in surveillance (Peters, 1979).

On the other hand, would a thorough understanding of the interrelations in the atmosphere and the relationship between the atmosphere and the oceans and land masses enable mankind to improve the climate? One could, in this connexion, consider the possibilities of eliminating severe weather conditions or perhaps even reversing the desertification process.

REFERENCES

SAC (1978). Satellite Telecommunications Experiment Project, Space Applications Centre, Indian Space Research Organization, Ahmedabad, India.

United Nations (1978). Agreement between the United Nations and Deutsch-Frauzösisches Fernmelde-satelliton-Projekt, ANNEX I, Description of ETS Station, Deutsche Forschungs-und Versuchsanstalt für Luft-und Raumfahrt V., Köln, Federal Republic of Germany.

Donoughe, P.L., Hunczak, H.R. and Gurski, G.S. (1978). Communications Technology Satellite - United States Experiments and Disaster Communications Applications, NASA Technical Memorandum 79109, Lewis Research Center, Cleveland, Ohio, USA.

Robinove, C.J. (1975). Worldwide Disaster Warning and Assessment with Earth Resources Technology Satellites. U.S. geological Survey, Project Report (IR) NC-47.

Schanda, E., Wyssen, D., Meylan, P. and Morzier, C. (1979). Soil Moisture Estimation by Correlated Ground-Based and Seasat Microwave Observation. Symposium on the Contribution of Space Observations to Water Resources Studies and the Management of these Resources, COSPAR XXII Plenary Meeting, Bangalore, India.

Peters, B. (1979). Invited Paper, Vikram Sarabhai Symposium on Space and Development, COSPAR XXII Plenary Meeting, Bangalore, India.

IAA 79-A29

THE APPLICATION OF COMMUNICATIONS SATELLITE TECHNOLOGY TO EMERGENCY MEDICAL SERVICES

Jerry Freibaum*, Cecil D. Burge[+]

Nearly 40 percent of the 25,000 daily medical emergencies in the United States occur in rural, wilderness or isolated areas. The probability of survival of the patient is directly affected by the availability of dependable, all-weather communications. Such communications are lacking in many areas of the world, including the United States.

An experiment is being conducted to determine the value and cost effectiveness of mobile communications via satellite for emergency notification, vehicle dispatch, and two-way voice and biomedical telemetry between paramedics and hospital physicians. Applications to rural, wilderness, and offshore recreational and oil rig areas are emphasized. The experiment described in this paper is being jointly conducted by the National Aeronautics and Space Administration (NASA), the Southern Regional Medical consortium (SRMC), and the National Highway Safety Transportation Administration (NHSTA).

This and other experiments, together with improvements in technology and a radio frequency allocation for a Land Mobile Satellite Service to be proposed at the 1979 World Administrative Radio Conference will facilitate planning for a future mobile satellite capability. The use of satellites as extensions of existing terrestrial systems should result in a major improvement in emergency medical care to many under-served areas of the world.

Keywords: Communications satellites; mobile communications; land mobile satellite services; emergency medical services; emergency response communications; disaster communications; rural communications; thin route communications; narrowband communications; 1979 World Administrative Radio Conference.

* *NASA Headquarters, Fourth and Maryland Ave., S.W., Washington, D.C. 20546*

\+ *University of Southern Mississippi, Hattiesburg, Mississippi 39401*

INTRODUCTION

In many areas of the world, particularly those which are largely rural or geographically remote, adequate health care services are not readily available. The maldistribution of health care professionals has left many non-metropolitan areas medically underserved with respect to general and specialized methods of treatment. Consequently, these areas experience higher mortality and morbidity rates in certain illnesses than do metropolitan areas. This is particularly true when expedient treatment is essential. For example, in a recent year, almost seventy percent of the deaths in the United States due to highway trauma were found to occur in rural areas. Coronary heart disease accounts for several hundred thousand deaths annually in the United States and by no means confines its impact to metropolitan areas. (Table 1)

TABLE 1 Principal Acute and Traumatic Conditions for Which Prompt Assistance is Critical to Life Preservation.

TYPE OF CONDITION	NUMBER OF CASES (injuries)	NUMBER OF DEATHS
Severe Injury from Vehicular Accidents (auto, aircraft)	1,900,000	50,000
Cardiac	600,000	352,000
Severe Burns	2,500	900
Drowning & Exposure	25,000	7,000
Poisoning (Ingestion Reptile & Insect Bite)	18,000	6,300
Home Accidents	3,600,000	24,000
Work Related Accidents	2,300,000	13,000

In these instances of patient crises, emergency medical services (EMS) are generally utilized. A decade ago the emphasis in emergency medical care was Basic Life Support (BLS). This concept stressed the transportation of the emergency patient as rapidly as possible to the nearest medical facility. Today, with the advent of better portable instrumentation, the emphasis has shifted to Advanced Life Support (ALS). The ALS concept stresses treatment and stabilization of the critically ill patient at the scene followed by transportation to a place of definitive medical care.

Dependable, all weather communications are essential to this process and directly improve the survival probability and morbidity of the patient. The reporting of emergencies and dispatching of ambulances is clearly dependent on communications, and fatality rates are highly dependent on response times. (Fig. 1 and Table 2)

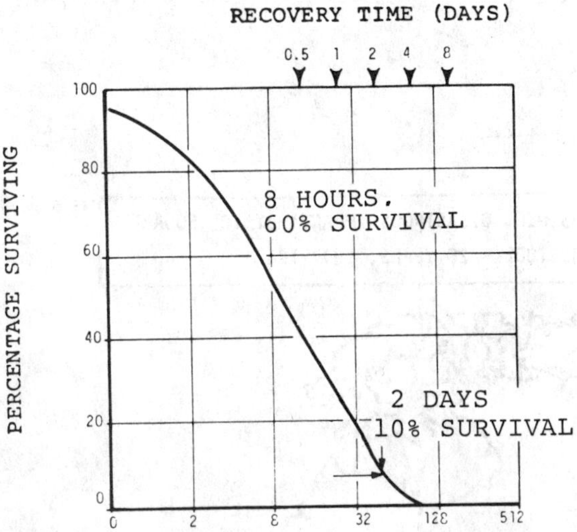

Fig. 1 Survival as a function of recovery time for search and rescue

TABLE 2

Approximate Time Needed to Limit Mortality to 50%

Severe Injury from Vehicular Accidents	Hours
Cardiac	**See note**
Severe Burns	Hours
Drowning	Minutes
Exposure	Hours
Poisoning	1 Hour
Home Accidents	Hours
Work-Related Accidents	Hours

NOTE: The 50% mortality delay times are averages drawn from a wide number of sources. Discussions with the Acadian Ambulance Service indicate that for the Cardiac entry there is a 3:1 increase in mortality if the time to care exceeds 40 min. More precise case histories are being collected.

Moreover, the ALS concept requires reliable, high quality links between emergency physicians at a command hospital and paramedics with the patient in the field. With two-way voice and telemetry of vital signs such as EKGs', physicians are able to effect diagnosis and immediate treatment through the paramedic. By shifting the legal responsibility for treatment from the paramedic to the physician, communications makes it possible for paramedics to practice defibrillation, fluid replacement, and drug therapy. The value of this life saving concept to severely injured or ill patients has been dramatic.

Patient survivability is similarly highly dependent on reliable, all weather communications with respect to other emergencies such as natural and man-made disasters and search and rescue operations. Emergency medical service needs related to natural disasters are not unique to any one country (Figures 2a and 2b)

Fig. 2a

AVERAGE YEARLY NUMBER OF DEATHS PER "DISASTROUS IMPACT" BY 10° SQUARES OF LATITUDE & LONGITUDE - 20 YEARS, 1947-1967

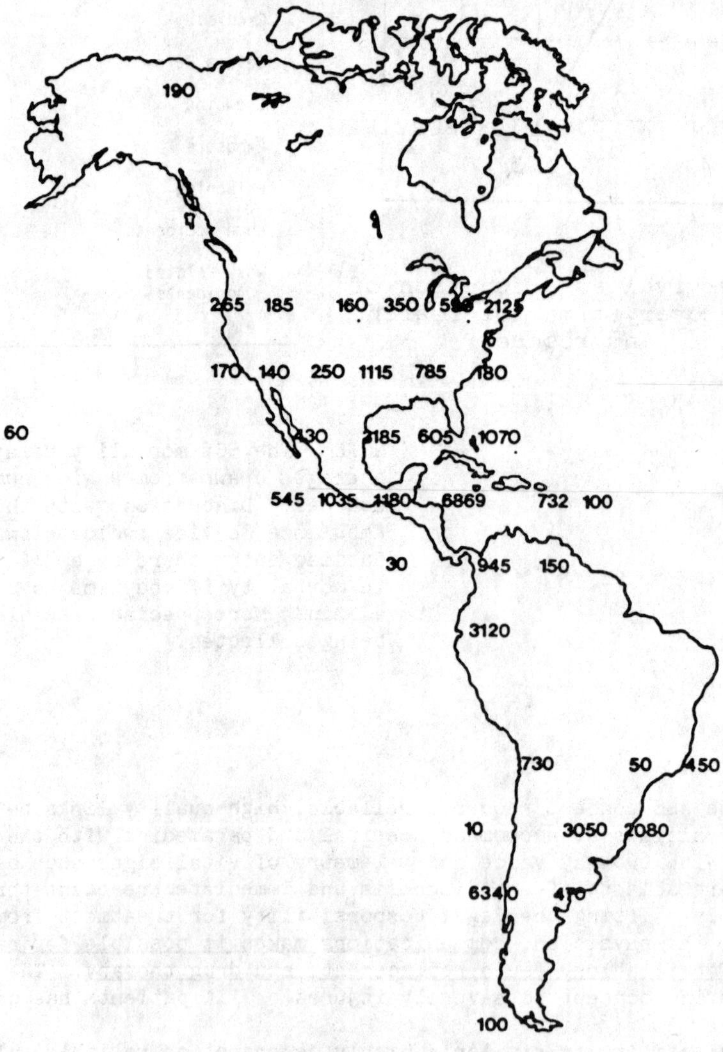

Fig. 2b

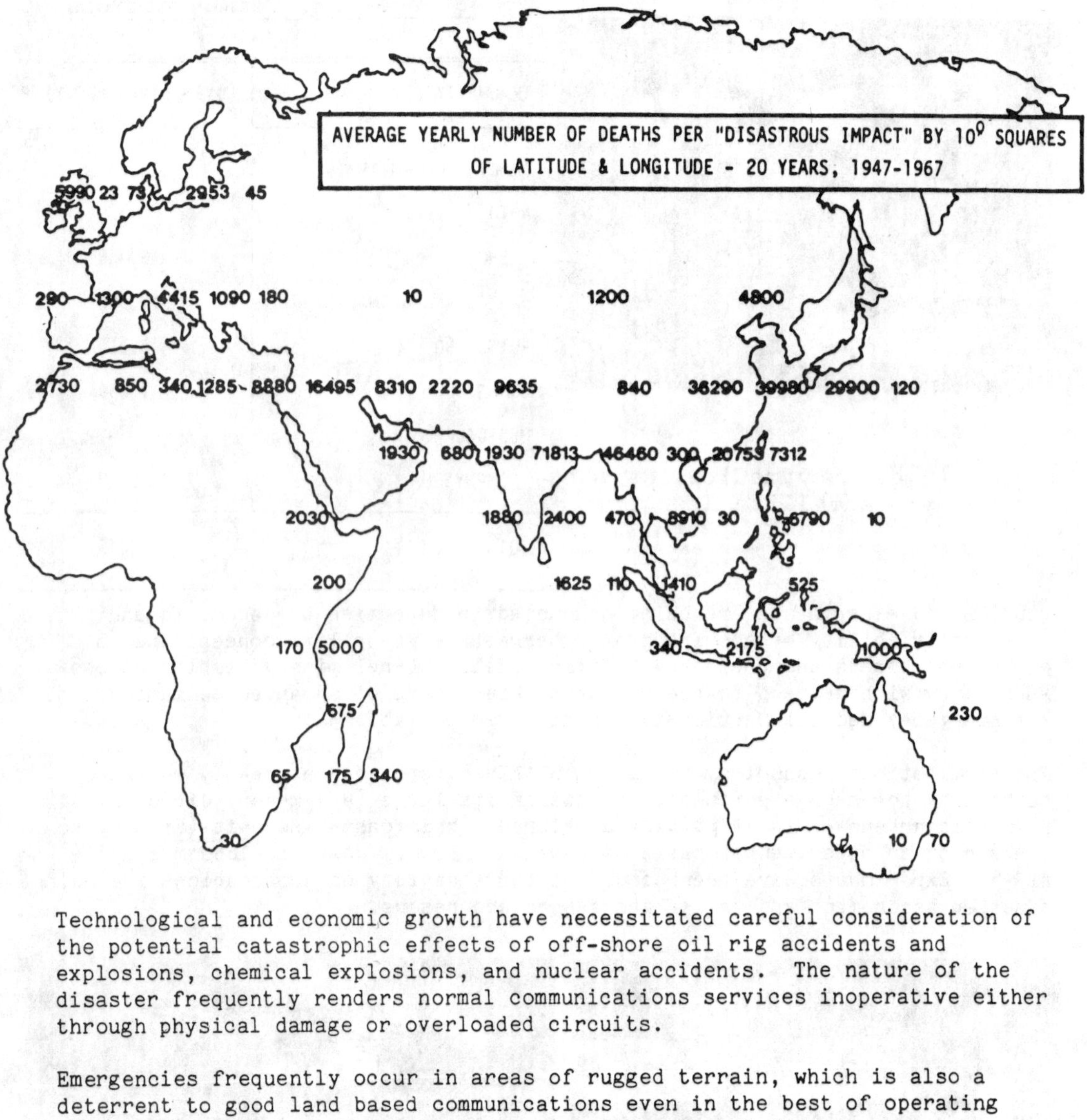

Technological and economic growth have necessitated careful consideration of the potential catastrophic effects of off-shore oil rig accidents and explosions, chemical explosions, and nuclear accidents. The nature of the disaster frequently renders normal communications services inoperative either through physical damage or overloaded circuits.

Emergencies frequently occur in areas of rugged terrain, which is also a deterrent to good land based communications even in the best of operating conditions. Emergencies occur in remote and/or isolated places where conventional communication facilities may not be economically warranted. Nearly 40 percent of the 25,000 daily medical emergencies in the United States occur in rural, wilderness, or isolated areas.

Satellites are capable of extending terrestrial communications and providing instantaneous communications for emergency, medical, and other applications to many of the non-urban areas where it may not be technologically, economically, or environmentally practical to use conventional systems (Fig. 3).

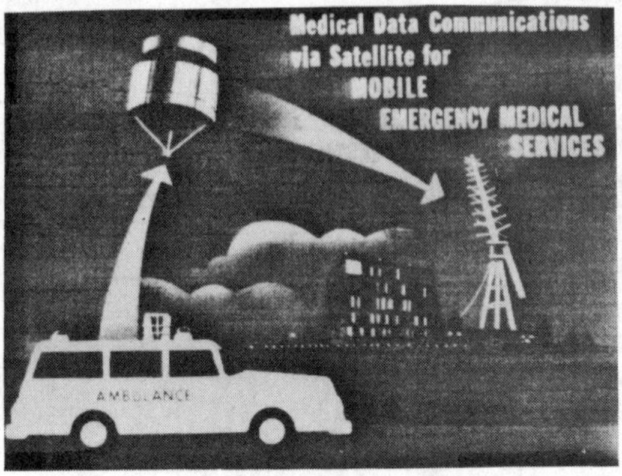

Fig. 3 Emergency medical service via satellite (concept)

TABLE 3
Estimate of Annual Benefits to United States from Advanced EMS Communications

TRAUMA TYPE	LIVES SAVED PER YEAR
CARDIAC, ARRHYTHMIA	13,296
AUTO	
SHOCK	26,356
HEAD AND SPINAL	14,996
POISONING	2,310
BURNS	1,059
MEDICAL	
EMPHYSEMA	1,428
TOTAL	59,445

Studies and experiments are being conducted to determine the economic and technical viability of the integrated terrestrial satellite concept and to assess user needs and projected traffic models. (Preliminary results of one such study with respect to the potential usefulness of advanced communications for Emergency Medical Services are summarized in Table 3).

The Applications Technology Satellite 6 (ATS-6) provided a greatly enhanced capability for such experiments because of its large (9.1 meter) directable parabolic antenna. In 1976, NASA developed a briefcase-size unit (Fig. 4) to operate on an experimental basis at L-band (1550 MHz down and 1650 MHz up) with ATS-6. Experiments have been conducted for a variety of applications including disaster assessment and relief and search and rescue.

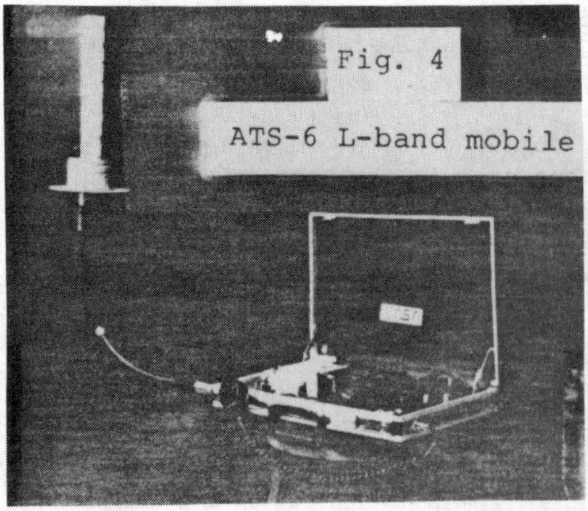

Fig. 4
ATS-6 L-band mobile

FIG. 5
ATS-3 VHF mobile

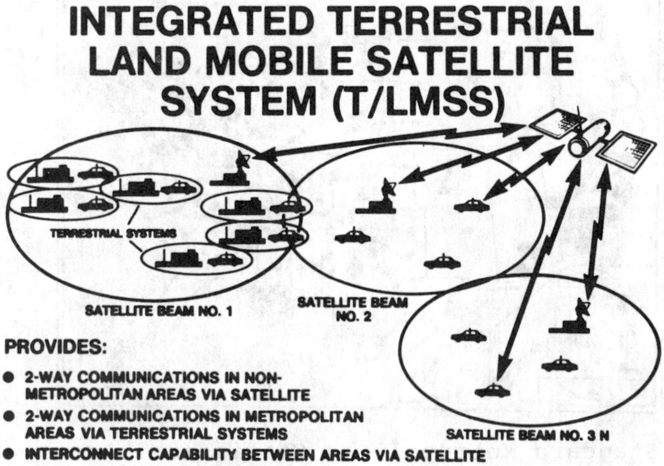

Fig. 6 An integrated land mobile system (terrestrial and satellite)

Units of similar size were developed by NASA's Space Technology Laboratory (NSTL) to operate with ATS-3 in the VHF band (Fig. 5). These units are being used in an emergency medical services experiment being conducted in Mississippi, Louisiana and Alabama by the Southern Regional Medical Consortium in conjunction with NASA and the National Highway Traffic Safety Administration.

Study considerations have included the technology needed for accommodating large numbers of thin-route, low-volume, diverse, commercial and public service users and for compatibility with existing and planned terrestrial systems.

An integrated, wide-area, land mobile system employing both terrestrial and satellite systems is depicted in Figure 6.

In areas not covered by terrestrial systems, direct communications between or with mobiles through a satellite is illustrated. A fixed earth station may perform the "base station" function within each satellite beam. Terrestrial and satellite systems can be interconnected between and within these service areas to serve a variety of applications such as: mobile telephone and paging; automatic dispatch; public safety, including: Federal, state and local law enforcement; disaster assessment and relief; search and rescue communications operations; emergency medical services; environmental and hazardous material control; and National Park management and protection.

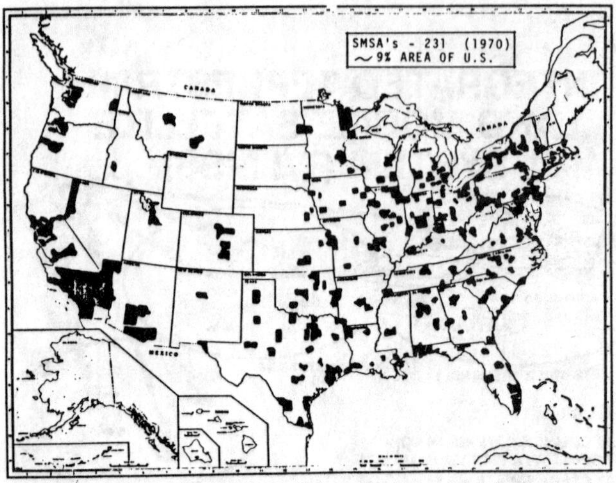

Fig. 7 Standard Metropolitan Statistical Areas of the United States and Puerto Rico: 1970

A large number of countries have population distributions, geographic characteristics, or economic, political or health considerations that lend themselves to land mobile communications satellite services. Using the United States as an example, this is depicted in Figure 7 which shows the scattered noncontiguous nature of the Standard Metropolitan Statistical Areas (SMSA) (shaded areas). Approximately 60 percent of the population is located in these 231 areas which comprise only 9 percent of the country's land mass.

It is in these SMSA's, each one with a population of at least 50,000, that terrestrial land mobile cellular systems are most likely to be established within this century.

In the rural areas, covering thousands of square miles, satellites could be used to extend terrestrial cellular systems. The need for and economic viability of this extension is presently under study. The extent and adequacy of conventional radiotelephone dispatch and paging operations in these rural areas are also under study.

At the present time, the provision of emergency medical services at the Advanced Life Support (ALS) level is dependent upon the installation of a terrestrial radio system for mobile to hospital communication. In urban areas, where the service demand is significant relative to the capital cost for a terrestrial communications system, the ALS concept is well developed. In rural areas, large investments are required for various coverage extension mechanisms (i.e., repeating tower units) in order to link field teams with the urban center hospital. In the more remote or wilderness areas where the terrain is rugged, these investments could be very large. Without some type of innovative augmentation cost, the coverage, reliability and environmental problems of existing terrestrial communications technology may thus limit the implementation of an advanced level of EMS care to urban/suburban areas.

In an effort to assess the demand, effectiveness, and benefits of an integrated satellite/terrestrial service the Southern Regional Medical Consortium (Fig. 8), with funding from NASA and the Department of Transportation, undertook an experiment utilizing the Applications Technology Satellite 3 (ATS-3) for emergency medical communications in three different EMS systems in the southeastern United States.

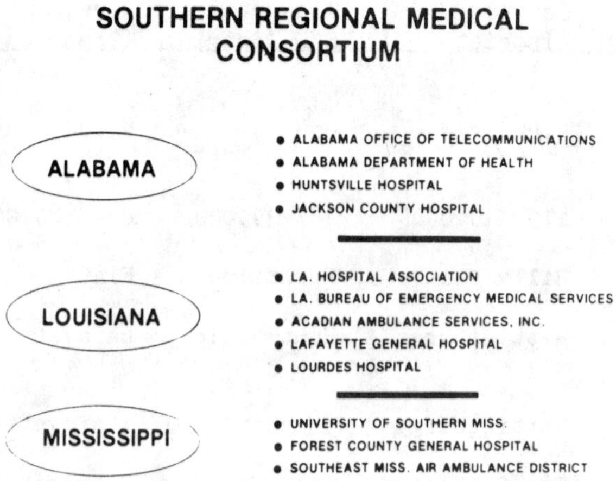

Fig. 8

Description of the Experiment Locales

Table 4 enumerates the salient characteristics of the EMS systems where the experiment is being conducted. The three EMS regions reflect diversity in terrain, financial structure, and level of previous experience with advanced life support techniques.

The Acadian Ambulance Service (AAS), a privately owned firm in southwestern Louisiana, is depicted in the table in terms of its traditional onshore role and also its offshore role. In the offshore environment, AAS responds to the emergency medical needs of about 60,000 petroleum workers.

While the three systems display diversity in many aspects, they also display similar communications requirements. The medical command hospitals possessing the only adequately staffed emergency departments within each region are all centralized. Hence, the implementation of a comprehensive pattern of ALS care throughout each district will require a communications network providing physician-to-paramedic links at ranges approaching 100 miles. All three EMS systems transport critically ill patients to hospitals as far as 100 to 200 miles from their base of operations.

Table 4: Summary Description of Project Locales

Item	System: Southeast Mississippi Air Ambulance Service District	North Alabama EMS District	Acadian Ambulance Service (Onshore)	Acadian Ambulance Service Offshore)
Service Area (square miles)	4,600	650	8,600	N/A
Population Urban/Rural	170/110,000	147/39,000	604/500,000	60,000 max.
Terrain	Hilly/Flat	Mountainous	Flat	Flat
Urban Center City	Hattiesburg	Huntsville	Lafayette LA	Lafayette LA
Medical Control Hospitals	1 (Hattiesburg)	1 (Huntsville)	3 (Lafayette)	3 (Lafayette)
Financial Structure	Tax Supported Public Utility	Provider Auxiliary	Subscription	User Pair
Central Dispatch System	Yes, VHF (155 MHz)	No	Yes, VHF (155 MHz)	N/A
Urban ALS/UHF Sys.	Yes	No	Yes	N/A
Existing ALS/UHF Range	10 miles	N/A	15 miles	N/A
Desired ALS/UHF Range	75 miles	50 miles	100 miles	250 miles
Patient Transfer Range (Maximum)	125 miles	150 miles	225 miles	N/A

While the onshore operations of the three are common to most other EMS systems, the offshore activity of the AAS dispatches paramedic teams via helicopter to oil platforms requesting aid. Enroute to and from a platform the helicopter, at its operational altitude, can communicate with shore. At lower altitudes (usually below 500 feet), and on an oil platform, communications is lost. From a medical standpoint the critical need is a link between the paramedics on the platform to a command hospital. This permits the patient to be stabilized prior to transport. (Fig. 9)

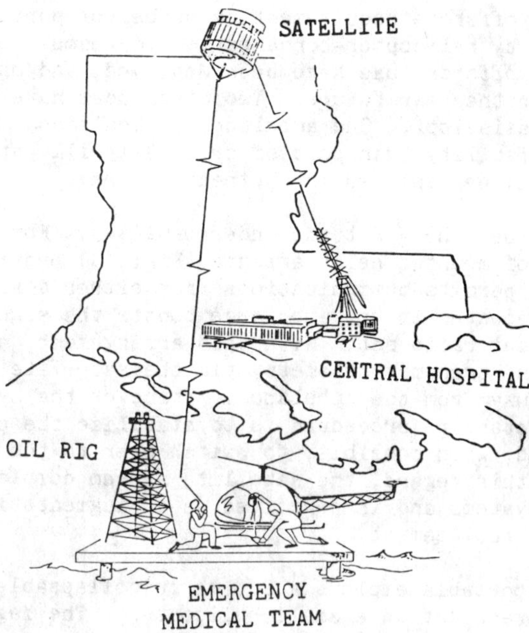

Fig. 9
Off-shore, oil rig
EMS communications

Description of the Equipment

A number of ambulances sufficient to meet the medical demands of selected rural areas beyond existing terrestrial UHF range are being equipped with VHF satellite transceivers (Fig. 10) to evaluate the application of mobile satellite communications in an emergency medical role.

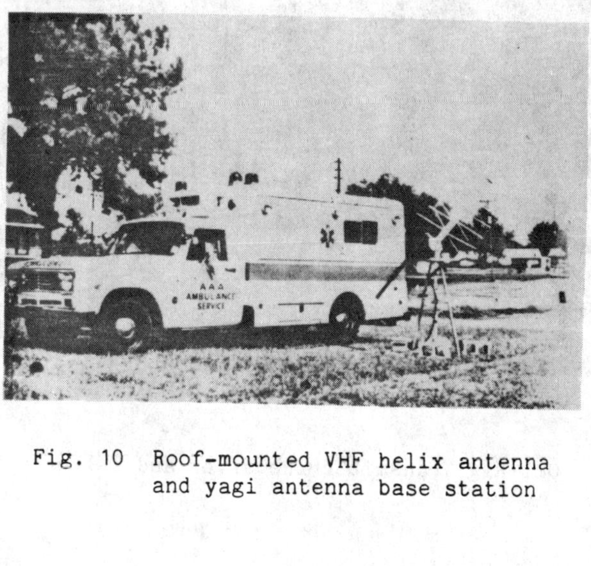

Fig. 10 Roof-mounted VHF helix antenna and yagi antenna base station

The base station antenna is shown to the right of the ambulance in Fig. 10.

To handle the offshore requirement, a number of portable terminals are being built for use by helicopter-borne paramedic teams. Onshore, three portable terminals for offshore use have been deployed, and one additional portable will be deployed in the near future. Two ambulances have been equipped in Alabama and two in Mississippi. One ambulance in Louisiana is being used exclusively for the interfacility transport of critically ill patients. Four additional ambulances will be deployed in the near future.

All terminals use the VHF transponder on ATS-3. For the ambulances, a compressed roof mounted helix antenna (Fig. 10) provides an omnidirectional pattern which permits communications from either a stationary or moving vehicle. A repeater in the ambulance boosts the signals of the paramedic's portable medical radio set. With this arrangement, a paramedic team can communicate and transmit telemetry via the satellite from a patient located outside and away from the ambulance. Whenever the system is employed with a patient, the standard procedure is to stabilize the patient using ATS-3, transport, and, when possible, to switch over to the terrestrial UHF/ALS network. In this regard, the satellite system complements the existing terrestrial systems and is perceived as an augmentation of the present system rather than a replacement.

The offshore portable employs a 2.6 meter collapsable Yagi antenna (Fig. 11) for ease of transport in a small helicopter. The Yagi in Fig. 11 replaced the antenna shown in figure 12 whose performance was considered marginal. Under

Fig. 11

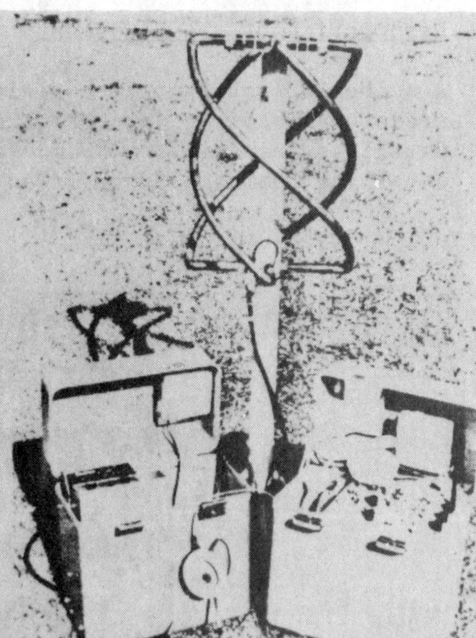

Fig. 12

Oil Rig Transmitter/Receiver and Telecare Unit

actual emergency conditions the antenna can be deployed in under two minutes. The use of the portable transmitter/receiver shown on the left, (Fig. 12) used with the medical sensors (Telecare Unit), on the right, permits telemetry and voice communications between paramedics on the oil platforms and the physician onshore. The equipment's light weight and mobility are illustrated in Fig. 13.

To support the mobile and portable terminals, dual base stations were installed at the command hospitals and dispatch centers.

A comprehensive data collection effort has been initiated. The analysis of this data will provide an insight into questions regarding the demand, cost, and benefits of mobile satellite communications in an emergency medical role. A summary of salient experiment results to date follows.

The Demand: Toward a Realistic Traffic Model

An important consideration in evaluating the need for new mobile satellite services in the emergency medical arena is how much the system will be used. While the ultimate traffic model must provide a designer with the needed channel capacity, a starting point, in this case, is the determination of the number and types of emergency calls to be expected from a given population. Beyond that, of course, are the methodologies and channel time required for each type of incident. With the data system now in place it is possible to account for each ambulance run within the study area and, from this, gauge the communications demand.

Prevailing medical opinion at the onset of the experiment held that cardiac victims, trauma, poisoning victims (including drug abusers), and burn victims would all benefit from an ALS level of care. The bulk of these were expected to be coronary cases and highway trauma victims. To the requirement for ALS communications was added the use of mobile satellite communications during major disasters. In Mississippi, the hurricane is a frequent problem. In Louisiana, both hurricanes and hazardous materials accidents are anticipated.

Table 5 details an incident distribution of an annual demand for advanced EMS communications in the Mississippi and offshore Louisiana. The projected incidents are based on population prorations of National Institutes of Health data for the entire country or, in the case of Louisiana, historical data. The "measured" columns are annual extrapolations based on data for the first four months of 1979. These data account for seriously ill patients by category that were or should have been handled with ALS communications.

Fig. 13 Demonstration of mobility

Table 5: Annual Demand for Advanced LMS Communications: Incident Distribution

Service Area

Type of Illness/Incident	Mississippi (Projected)[1]	Mississippi (Measured)[3]	Louisiana Offshore (Projected)[2]	Louisiana Offshore (Measured)[3]
Cardiacs	470	517	6	0
Cardiac Deaths	268	268	3	0
Trauma/Shock	100	153	36	8
Trauma/Head & Spinal	8	12	6	0
Poisoning	5	6	0	0
Burns	5	6	0	0
Critical Transfers	12	114	0	N/A
Manmade Disasters	0	0	2	1
Natural Disasters	0.33	0	0.33	0
Disaster Standbys	0	1	12	6

1. Prorated on a population basis from national data.
2. Extrapolated from four months (1979) data to annual basis.
3. AASI estimates from CY 1978 data.

From Mississippi data the following trends emerge:

(1) The primary driver for ALS communications is the coronary patient. During calendar year 1978, physicians in the Mississippi EMS region under study admitted 955 patients to coronary care units. While almost half of these bypassed the ambulance system, the cardiac load was the predominant source of critical EMS patients.

(2) The trauma load seems consistent with the projection but bears considerably more investigation.

(3) A surprising feature of the data collected to date is that critical transfers appear to be the second leading load behind cardiacs in rural EMS. Because the level of definitive care is quite low at rural hospitals, seriously ill patients must, of necessity, be transferred to regional facilities. Neonates, cardiacs, and stroke victims predominate. Forty-four (44) extremely critical cardiac patients were transferred in calendar year 1978.

(4) The use of the telephone by paramedics in the patient's home has been found to be minimal. The paramedics prefer to work with patients, if possible, in the vehicle.

The data from the Mississippi area strongly indicate that the patient loads for ALS communications are significant for rural areas; that the primary driver is the cardiac case, and; that the long distance critical transfer is a secondary driver. It should be kept in mind, of course, that forty to fifty percent of the patients reflected in Table 5 fall under existing terrestrial coverage.

The Benefits: Results to Date

Table 6 is a summary of incidents with the experimental system at all three locales through May 28, 1979. An "incident" in this table is a call for service that requires deployment of satellite communications terminals. A patient "encountered" is any patient causing an "incident". A patient "served" is one whose condition occasioned use of satellite communications. A patient "salvaged" represents one who survived a life threatening crisis and arrived alive at a medical facility. A "standby" is a deployment that does not result in satellite traffic.

Table 6: Summary of Incidents with Experimental Mobile Satellite System: January 1, 1979 to May 28, 1979

	Location			
Item	Mississippi	Louisiana	Alabama	Totals
Incidents	13	8	2	23
Patients Encountered	14	6	2	22
Patients Served	8	3	2	13
Standbys	1	2	0	3
Failures with Patients	6	0	0	6
Cardiac Encountered	10	2	1	13
Traumas Encountered	2	0	0	2
Other Illnesses Encountered	2	1	1	4
Voice Only (Patient Successes)	4	2	2	8
Voice & EKG (Patient Successes)	3	1	0	4
Patients Salvaged	2	0	1	3

The patient loads here are extremely small in view of Table 5. The explanation for this is twofold. First, when the hardware was initially installed a variety of problems and equipment failures occurred. This temporarily diminished the confidence and reliance of practitioners on the system. Second, the installation of an ALS communications capability at smaller rural hospitals revealed institutional impediments to its use. Paramedics, in some cases, when confronted with seriously ill victims retrogressed to a Basic Life Support (BLS) routine. As a result, modified protocols, paramedic refresher courses, and the use of registered nurses on ambulances have been initiated to solve this problem. The reasons for the reported small number of offshore incidents are being assessed. The data collection system reveals that in onshore rural areas there is no lack of patients.

Of the three patients whose lives were apparently saved two were transfers who had cardiac crises enroute. One was an emphysema victim who experienced a cardiac incident. All patients were handled at distances exceeding 40 miles from the command hospital. One rough measure of the economic benefit of ALS communications is to multiply the "lives saved" by some established value per life. Figures ranging from $24,000 (2) to $287,000 (3) have been published. The economic benefits of ALS communications in trauma cases is more complicated to determine since resulting residual impairment may create significant rehabilitative costs to society for survivors. One goal of this evaluation is to examine these benefits in microeconomic details for a sample of patients. This work is currently ongoing.

Some qualitative benefits of the experimental system have been the creation of institutional agreements for sharing scarce medical resources and the refinement of existing protocols. The sophistication of rural emergency medical department personnel has increased as a result of adaptive pressure from the technological advances. Because of this, public awareness will force the institution of some type of limited terrestrial ALS system around rural hospitals after the completion of this experiment.

The Cost of Terrestrial Alternatives

It has been argued that terrestrial technology can provide wide area coverage for rural EMS communications. A coverage extension mechanism employing strategically located repeating towers linked by microwave, leased phone lines, or the dial network through radio-telephone switches is required. As part of the evaluation effort a cost model of terrestrial systems based on these three approaches and driven by medical incidence data has been developed. The most cost effective terrestrial approach appears to be the radio telephone switch system (RTSS). These models are being refined with cost data from the three cooperating EMS systems.

The microwave system employed in Mississippi covers a small area beyond the city of Hattiesburg. The EMS communications here as in many rural areas is a dedicated one. Although interagency sharing is desirable, the reality is that fire, law enforcement, and medical services have separate systems. The cost of providing advanced ALS communications for the rural fringe area under the present terrestrial system approaches $2,000 per channel hour with the medical incidence rate measured to date. A key data item in all these analyses is the medical incidence rate for a rural population. The acquisition of this demand data in terms of illness number, type and communication usage is continuing.

Land mobile satellite services of the type described in this paper are not yet available for general use. The experimental satellites that have been used (ATS-3 and ATS-6) are operating far beyond their design lifetimes and no comparable commercial satellites have been developed for this kind of service. It is expected that any frequencies to be allocated for Land Mobile satellite services by the 1979 World Administrative Radio Conference (WARC) in Geneva will probably be different from those presently being used experimentally.

In anticipation of further experimentation and the possible implementation of land mobile satellite services, the United States is proposing at the WARC to allocate the 806-890 MHz band for "mobile-satellite service for use and development of systems using space radiocommunications techniques."

If the final analysis of these U.S. experiments and studies by the U.S. and other countries demonstrates that satellite mobile communications are economically feasible and desirable, frequencies for this service would be available. NASA estimates that approximately 20 megahertz in the 800 MHz band would be needed for such service. Eight or more antenna spot beams might be used in an initial system.

This capability has a potential for inclusion in a proposed narrowband experiment being planned by NASA for the mid or late 1980's. This development would emphasize thin-route, low-volume traffic requirements and could include a combination of fixed and mobile services to be provided by the private sector. Government and non-government users, many with diverse needs and problems can be addressed with a common communications system. This commonality of use is probably crucial to the economic viability of the system.

Summary

A significant portion of the world's emergency medical needs occur in rural or isolated areas. In life threatening illnesses, the patient's survival probability and morbidity are directly affected by the availability of dependable, all weather two-way mobile communications. In other emergencies, such as natural and man-made disasters and search and rescue operations, victim survivability is similarly highly dependent on reliable, all weather mobile and portable communications. Many of these incidents occur in places where conventional communications facilities are economically, technically, or environmentally impractical. Satellites are capable of extending or augmenting terrestrial communications for service to these areas.

NASA is conducting studies and experiments to determine the economic and technical viability of this augmentation concept and to assess user needs and projected communications traffic models. A mobile communications experiment using an experimental satellite, ATS-3, is helping quantify the requirements for advanced medical communications in rural areas. The physician-paramedic team is linked by two-way voice and telemetry through ATS-3. Results thus far have been very positive particularly for cardiac and critical patient transfers. The data base from this and other related experiments and studies will be used to help conceptualize an integrated terrestrial-satellite communications system for thin-route, narrowband communications applications. Since preliminary assessments indicate potential great need for and value in reaching underserved areas via satellite, the United States is proposing a UHF frequency allocation for a Land Mobile Satellite Service at the 1979 World Administrative Radio Conference.

REFERENCES

Burge, C.D. (1977). Some potential applications of a public service satellite in health care delivery.

Burge, C.D., F. K. Nagurney, O. Salguero (1979). Application of satellite communications in emergency medical service systems: evaluation methodology (interim report).

CCIR Documents: (1/18/79). USSG/BC/828 (Rev. 1) CCIR Study Group SPM, Geneva Doc. 8/452. Preferred technical characteristics for the land mobile-satellite service (draft new question); (1/19/78). Doc 8/480. Technical considerations for a land mobile satellite service operating in Band 9; (5/17/78). Doc 8/26 Land mobile satellite service considerations.

Consulting Communications Engineers, Inc. (1976). A complete mobile radio market study on the 900 MHz high capacity system.

ECON, Inc. (1979). A study of the benefits and cost-effectiveness of satellite-aided communications for emergency medical services, fighting forest fires and pickup and delivery trucking applications.

ECOsystems (1979). Draft report. The impact of the mobile telephone cellular system on the mobile communications satellite market.

Fairchild Space and Electronics Company (1977). Communications systems technology assessment study (vol. I and II).

Frost & Sullivan (1976). Communication satellite systems market in developing countries for the period 1976-1990 (2 vols.).

Hewett, K. (1969). A pilot survey of global natural disasters in the past 20 years. Paper No. 11. University of Toronto.

MITRE Corp. (1979). Satellite-based law enforcement mobile communications (a cost benefit effectiveness study).

Mundo, C., L. Tami, and G. Larson (1974). Final reports program plan for search & rescue electronics alerting & locating system.

World Health Organization Report (Jan. 5-7, 1977) Geneva.

Note:

Efforts were successful in obtaining almost world wide UHF allocations at the 1979 World Administrative Conference for Land Mobile Satellite Services for domestic applications. World wide Mobile Satellite allocations for distress and safety at L-Band were also obtained.

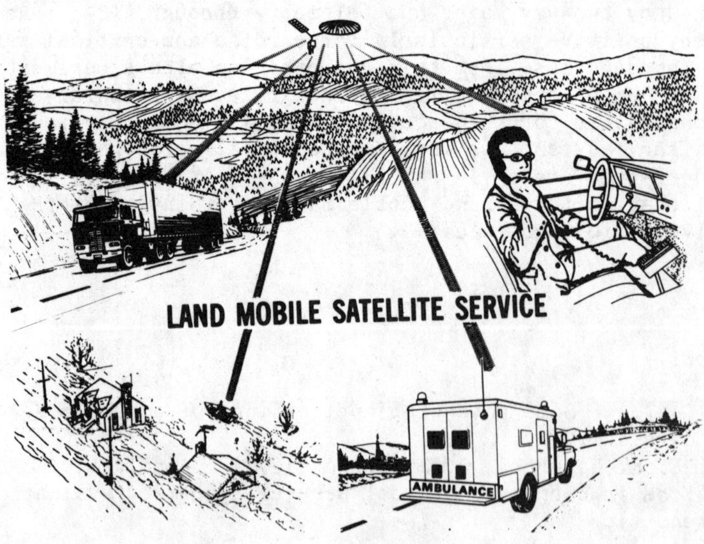

IAA 79-A28
(AAS 79-331)

THE FRENCH LOCAL USER TERMINAL FOR THE SARSAT DEMONSTRATION AND EVALUATION PHASE*

H. Castetbert, C. Lerr,
M. Midroit, M. Winterholer[+]

The Sarsat program, a cooperative program between USA, Canada and France, is intended to evaluate a search and rescue system by satellite. France participates in this program and provides the space-borne processor of the 406 MHz emergency calls.

For the evaluation and demonstration phase, the CNES (Centre National d'Etudes Spatiales) is studying a Local User Terminal (LUT). The 406 MHz on-board processed data will be processed in the LUT to detect and locate the 406 MHz emergency calls from distress platforms.

The anticipated performances of the Local User Terminal and its characteristics to be evaluated during the demonstration and evaluation phase are presented.

Keywords: Search; rescue; Argos; Sarsat; Eole; ELT; EPIRB; LUT; COSPAS; Sargos.

---

* *This paper appears in full in Volume 39, AAS Microfiche Series.*

+ *Centre National d'Etudes Spatiales, Centre Spatial de Toulouse, 18 avenue Edouard Belin, 31055 TOULOUSE CEDEX, France*

IAA 79-A31
(AAS 79-333)

WORLDWIDE DISASTER AND RESCUE RESPONSE EMPLOYING SPACE SYSTEMS*

E.E. Anderson+

A briefing is given concerning the functions and responsibilities of the United Nations Disaster Relief Office (UNDRO) which came into existence in March 1972. Emphasis is placed on the need for a means of rapid communication from any given disaster area and also for accurate information. This responsibility rests with the government of the country concerned. This implies the need for a good communication network. It also requires forthrightness on the part of the government of the country affected. Other deterrences to rapid deployment of emergency relief, such as bureaucratic procedures including immigration, customs, overflight permission requirements and other problem areas are presented.

* *This paper appears in full in Volume 39, AAS Microfiche Series.*

+ *General Anderson is Director, Relief Coordination, United Nations Office of Disaster Relief Coordination, Geneva, Switzerland*

IAA 79-A33
(AAS 79-334)

COSPAS PROJECT - A SATELLITE-AIDED EXPERIMENTAL SYSTEM FOR SAR APPLICATIONS*

Y.G. Zurabov, L.S. Pcheliakov, V.A. Bogdanov[+]
I.S. Bronitsky[++]

A brief description is given of the radio facilities installed on-board Soviet ships. Experience shows that conventional radio aids are often of poor efficiency in distress situations. Satellite approach is discussed, and a COSPAS system is described which is a part of a joint experimental satellite-aided search and rescue project designated as COSPAS/SARSAT project (USSR/Canada, France, USA). The COSPAS system spacecraft will be equipped with a two-channel receiver to receive "simultaneously" up to 20 signals of EPIRB-406 MHz. The EPIRB signals will be partly processed on-board the spacecraft and then will pass to a down-link transmitter for immediate transmission and to the spacecraft memory unit for further read-out when the spacecraft is within the view of the USSR LUTs. After complete processing of the signals the data needed for arrangement of SAR operations will be transmitted to SAR services as appropriate.

Keywords: Search and rescue; COSPAS/SARSAT project; Local User Terminal (LUT); processing aboard the satellite; spacecraft memory; experimental EPIRB-406 MHz.

* *This paper appears in full in Volume 39, AAS Microfiche Series.*

+ *V/O "Morsviazsputnik", Moscow, USSR*

++ *Merchant Marine Central Research Institute, Leningrad, USSR*

APPENDIX A

This appendix contains abstracts or summaries of all available papers presented at Space Safety and Rescue Symposia held 1976 to 1978. Many of these papers appear in full in Volume 40, AAS Microfiche Series, a supplement to this volume. These abstracts or summaries appear in this appendix in numerical order.

IAA 76-A01

CREW SAFETY*

Donald K. Slayton+

Crew safety is not a discrete function performed by specialists at a specific time and place in a program. It is embodied in the total efforts of everyone involved in a project from concept through completion. Factors associated with improving safety at each phase of development are discussed in varying degrees of detail. Mission success and crew safety carry almost identical definitions in manned space flight.

* *This paper appears in full in Volume 40, AAS Microfiche Series.*

\+ *Dr. Slayton is associated with the National Aeronautics and Space Administration, Johnson Space Center, Houston, Texas 77058*

IAA 76-A04

RECOVERY FROM A TUMBLING CONDITION IN SPACE*

G. S. Canetti+

The Space Division of Rockwell International has examined means for arresting the motion of an out-of-control tumbling manned spacecraft. Twenty-three different concepts for checking the tumbling by using a rescue Space Shuttle have been evaluated. The most practical of these approaches were a stick-on rocket concept and a water-stream concept. Analysis of these two methods, which are described in this paper, shows that they are feasible and can be used to stop the worst-case tumbling considered and allow rescue of the crew with one Shuttle flight.

* *This study was performed for the National Aeronautics and Space Administration, Johnson Spacecraft Center, under Contract NAS9-12004. This paper appears in full in Volume 40, AAS Microfiche Series.*

+ *Project Engineer, Space Division, Rockwell International, 12214 Lakewood Blvd., Downey, California 90241*

IAA 76-A05

LUNAR ESCAPE SYSTEMS FEASIBILITY STUDY*

J. O. Matzenauer[+]

On Apollo lunar missions, the safe return to orbit of the two landed astronauts depended on the proper functioning of the single-engine ascent stage of the lunar module. Among many options considered by NASA during early Apollo development were alternative ways of rescuing the astronauts if they became stranded on the moon. Investigated in the study, which was funded by NASA's Langley Research Center, were the feasibility and characteristics of a simple, "bare-bones" flying platform that would provide emergency escape to orbit. This concept was found to be feasible and provided program planners with another potential option for increasing the safety of lunar exploration.

* *This paper appears in full in Volume 40, AAS Microfiche Series.*

+ *Study Manager, Space Division, Rockwell International, Downey, California 90241*

IAA 76-A06

SPACE RESCUE AND OTHER SPACE OPERATIONS FROM EXISTING AIRSTRIPS*

Robert Salkeld+

This paper explores the possibility that space rescue and other missions which impose special requirements for quick reaction and operational flexibility, can be conducted from existing airfields. It is shown how the C-5 aircraft with modest modification, could serve as a mobile launch platform for a small single-stage-to-orbit (SSTO) shuttle, and that a somewhat larger SSTO could operate directly from a heavy-duty runway carrying its own integral rolling gear. It is suggested that this vehicle, air-launched from a new large subsonic aircraft, could orbit sufficiently large payloads to open up this mode of operation as a potentially promising alternate approach to low-cost high-volume space transportation. It is recommended that this concept receive increased attention.

---

* *This analysis was conducted in support of this paper by Lockheed-Georgia Company, under the direction of Dr. H.S. Sweet. This paper appears in full in Volume 40, AAS Microfiche Series.*

+ *System Development Corporation, 2500 Colorado Avenue, Santa Monica, California 90406*

IAA 76-A09

USE OF WATER SPRAYS IN SPACE RESCUE AND RETRIEVAL OPERATIONS*

M. H. Kaplan [+], D. C. Freesland[++]

Recent experiments involving liquid jets exhausting into a vacuum have led to significant conclusions regarding techniques for detumbling and despinning spacecraft during retrieval and rescue operations. A fine water spray directed toward a tumbling or a spinning object may quickly form ice over its surface. The added mass of water will absorb angular momentum and slow the vehicle. As this ice sublimes it carries momentum away with it. Thus, a complete detumble or despin is possible by simply spraying water at a disabled or spinning vehicle. Experimental and analytical results are presented on performance and physical properties. Although these are of a preliminary nature, the results are quite promising. Example situations are considered to illustrate potential applications.

---

* *Work supported by NASA Grant NSG-7078.*
This paper appears in full in Volume 40, AAS Microfiche Series.

+ *Associate Professor of Aerospace Engineering, The Pennsylvania State University, University Park, Pennsylvania 16820*

++ *Graduate Assistant, The Pennsylvania State University, University Park, Pennsylvania 16820*

IAA 76-A11

SPACE SHUTTLE PROGRAM SAFETY OVERVIEW*

N. E. Brown+

Safety systems, techniques and provisions unlike those previously employed in United States space vehicle development are being implemented during the Space Shuttle Program to ensure both crew and vehicle safety. Concurrent with ground-based flight simulations, computer-equipped aircraft--designated Shuttle Training Aircraft and programmed to simulate Orbiter non-powered flight profiles from 10,700 m. (35,000 ft.) to touchdown--are used to train Shuttle pilots. Prior to initial orbital flights, the first Orbiter vehicle will be released at 6,000 m. (20,000 ft.) from atop a Boeing 747 aircraft for Shuttle approach and landing tests and pilot training. Safety systems, hardware and procedures are provided for all phases of Shuttle operation, pre-launch through touchdown: launch pad crew escape systems; crew ejection capability through 23,000 m. (75,000 ft.) on early flights; and Orbiter intact abort for identifiable vehicle launch system failures. Once on orbit, safety equipment and contingency reentry options are available in the event of fire, equipment malfunction or loss of cabin pressure. Orbiting crewmembers can be rescued from a disabled vehicle on operational flights via a rescue Orbiter using onboard personnel extravehicular equipment, remote manipulator system and autonomous manned maneuvering units. Upon reentry, both primary and alternate landing sites are available, contingency ground egress systems provided, and emergency Orbiter ditching safety subsystems incorporated. An overview of the major safety aspects for each Shuttle operational phase is presented.

* *Shuttle safety information presented was current at manuscript preparation and is subject to change as development progresses. This paper appears in full in Volume 40, AAS Microfiche Series.*

+ *McDonnell Douglas Corporation, 16915 El Camino Road, Houston, Texas 77058*

IAA 76-A12

ANNUAL SURVEY OF RECOVERY CAPABILITIES*

Gloria W. Heath+

Studies undertaken to assure the safe return of astronauts have extended the capabilities of terrestrial rescue forces to respond to the needs of survivors of accidents or natural disasters generally. They have contributed, as well, to alleviating the causes of accidents and to diminishing their deleterious results.

The protection of occupants of space vehicles returning under contingency conditions has, in turn, drawn on the results of research un-allied to space program requirements.

The SURVEY reports advances that have been made since the inception of the Space Rescue and Safety Symposia in 1968 and considers further improvement that may be realized in recovery performance in situations of distress.

---

* *This paper appears in full in Volume 40, AAS Microfiche Series.*

+ *Director, SAR-ASSIST, Inc., One Island Lane, Greenwich, Connecticut 06830*

IAA 76-A13

SURVEY OF SPACE FLIGHT SAFETY SYSTEMS
SIXTH SUPPLEMENT
SURVEY PERIOD JULY 1975 - JUNE 1976*

Henry D. Wolf+

This paper surveys and assesses the space flight safety developments and achievements of the last 12 months as reported in the published literature dealing with space escape, rescue and survival, hazards of working in space, and experiment safety. Trends and research gaps are identified and their impact on future space efforts evaluated.

Significant open research areas identified in this survey include (1) the need for safety evaluations of material processing in space, (2) study of the impact on human performance of pulsed and long-duration magnetic fields, and (3) understanding of the transition and equipment carryover from Space Laboratory to Space Station.

---

* *This paper appears in full in Volume 40, AAS Microfiche Series.*

+ *Grumman Aerospace Corporation, Bethpage, New York 11714*

IAA 77-A31

APPROACH AND LANDING TEST PROJECT*

Donald K. Slayton, Thomas U. McElmurry[+]

The National Aeronautics and Space Administration is taking advantage of an early opportunity to investigate Orbiter airworthiness and systems performance by flying a program of mated Orbiter/Shuttle Carrier Aircraft and Orbiter free flight tests to the Rogers dry lakebed at Edwards Air Force Base, California.

The Approach and Landing Test Project was established to accomplish this flight test program, operating out of facilities at Dryden Flight Research Center and landing on Rogers Dry Lake. This paper discusses the flight testing accomplished and planned under this effort.

---

* *This paper appears in full in Volume 40, AAS Microfiche Series.*

+ *NASA, Johnson Space Center, Houston, Texas 77058*

IAA 77-A32

NON-SPECIFIC TRAINING AS A FACTOR TO ENSURE SPACE FLIGHT SAFETY*

G.M. Kolesnikov, N.V. Krylova
V.A. Popov, I.B. Solovyova+

SUMMARY

The safety of a spaceflight depends upon the operational flight control, the efficiency of the crew, and the ability of the crew to make correct and appropriate decisions by themselves, and upon their ability to act in a situation of emotional stress.

The efficiency of the training of cosmonauts to act in a situation of emotional stress depends upon the ability to simulate stress factors on earth and upon the ability of the cosmonaut to react to a stressful situation. Analysis of available data shows the necessity for more extensive use of highly developed modes of real stress factors, for training to perform particular tasks and for training to act in case of emergency.

As an addition to our previous publications the problems considered here include developing a particular mode of operative activity, for example, in parachute jumps, the possibility of supplying an operator with adequate information during a jump, and the ability of a parachute jumper to accept it and analyze it.

In these experiments, a process of adaptation of man to the activity under emotional stress has been analyzed and the activation of attention and operative thinking in performing a given task in parachute jumps has been studied. Perspectives of increasing spaceflights are closely connected with the intensification of working activity of a cosmonaut. In this connection, the problems of spatial orientation gain a new psychophysiological and psychological feature.

The work done shows that the problem of spatial orientation is closely connected with the efficiency and reliability of an operator's activity in a spaceflight.

Along with the investigation of physiological characteristics of motion sickness more and more attention is being paid to the study of those changes in the functional systems of an operator which are responsible for the efficiency and reliability of his activity.

Available materials permit us to think that it is possible not only to forecast an operator's activity during motion sickness, but also to determine new ways of improving the system of selecting and training cosmonauts.

---

* *Only a summary of this paper was available for publication. It also appears in Volume 40, AAS Microfiche Series.*

+ *Institute of Psychology, Academy of Sciences, USSR*

IAA 77-A33

SYSTEM SAFETY IN MANNED SPACE FLIGHT*

Jerome B. Hammack[+]

As NASA is nearing completion of the space shuttle approach and landing test phase and the first orbital flight is within 2 years, it is appropriate to discuss those system safety techniques utilized for manned space flight safety.

NASA manned space flight system safety today is the result of the evolution of techniques that had their beginnings in the original manned space flight program; that is, Project Mercury. Although at the time Project Mercury was conceived it was considered a complex program, actually it was much less demanding than manned space flight programs that followed. The total safety effort, therefore, had to continually evolve with the manned space flight programs. This evolution has resulted in the present system that includes distinct safety requirements, identification of responsibility for program safety, application of independent safety analyses, special safety panels, and management visibility of the safety of a program. This paper outlines the scope of system safety in manned space flight today in terms of the organization of the safety effort, hazard identification and resolution, safety program integration, risk management visibility, and finally, payload safety.

---

* *This paper appears in full in Volume 40, AAS Microfiche Series.*

+ *National Aeronautics and Space Administration, Lyndon B. Johnson Space Center, Houston, Texas 77058*

IAA 77-A34

THE SYNTHESIS OF CONTROL ALGORITHM FOR MANEUVERING A SPACE SHUTTLE VEHICLE*

A.A. Zhevnin+

The space maneuver of a moving Space Shuttle (SS) in the landing phase from a height of several hundred meters to zero is considered.

The main objective for Space Shuttle is to get the proper final approach flight conditions.

The unknown disturbing factors (wind, asymmetric configuration rate) deflect the Space Shuttle from the vertical reference plane corresponding to the lateral direction of the runway. For this reason vehicle stabilization is achieved through aileron and rudder control.

Space Shuttle motion is described by nonlinear differential equations. Disturbing factors are considered to be nonobservable but they affect the current state x(t) of SS.

Control system synthesis is based on works by N.N. Krasovsky, A.B. Loe, and L. Marcus.

According to this approach, rapid calculation of the programmed control on the base of current state x(t) is equivalent to the feedback control.

This nonlinear time-variable feedback defines the programmed flight with initial and final conditions.

The equations for the current angles of attack and role were also found as lateral and angular motion are linked.

Elevator, aileron and rudder control laws were derived in analytical form from the angular equations of motion.

Created algorithms require a few arithmetic operations and it is therefore convenient to use such algorithms in a real-time guidance system.

* *Only an abstract was available for this paper. It also appears in Volume 40, AAS Microfiche Series.*

+ *Academy of Sciences, USSR*

IAA 77-A35

SURVEY OF SPACE FLIGHT SAFETY SYSTEMS
SEVENTH SUPPLEMENT
SURVEY PERIOD: JULY 1976 -- JUNE 1977*

Nelson E. Brown+

In this paper, published documentation encompassing space flight safety developments for the period July 1976 through June 1977 is surveyed and summary reports provided for 28 manuscripts. As the seventh of a series, this edition places emphasis on safety activities conducted by NASA during the development of the Space Shuttle Program prior to the availability of flight-ready systems and vehicles. Activities include systems safety assessments, hazards identification, simulations, testing and crew training, procedures development and guidelines to enhance crew and vehicle safety. Each space component, system and project is subjected to rigorous paper analyses and testing, and crewmembers are familiarized with the functions and trained in the operations prior to initial flight. The enormous quantity of documentation resulting from analyses and assessments of Space Shuttle and payload safety aspects represents extensive efforts in developing a vehicle that should surpass all previous spacecraft in overall flight safety. A sampling of such documentation is summarized.

Synopses of several published articles addressing crew escape and rescue, systems safety, vacuum exposure and safety hardware/systems developments are provided.

* *This paper appears in full in Volume 40, AAS Microfiche Series.*

+ *McDonnell Douglas Corporation, 16915 El Camino Real, Houston, Texas 77058*

IAA 77-A36

DESIGNING PAYLOADS FOR SAFETY ON THE SPACE TRANSPORTATION SYSTEM*

G. S. Canetti[+]

Payloads using the Space Transportation System (STS) (which consist of the Space Shuttle, Spacelab, Interim Upper Stage and other carriers) must meet STS safety requirements. These have been established by NASA and are intended to protect the STS and other payloads flying in the STS, while allowing maximum design flexibility to the payload designer.

Examples are given of the main safety concerns for different classes of payloads and how they can be resolved. These payloads include satellites, Spacelab experiments, and small self-contained payloads. In addition, structural integrity and caution and warning are discussed as general safety concerns.

It is concluded that safety for STS payloads can be implemented simply, easily, and at low cost.

* *This paper appears in full in Volume 40, AAS Microfiche Series.*

+ *Space Division, Rockwell International, 12214 Lakewood Blvd., Downey, California 90241*

IAA 77-A37

PRINCIPLES TO DEVELOP A MODEL OF DANGER IN SPACE*

G.T. Beregovoy, G.P. Shibanov

V.I. Yaropolov, I.I. Baranetsky+

SUMMARY

Results of theoretical analysis and an impressive volume of factual data give us a better understanding of danger resulting from major irregularities in spaceflight and can provide the basis for developing a model of danger in space (MDS). MDS lays the foundation for spaceflight safety theory (SFST). It explains the inner mechanisms of danger in space, their qualitative and quantitative characteristics, the dynamics of their development during the spaceflight and develops a uniform scheme for crew safety analysis of piloted space ships (SS).

Theoretical analysis of SFST works shows that they usually utilize this or another model of danger. Despite significant differences between the models, they all essentially represent different realizations of a more general MDS. Although the general MDS reflects the main characteristics of particular ones we should bear in mind that there is no reciprocal correspondence between them, and the general MDS obtains new characteristics which differentiate it from a mere aggregate of particular models. This is why the authors of the report to the XXIVth Congress of the IAF offered the logical structure of danger which, when adequately developed, allows the establishment of interrelations between particular models, which significantly alter the model characteristics in general.

The report deals with the problems of the conceptual basis of the model, the degree of model adequacy when applied to real spaceflight conditions and fields of its application. The report gives the following principles of the model development:

1. Accounting of the whole spectrum of danger, which includes not only emergencies, but also unexpected situations as preconditions to emergencies

2. Strict mathematical description of known and revealed regularities of danger origin in the process of spaceflight that takes into account statical and dynamical factors, determined and causal characteristics, origin in or external to station for processes in the crew-SS-environment system

---

* *Only a summary of this paper was available. It appears also in Volume 40, AAS Microfiche Series.*

+ *Intercomos Council, USSR Academy of Sciences, Leninsky Prospect 14, 117901 Moscow V-71, USSR*

3. Flexibility of the model that provides its structure alterations depending on the tasks being solved, types of SS and main characteristics of the spaceflight

4. Applicability of the model to a broad spectrum of initial conditions and data

5. Possibility of further model improvements taking into account new spaceflight data.

In conclusion, we should like to note that MDS development would promote not only the degree of reliability of safety analysis results, but also unanimity of understanding and approaches to problems being solved by specialists in the field of spaceflight safety in different countries of the world.

IAA 77-A38

SHUTTLE REMOTE MANIPULATOR SYSTEM
SAFETY AND RESCUE SUPPORT CAPABILITIES*

Jeri W. Brown[+], Graham D. Whitehead[++]

A Remote Manipulator System being developed to enhance payload applications for the Space Shuttle Program represents a significant advancement in man's capabilities in space and in his progress to more effectively utilize the "weightless" environment. The Remote Manipulator System (RMS), an Orbiter-attached manipulator arm, will be used to deploy scientific payloads from the Shuttle Orbiter payload bay and also retrieve free-flying satellites from earth orbit. The satellites may be serviced on-orbit with the aid of the RMS and released again into space or stowed in the payload bay for return to earth.

The capabilities inherent in the RMS will be exercised in operations ranging from handling large, massive payloads to inspecting the Orbiter exterior for safe reentry status. The manipulator will also assist in extravehicular activities--including crew rescue from a disabled vehicle. Rescue support primarily involves the transfer of crewmen requiring aid from the disabled earth-orbiting spacecraft.

Although remote manipulation is relatively new to man's repertoire in space, it is imperative that RMS operations do not compromise the safety of Shuttle flight crews or the structural integrity of the reusable Shuttle Orbiter and Orbiter payloads. To ensure non-hazardous RMS operations, safety has been and continues to be an integral element of the design and development programs since the earliest concepts were conceived. The manner in which safety is being integrated into the RMS and crew rescue support capabilities are the subjects of this paper. A description of the system is initially presented for reader familiarization.

---

* *This paper appears in full in Volume 40, AAS Microfiche Series.*

+ *NASA, Lyndon B. Johnson Space Center, Houston, Texas 77058*

++ *Dr. Whitehead is associated with Spar Aerospace Products Ltd., Toronto, Ontario, Canada*

IAA 77-A39

RADIATION PROTECTION FOR MANNED ORBITAL STATIONS*

N.N. Gurovsky, E.E. Kovalev, V.M. Petrov+

SUMMARY

Radiation exposure aboard an orbital station strongly depends on spaceflight conditions (orbit parameters, mission duration, thickness of radiation shielding, etc.). A new approach to the estimation of radiation hazard is proposed which is based on the concept of risk of exceeding the design level of radiation established by a competent authority. A comparison of the risk levels under various flight conditions permits evaluation with reasonable accuracy of the radiation hazard for different space missions as well as the efficiency of protective measures. Thus, at an orbit inclination of 71°, mission duration of one year and manned modules shield thickness of 1 $g.cm^{-2}$ of aluminum, an increase in the orbital altitude from 360 to 600 km will result in an increase in the risk of exceeding the established design level of radiation for the bone marrow (150 rem) from 1 to 3% and for the eye lens (225 rem) from 15 to 30%. These estimations show that an additional shield thickness is required for the above flight parameters. When some time during the flight there is a danger of an impending major proton flare, the crew should move to the radiation shelter and stay there as long as the radiation situation outside remains dangerous. Then, if the radiation shelter shield thickness is 4 - 5 $g.cm^{-2}$ of aluminum the risk of exceeding the design radiation level lowers for the bone marrow to 0.05% at an orbital altitude of 360 km and to 0.3% at 600 km, and for the eye lens to 0.2% and 1% at the respective altitudes. When the shelter shield is less thick, an emergency descent might be considered necessary. At an orbital altitude of 600 km and the shield thickness of 1 $g.cm^{-2}$ the risk of an emergency descent is about 1%. This approach is expected to be used in the operational radiation protection service for extended space missions to be flown by the USSR in the future.

* *Only a summary of this paper was available for publication. This summary appears also in Volume 40, AAS Microfiche Series.*

+ *Academy of Sciences, USSR*

IAA 77-A41

THE PROBABILITY OF DECOMPRESSION SICKNESS AS A RESULT OF GOING OUT IN A PRESSURE SUIT FROM A SPACECRAFT WITH AN ATMOSPHERE CLOSE TO THAT ON EARTH*

A.S. Barer, L.G. Golovkin, S.N. Filipenkov,
I.N. Chernyakov, A.A. Sheikin+

The retention of an atmosphere close to that on Earth in spacecraft and stations is of physiological importance especially during the course of prolonged spaceflights. In this case the transition to reduced pressure in a pressure suit may be dangerous and lead to the development of decompression sickness (DS). The present investigation was stimulated by the supposition that for the solution of this problem one could not use the literature data concerning the simulation of the mentioned conditions without a pressure suit.

The data from 536 experiments performed by different authors for the past 15 years are analyzed. About 200 subjects were tested. The test subjects were selected in accordance with their decompression resistance.

In some cases DS symptoms can disappear spontaneously or maintain themselves in light forms which do not prohibit the performance of a number of working tasks. This will be illustrated in the report using quantitative data. Also the relationship between frequency and severity of the mentioned symptoms versus experimental duration and the intensity of physical work will be presented.

---

* *An extensive summary of this paper appears in Volume 40, AAS Microfiche Series.*

+ *Academy of Science, USSR*

IAA 77-A42

ADVANCED VEHICLE CONCEPTS FOR EARTH-ORBIT TRANSPORTATION AND RESCUE*

Rudolph C. Haefeli+

The advent of the Space Shuttle era of the 1980-1990 time period is stimulating international interest in using space for scientific and earth resource applications. Beyond the 1990 time period, the future potentials of space use will continue to be realized, and their level of achievement could be significantly enhanced by development of even more efficient, lower-cost earth-orbit transportation. Advanced transportation systems that could evolve using complete reusable single-stage-to-orbit (SSTO) vehicles are described in this paper. These vehicles would use the newest technology capabilities then available to achieve lightweight thermo-structures, efficient rocket propulsion, sophisticated avionics, and largely automated ground operations. These SSTO concepts could provide means for safety and rescue with fast reaction for both normal and emergency operations, and could be the same basic vehicles built and used for the transportation of personnel, cargo and equipment to orbit, readily modified for use in rescue. This paper will discuss specific concepts for SSTO vehicles based on advanced technology capabilities projected for the 1995-2025 era.

---

* *This paper appears in full in* *__Volume 40, AAS Microfiche Series__.*

\+ *Martin Marietta Corporation, Denver Division, Box 179, Denver, Colorado 80201*

SURVEY OF RECOVERY CAPABILITIES*

Gloria W. Heath[+]

This survey is presented annually as a part of the Space Rescue and Safety Symposia. For this 1977 survey see Worldwide Disaster/Rescue Response (IAA 77-A81)

Presentations at previous sessions considered classic and innovative rescue response - initially by dedicated response organized by the launching nation to recover its re-entering manned spacecraft. Subsequently, the SURVEY paper assessed the extent to which the space rescue system could draw on capabilities of the existing international cooperative rescue system, organized to respond to aviation and maritime craft in distress, rather than to duplicate such capabilities. The 1975 SURVEY paper further advocated that space rescue planning systems establish liaison with the terrestrial rescue response system, observing that the lack of such liaison could cause rescue response confusion and the consequent placing of rescuers and survivors at risk.

Placing the development of earth-surface location systems utilizing space-based elements for spacecraft recovery, the SURVEY pursued application of such systems to search and communication aspects of terrestrial rescue response.

The 1977 SURVEY is further expanding its purview not only to the tasks of locating and responding to terrestrial "transportation" emergencies but also to placing this portion of the rescue problem in an encompassing, worldwide, distress-assistance perspective that includes applications of space communications and remote sensing technology to disaster occurrences.

* *Only the above abstract was available for this paper. For additional material see IAA 77-A43A and IAA 77-A81 in this volume and in Volume 40, AAS Microfiche Series.*

+ *SAR-ASSIST, Inc., One Island Lane, Greenwich, Connecticut 06830*

IAA 77-A43A

WORLDWIDE DISASTER AND RESCUE RESPONSE EMPLOYING SPACE-BORNE SYSTEMS*

I.H.Ph. Diederiks-Verschoor+

This discussion is directed to statements of fundamental challenges and voids in current disaster/rescue response methodologies, analysis of the potential contribution of space-borne systems to improvement in disaster/rescue response performance (land, sea, air), and speculation on short-range and long-range changes in local, national, regional, and international arrangements that should come about as disaster/rescue response systems seek to incorporate space-borne elements. Topical areas in which the Discussants are experienced include the organization of national and international disaster/rescue response, the conduct of varied, specialized aspects of land, sea or air response missions, and the development of space systems directed to receipt of information that a distress situation requiring response exists, location of the distress site, assessment of conditions at the distress site, relay of distress communications and coordination of rescue/relief response, issuance of warning of impending disaster to threatened populations as well as for disaster prevention measures.

This presentation outlines space law and maritime law conventions and suggests prospects for future search and rescue with reference to specific problems such as use of radio frequencies for spacecraft and legal responsibilities of sovereign states.

---

* *Additional material on this presentation appears in* *__Volume 40, AAS Microfiche Series__*.

\+ *Dr. Diedriks-Verschoor is President of the International Institute of Space Law.*

IAA 77-A73

A SPACE-BASED PUBLIC SERVICE PLATFORM FOR TERRESTRIAL RESCUE OPERATIONS*

Ross Fleisig+, Joel Bernstein++

David C. Cramblit+++

The space-based Public Service Platform (PSP) is a multibeam, high-gain communications relay satellite that can provide a variety of functions for a large number of people on earth equipped with extremely small, very low cost transceivers.

This paper describes the PSP concept, the rationale used to derive the concept, the criteria for selecting specific communication functions to be performed, and the advantages of performing such functions via satellite.

The discussion focuses on the benefits of using a PSP for natural disaster warning; control of attendant rescue/assistance operations; and rescue of people in downed aircraft, aboard sinking ships, lost or injured on land.

---

* *This paper appears in full in Volume 40, AAS Microfiche Series.*

+ *Fellow AAS, Associate Fellow AIAA, Corresponding Member IAA, Grumman Aerospace Corporation, Bethpage, New York 11714*

++ *Member IEEE, Grumman Aerospace Corporation, Bethpage, New York 11714*

+++ *Alabama Society of Professional Engineers, George C. Marshall Space Flight Center, NASA, Huntsville, Alabama 35812*

IAA 77-A75

DISTRESS DETECTION, LOCATION, AND COMMUNICATIONS USING ADVANCED SPACE TECHNOLOGY*

W. E. Sivertson, Jr.+

This paper briefly introduces a concept for low-cost, global, day-night, all-weather disaster warning and assistance. Evolving, advanced space technology with passive radio frequency reflectors in conjunction with an imaging synthetic aperture radar is employed to detect, identify, locate, and provide passive communication with Earth users in distress.

This concept evolved from a broad NASA research on new global search and rescue techniques. Appropriate airborne radar test results from this research are reviewed and related to potential disaster applications. The analysis indicates the approach has promise for disaster communications relative to floods, droughts, earthquakes, volcanic eruptions, and severe storms.

---

* *This paper appears in full in Volume 40, AAS Microfiche Series.*

+ *NASA Langley Research Center, Hampton, Virginia 23665*

IAA 77-A81

WORLDWIDE DISASTER AND RESCUE RESPONSE*

Discussion Perspective

Gloria W. Heath+

Life-saving efforts conducted to rescue persons endangered as a result of land, sea, or air emergencies and to bring relief in the wake of disasters were impeded by the technical infeasibility of overcoming a number of formidable challenges - prior to the advent of space-borne systems for communication, remote sensing of the earth's surface, and weather surveillance.

Advances in space technology have now reached the stage where the potential for realizing improvements in worldwide rescue/relief response is being evaluated jointly by space agencies and rescue/relief offices - notably, the possibility of issuing warning of impending disaster to threatened populations, receiving early information that an emergency situation exists and its precise whereabouts, coordinating response by diverse organizations and rescue personnel, and assessing physical aspects of the distress site both before-the-fact so that long-term preventive measures may be taken and after-the-fact for assessing damage, escape routes, and the like.

Such initial evaluations may well demonstrate clear advantages to employing space-borne elements in one or another aspect of response systems. However, it is suggested that evaluations such as the following - beyond solely technical feasibility - should be carefully weighed:

* *This paper appears in full in* *Volume 40, AAS Microfiche Series.*

+ *SAR-ASSIST, Inc., One Island Lane, Greenwich, Connecticut 06830*

- Can equivalent results be obtained in other (less costly or "disruptive") ways?
- Can orderly arrangements be effected for use of the new communications or remote sensing capabilities (and within what time-frame)?
- Will the near-term incorporation of space-borne elements in national or regional rescue response stand in the way of the long-term achievement of a global response system (considerations of radio frequency, earth-surface user equipment)?
- Will incorporation of space-borne elements in global systems rule out the near-future participation in the system of those presently having access to cooperative, international rescue assistance? What significance will this bear with respect to achievement of a unified, global response system?

Finally, where clear advantage is offered after considering these further evaluations, it is suggested that thorough understanding of geographical and organizational differences in rescue/response resources and exposures is essential in order to plan for transition to the incorporation of new methodologies in worldwide rescue/relief response systems.

The purpose of this paper and the WORLDWIDE DISASTER/RESCUE RESPONSE Session - XXVIIIth IAF Congress, Current Events Session III - is to offer perspective for the making of such assessments and to provide a forum for the phrasing of international initiatives to incorporate promising space-borne elements into the framework of improved global response organization.

IAA 78-A55

SAVER—A POTENTIAL SPACE EMERGENCY RETURN SYSTEM*

G.S. Canetti[+]

The SAVER—System for Astronaut Variable Entry and Rescue—is a concept for an escape capsule for returning an astronaut from space in an emergency. It employs a large inflatable balloon as a drag device. Deformation of the balloon by aerodynamic forces during entry allows the drag, and hence the peak deceleration, to be modulated to an acceptably low value. The device can also be used for low-altitude recovery and landing.

* *This paper appears in full in Volume 40, AAS Microfiche Series.*

+ *Space Systems Group, Rockwell International, 12214 Lakewood Blvd., Downey, California 90241*

IAA 78-A56

ON MAN'S ADAPTATION TO THE OPERATOR'S WORK UNDER STRESSFUL CONDITIONS OF SPACE FLIGHT*

G.T. Beregovoy, N.V. Krylova, I.B. Solovyova+

Space flight safety is equally dependent on the reliable functioning of various spacecraft systems, and the correct and efficient performance of the crew. The latter is associated with their preparedness for the mission as well as their adaptation to adverse space flight effects. The reliability aspect of the crew-members at the stage of their adaptation involves a number of components:

- o Variations in the man's activity under the conditions of psychogenic effects (danger, risk, independence of actions and responsibility for their results, etc.)
- o Formation at the training stage of man's operational stability as a capacity for emotional isolation (abstraction) from the environment
- o Development of methods for training operations to perform complex work under the stressful conditions, etc.

In order to establish peculiar features of the emotional perception of stress-effects at the adaptation stage, stress effects on the man's performance depending on the task complexity, motivation level, etc., we have carried out experiments in real stress parachute jumps.

An examination of the man's emotional state during parachute jumps shows that the major effects are the unusual conditions, independent actions, certain risk, and limited time. The fact that the operator realizes his responsibility and potential hazard of the jump makes it possible to use it as a model of space crew activity.

* *This paper appears in full in Volume 40, AAS Microfiche Series.*

+ *Institute of Psychology, USSR Academy of Sciences, Moscow, USSR*

In our presentation of the XVIIth IAF Congress (October 1976) we demonstrated that while doing a parachute jump, man is able to perform certain elements of operator's work. This communication presents data from further investigations. Specific procedures were selected with respect to the psychological content of the simulated work. The complexity of the tasks to be performed and conditions of their fulfillment depended on the level of operator's training.

IAA 78-A57

RISK ASSESSMENT PROCESS AS APPLIED TO THE SPACE SHUTTLE*

R.L. Peercy, Jr.,+

This paper summarizes risk identification activities associated with system safety on the Space Shuttle vehicle as a pattern for their use on other complex, multiple-element programs utilizing emerging technologies. A building-block approach beginning first with safety analyses for each element and later with multiple element analyses is described in some detail. Examples of hazards that were identified and eliminated are provided. From a safety point of view, it is of interest to note that the solutions to element hazards often may become counter-productive when viewed from the total system perspective.

* *This paper appears in full in Volume 40, AAS Microfiche Series.*

+ *Professional Engineer-Safety, Space Systems Group, Rockwell International, 12214 Lakewood Blvd., Downey, California 90241*

IAA 78-A59

FLIGHT SAFETY ASPECTS OF ASTRONAUT TETHER DYNAMICS*

Alfred I. Sibila+

An experimental program was conducted to determine the motion of an astronaut upon reaching the end of a safety tether. The situation simulates an astronaut with a disabled maneuvering unit having a resultant velocity drifting away from a space station. The program was conducted using a frictionless platform to represent "0"-g flight. The test subject was in a pressure suit and was equipped with a maneuvering unit mock-up. Tests were performed in 3 degrees-of-freedom in both the vertical position (representing yawing motions) and in a horizontal position (simulating pitching motions). The subject was propelled away from the spacecraft with velocities up to 1.5 meters per second, and the resulting motions noted when the umbilical tether became taut. Investigations were made to assess the capability of the astronaut to prevent rotational motions by manipulating the tether as it became taut. A separate series of tests were also performed with the subject propelled toward the spacecraft to determine the ability of the astronaut to "fend off" and grasp hand holds.

The results showed that acceleration on reaching the end of a tether even at 1.5 meters per second were relatively mild, but the rotations and tether wrap-around were severe in some situations. It was learned that the astronaut could be trained to manipulate the tether to prevent, or at least greatly ameliorate, rotational motions. Motion picture records of the maneuvers were made over a grid painted on the platform, but attempts to correlate these with computed motions were not successful. This is attributed to camera angles that did not lend themselves to easy analysis without extensive computer programming and to hysterisis effects in the load-deflection characteristics of the tether. In the simulated space station encounter tests it was found that the astronaut could easily arrest linear and rotational velocities by proper use of either hands or feet, or both.

Overall, astronaut tether dynamics and space station encounters up to a velocity of 1.5 meters per second do not appear to present serious safety problems.

Keywords: Tether, space flight safety, simulation, EVA, space station.

---

* *This paper appears in full in Volume 40, AAS Microfiche Series.*

+ *Vought Corporation, An LTV Company, Dallas, Texas 75265*

IAA 78-A60

SALYUT 6 EXTRAVEHICULAR SEMI-RIGID SPACE SUIT*

G.I. Severin, V.I. Svertshek,
I.P. Abramov, A.Yu. Stoklitskiy[+]

A semi-rigid space suit with integrated self-contained life support system was tested during the *Salyut 6* orbital station flight for EVA purposes for the first time in the history of space exploration.

The space suit has an original structural arrangement ensuring a number of advantages: easy and quick donning and preparation for egress; compactness, good mobility and high reliability, multi-purpose role, working comfort conditions.

The egresses into space out of the *Salyut 6* station performed by G.M. Grechko and Yu.V. Romanenko on the 20th of December 1977 and by Cosmonauts A.S. Ivanchenkov and V.V. Kovalyonok on the 29th July 1978 demonstrated the correctness of the chosen arrangement and high suit performances.

Keywords: *Salyut 6* orbital station, semi-rigid space suit, suit design, self-contained life support system, hard-torso - "curaiss", water-cooled garment, suit thermal control system.

---

* *This paper appears in full in Volume 40, AAS Microfiche Series.*

+ *Intercosmos Council, U.S.S.R. Academy of Sciences, Leninsky Prospect 14, 117901 Moscow V-71*

IAA 78-A61

ANNUAL SURVEY OF SPACE FLIGHT SAFETY SYSTEMS: 8th SUPPLEMENT
SURVEY PERIOD: JULY 1977--JUNE 1978*

Nelson E. Brown+

Ensuring the safety of crew members onboard a spacecraft, the space vehicles during each phase of operation and personnel on the ground during vehicle preparation, launch and return involves an extensive number of safety projects, systems and innovative ideas. This paper, an annual survey, reports on safety developments and achievements as published in literature dealing with space rescue, crew and vehicle safety, systems safety assessments and systems testing. An overview of the characteristics and applications of the Space Shuttle Manned Maneuvering Unit is provided and the results of the Shuttle Approach and Landing Test Program are summarized.

Enhancement of crew safety as a result of Extravehicular Mobility Unit evaluations is discussed and several major safety concerns of the Shuttle program are identified. The paper concludes by reporting open research areas considered significant in providing adequate safety for perhaps a not-too-distant major space undertaking -- space colonization.

---

* *This paper appears in full in Volume 40, AAS Microfiche Series.*

+ *ORI, Incorporated, 1400 Spring Street, Silver Spring, Maryland 20910*

IAA 78-A66

THE NORWEGIAN SEARCH AND RESCUE SERVICE*

Torbjørn Høydal+

The Norwegian Search and Rescue Service is responsible for vast desolated mountain areas and considerable areas in the North Sea, the Norwegian Sea and in the Arctic Ocean. To cover these areas, however, Norway has only limited resources, and this fact creates special problems. In order to solve these problems the Norwegian government has established a rather unique SAR system. The main principle is that the service has to be carried out by cooperation between a number of public and private agencies which have facilities that normally are utilized for other purposes, but which also have SAR capabilities. In order to utilize the existing resources as efficiently as possible, it is also agreed that the SAR organization has to be a fully integrated service capable of handling land, sea and air rescue operations. In principle satellite-aided distress and locating systems are of interest for the Norwegian Sar system. Such satellite systems will in all probability improve coverage, reduce lost time between the occurrence and detection of a distress incident and improve the detection and location of false alarms. In addition a satellite system is likely to improve the efficiency of SAR mission planning. The operational SAR coordinators in Norway are two specially established Main Rescue Coordination Centers and 54 Rescue Sub Centers situated at Police Headquarters.

---

* *This paper appears in full in Volume 40, AAS Microfiche Series.*

+ *Deputy Chief of Police, Main Rescue Coordination Center, Bøde Police Headquarters, 8000 Bøde, Norway*

IAA 78-A67

AN INTERNATIONAL SATELLITE-AIDED SEARCH AND RESCUE SYSTEM*

A.E. Winter+

W.N. Redisch, B.J. Trudell++

For the past several years both the United States and Canada have been studying the application of satellite technology as an aid to Search and Rescue (SAR). The work has included examination of SAR requirements and the investigation of various system concepts including both low-orbiting and geostationary types of satellites.

In the past these studies and experiments were conducted separately by both Canada and the United States. In the fall of 1976 a joint program was proposed to demonstrate a satellite-aided search and rescue system. This joint program calls for a flight demonstration of the system commencing in 1982 using suitably-instrumented operational satellites operated by the National Oceanic and Atmospheric Administration (NOAA). France has also decided to participate in this project. The satellite system under consideration would be capable of detecting and locating Emergency Locator Transmitters (ELTs) and Emergency Position Indicating Radio Beacons (EPIRBs) operating at 121.5 and 243 MHz as well as a new experimental ELT/EPIRB operating in the 406 and 406.1 MHz band authorized at the 1971 World Administrative Radio Conference (WARC) for distress alerting and location via satellite.

A brief review of past systems studies is presented. The necessity for any early system to operate with the existing ELT/EPIRBs in the field (over 160,000) has led to the present system concept which was developed for the near-term solution to the distress alerting and locating problem. It consists of polar-orbiting satellites at 850 kilometers altitude to detect and estimate the location of activated ELT/EPIRBs. The international system is described and the responsibilities of each participating country are outlined.

Finally, the system has been discussed in the Joint IAA/IAF Working Group on Global Satellite-Aided Search and Rescue and International participation is expected in the demonstration.

---

* *Only an abstract was available for publication from this presentation. It also appears in Volume 40, AAS Microfiche Series.*

+ *Telesat Canada, 333 River Road, Ottawa, Canada K1L 8B9*

++ *NASA/Goddard Space Flight Center, Greenbelt, Maryland 20771*

IAA 78-A68

DESCRIPTION OF THE DISTRESS RADIO CALL SYSTEM OF THE FEDERAL REPUBLIC OF GERMANY*

Walter Goebel+

To extend the surveillance area of distress radio calls to a full ocean, the use of a geostationary satellite is a practical answer to this requirement. As a matter of cost saving it is furthermore important to be compatible with the existing maritime satellite system of the next decade with very little additional provision. The second basic idea was to be cheap on the user side, which means to take the complexity out of the numerous equipments and put it in the ground station which exists in a very small number only.

Considering furthermore that vessels with a satellite station would have an antenna steering mechanism which is much more complex than a wire antenna of nowadays, a breakdown of the communication link via satellite is supposed to be more likely. In this situation and when the danger of an immediate ship sinking does not exist, which is the predominant situation, a means of communication is required which is completely self-contained and portable without any support from the ship. This was the basic idea of the distress keyboard sender which was developed for the capability of transmitting an individual message to the coastal station which may consist of position data, kind of distress, requested assistance or whatever has to be reported.

The transmit frequency is defined to be in the range 1600 MHz, which is very practical with respect to the size of the antennas for the distress equipment and which can be produced with high efficiency and low cost.

---

* *This paper appears in full in Volume 40, AAS Microfiche Series.*

+ *DFVLR, Oberpfaffenhofen, 8031 Wessling, West Germany*

IAA 78-A69

PERSPECTIVES ON WORLDWIDE DISASTER AND RESCUE RESPONSE EMPLOYING SPACE-BORNE SYSTEMS*

Annual supplement to SURVEY OF RECOVERY CAPABILITIES

G.W. Heath+

The potential for realizing improvements in worldwide distress assistance through the employment of space-based systems is of opportune interest to agencies bearing responsibility nationally and worldwide for disaster/rescue response - for the short-term concerns of organizing and conducting the immediate distress assistance mission and for the institution of long-range preventive measures directed to alleviating the results of disasters or of preventing their occurrence. Space-based systems of interest include those related to giving early warning of impending disaster or other emergency notifications, facilitating distress communications, establishing the geographical location of distress sites, and remote sensing of earth for detection and monitoring of incipient disasters and conduct of vulnerability analyses.

Based upon studies of actual missions conducted to assist endangered persons on land, at sea, aboard aircraft and spacecraft returning to earth under contingency conditions, impediments - both technological and institutional - to delivery of optimum assistance are identified. Features of space systems that offer promise of alleviating or removing technological impediments are elucidated. It is cautioned that evaluations beyond solely technological feasibility need to be considered, including those concerned with identification of realizable economies with respect to the financial bases for utilizing space systems. Further, an understanding of organizational differences in agencies that respond to distress occurrences - occasioned in part by differences in geographical situation - is essential in order to plan for orderly incorporation of space-derived methodologies into worldwide disaster/rescue response systems.

---

* *This paper appears in full in Volume 40, AAS Microfiche Series.*

+ *SAR-ASSIST, Inc., One Island Lane, Greenwich, Connecticut 06830*

APPENDIX B

This appendix contains three papers presented at the following symposia:

1. Second World Congress on Emergency and Disaster Medicine (Pittsburgh, Pennsylvania, May 31 - June 3, 1981)

2. Space Technology in Disaster Relief Symposium (University of North Carolina, Chapel Hill, North Carolina, October 27, 1981)

SHARED OBJECTIVES of the International Academy of Astronautics' Studies Committee on WORLDWIDE DISASTER & EMERGENCY RESPONSE EMPLOYING SPACE-BORNE SYSTEMS and the Participants and Cooperating Organizations of the SECOND WORLD CONGRESS ON EMERGENCY AND DISASTER MEDICINE

The objectives of your Congress (cited on page 4 of the Congress Program) are closely aligned with those of our Studies Committee. The fundamental concern we share is that of reducing human suffering in the wake of life-threatening natural forces, man-made disasters, or emergencies experienced in the course of daily life. The overarching objective is to reduce to a minimum a population's vulnerability to disastrous occurrences by anticipating exposures accurately and setting in place wise precautionary systems.

The contribution made by space-borne systems is essentially that they can provide accurate information rapidly, clearly, and dependably over wide areas.

Our Studies Committee organizes an annual Symposium held during the International Astronautical Congress. Among proposals introduced at these Symposia over the course of the last several years have been the following:

- Rapid visualization of earth-surface features to assess disaster scenes and aid in such tasks as planning evacuation and access routes - by means of small, simple, low cost equipment
- Issuance of tsunami warnings in discrete native dialects to remote island populations in the Pacific Basin region that are not served by conventional communications systems - a low cost, automatically operated system
- A model for evaluating radioactive debris safety thresholds following uncontrolled re-entry of nuclear-powered space craft
- Receipt of satellite alert warning as a catalyst for an international disaster response legal regime

Our Studies Committee is presenting a paper at your 1981 Congress - on Free Paper Session #5 - that draws your attention to two new space technology tools:

The first reports the experimental development and use in actual emergency medical occurrences of two-way satellite communications services compatible with existing terrestrial emergency medical response systems to bring mobile emergency medical services to sparsely populated areas - at lower cost and with greatly increased effectiveness than could be achieved by extending exclusively terrestrial systems.

The second offers a model for disaster medical planning that can predict the outcome of decision-making while planning is still in the formative stages and later can refine it during the actual stages of the disaster. Use of the planning model results in more effective patient management and more effective deployment of communications, transportation, and other resources.

The Studies Committee will study carefully the initiatives and Recommendations phrased at the termination of your Congress. Problems you have found difficult to solve or resolve that can be assisted by space-borne systems will be placed on our own agenda.

We invite formal representation of your constituency on our Studies Committee and we welcome informal exchange of information with individual investigators and practitioners of emergency and disaster medicine.

Worldwide Disaster & Emergency Response Employing Space-borne Systems
Studies Committee Chairman:
GLORIA W. HEATH
SAR-ASSIST, Inc.
One Island Lane
Greenwich, CT 06830 U.S.A. TELEPHONE: (203) 869-1322

Experiments in Land Mobile Satellite Communications Compatible with Existing Terrestrial Mobile Emergency Medical Response Systems:
Jerry Freibaum EC-4
Technical Consultation Service
Space & Terrestrial Applications
NASA Headquarters
Washington, DC 20546 U.S.A. TELEPHONE: (202) 755-8570

Methodology for Disaster Response Planning
Dr. Paul B. Richards Code 4103
Director, Fleet Medical Support Project
Naval Research Laboratory
Washington, D.C. 20375

Second World Congress on
Emergency and Disaster
Medicine Pitt. PA 5/31-6/3/81

SPACE-BORNE SYSTEMS IN SUPPORT OF EMERGENCY AND DISASTER MEDICINE

Gloria Heath/SAR-ASSIST, Greenwich, CT
Jerry Freibaum/National Aeronautics and Space Administration, Wash. DC
Paul B. Richards/Naval Research Laboratory, Wash. DC

THE PROBLEM

The quick reestablishment of communications is vital to all disaster relief operations, but simple effective means of doing so are usually lacking. The unpredictability of the type, time, and location of catastrophic or disaster events requires a communications capability which is mobile, versatile, and dependable under all conditions of weather and terrain. Similar needs are essential for emergency medical services and search and rescue operations. These are problems without national boundaries.

There is also a growing realization that current disaster response plans and preparedness measures in many countries are inadequate to cope with the casualties and damage from a catastrophic earthquake, and indeed from many "lesser" disasters. For those plans that do exist they may only be delegations of assignments and responsibilities from one organizational level to another, without instructions or plans on how to carry out those responsibilities.

Numerous countries and institutions, individually and collectively, have begun to address the problem through studies, experiments, technology and planning. Examples are described in the ensuing paragraphs.

DISASTER ASSESSMENT AND RELIEF EXPERIMENTS

At ages 14 and 13 years respectively, NASA's experimental Applications Technology Satellites (ATS-1 and 3) have been providing experimental emergency communications in the Western Hemisphere and South Pacific.

An example of the rapid reestablishment of emergency communications via satellite occurred in May 1980 at the Mt. St. Helens volcano eruption in the western U.S. An immediate need following the eruption was communications for search and rescue operations and disaster assessment. Local power and telephone lines were disabled, and range limitations of conventional mobile radio equipment in the remote and mountainous region posed serious problems so that emergency personnel were unable to communicate with each other. The day after the eruption a special communications jeep equipped with an antenna capable of sending and receiving signals via NASA's ATS-3 satellite was flown to the disaster site, allowing emergency personnel to communicate with each other, with search aircraft, and with the outside world.

In the Pacific, NASA's ATS-1 experimental communications satellite is being used for disaster relief, medical, and educational applications. Numerous islands are linked via ATS-1. When sea waves swept over Majuro in 1979 leaving 4,000 people homeless, ATS-1 was the principal communications channel used to coordinate relief operations.

Near tragic consequences were averted in Southeast Mississippi in 1979 with the help of a satellite transceiver equipped ambulance. A tornado, spawned by Hurricane Frederick, severely damaged the area and knocked out the emergency power system to the County Hospital. The hospital was totally non-functional. Consequently, some patients, critically dependent on electrical monitoring equipment at the time of the power failure, were put into a life threatening situation. The satellite transceiver equipped ambulance (part of a NASA emergency medical experiment) was the only means of communications in operation. It linked the hospital with State emergency officials through ATS-3 and an emergency power generator was quickly dispatched.

EMERGENCY MEDICAL SERVICE EXPERIMENTS

There are 25,000 daily medical emergencies in the United States; nearly 40 percent of them occur in rural, wilderness, or isolated areas. In an effort to improve emergency assistance in these areas and thus save lives, experiments were conducted by NASA through the Southern Regional Medical Consortium to determine the value and cost effectiveness of mobile communications via satellite for notification, vehicle dispatch, two-way voice communication and biomedical telemetry between paramedics and hospital physicians.

The utilization of a satellite communications link permits, under the medical practice statutes of many states, the paramedic to perform sophisticated life saving treatments including drug therapy, fluid replacement, electroconversion, and in some cases, tracheal intubation. When coupled with the provision for telemetry of electrocardiograms, this service can prove of immense benefit to many types of patients including cardiacs.

Two major problems being addressed are: response to trauma and cardiac cases among the 60,000 oil rig workers stationed as far as 200 miles offshore in the gulf of Mexico and land transfers of the critically ill who are as far as 150 miles from specialized medical treatment.

SEARCH AND RESCUE EXPERIMENTS

Since the mid 1970's there has been extensive interest in using satellites to detect and locate emergency transmissions from aircraft and ships in distress. This interest has led to an international joint venture in satellite-aided search and rescue. The partners in this experiment are Canada, France, the United States, and the USSR. Several other nations have expressed interest in participating in the project.

NEW TECHNOLOGY

In the U.S., both the private and public sectors have undertaken research and development to improve satellite communications and facilitate its application for public and commercial services. NASA, for example, is studying a proposed thin route, narrowband, communications program. Thin-route communications means service primarily to low population densities or non-metropolitan areas. Thin-route communications requires primarily inexpensive two-way voice, two-way data, slow-scan video and one-way video and two-way voice. A land Mobile Satellite in the 806-890 MHz band is part of the proposed experimental thin route communications program.

Discussions on a possible joint mobile-satellite project between NASA and the Canadian Department of Communications were initiated in April 1980. Regular meetings have been held since then with an agreement to continue with separate but coordinated definition and conceptional design studies during FY81. One of several spacecraft candidates being considered for a possible Canadian-U.S. initiative would provide full coverage of Canada and the U.S. by means of a 50 foot spacecraft antenna. Small, standard, hand held and commercially available, car mounted transmit/receiver mobile units using whip antennas are envisioned.

A METHODOLOGY FOR DISASTER RESPONSE PLANNING

Most disasters can be placed in one of two categories: (1) the population at risk has a warning period for evacuation or sheltering. Examples include hurricanes and forest fires. (2) the population at risk has essentially no warning period for evacuation or sheltering. Examples include earthquakes and aircraft accidents.

For the disasters wherin there is a warning period, disaster response planning should focus on the timely evacuation and or sheltering of the population residing in the threatened area, before the population is affected. For those disasters wherein there is no warning, planning should focus on the management of mass casualties - identification and disposition of the dead; search and rescue of the missing; emergency medical care and hospitalization of the injured; emergency sheltering, feeding and care of survivors.

A computer simulation model has been developed by P. B. Richards at the Naval Research Laboratory in Washington, D.C. that can be adapted as a planning vehicle in developing disaster response plans. The model was designed to simulate the medical treatment and evacuation of combat casualties, to forecast the consequences of decisions and procedures, to optimize the patient management process, to evaluate capabilities of specific support systems and to determine resource requirements.

Research is currently underway to identify the pertinent characteristics needed to simulate various disasters in order to test the applicability of this model as a planning tool. This research will specify the types and the accuracy of information needed, and hence will dictate the requirements to be placed on various communication and information sources, be they ground-based, air-borne, or space-borne.

SUMMARY

This decade should see the application of two new tools to help mitigate the effects of disasters and other emergencies. (1) space technology for more effective, affordable, wide area communications and (2) planning models for more effective deployment of communications, transportation and other resources.

REFERENCES

"An Evaluation of the Experimental Application of Mobile Satellite Communications in Rural EMS Systems: Demand, Costs, and Benefits," Prepared for NASA, Contract NAS13-95, by Univ. Southern Mississippi, C. Burge and F. Nagurney, December 1980.

"The Application of Mobile Satellite Services to Emergency Response Communications," J. Freibaum ICC Paper, 1980.

"The Application of Communications Satellite Technology to Emergency Medical Services," IAF Paper 79-A29, J. Freibaum and C. Burge, August 1979.

"Computer Model for Simulation of Emergency Medical Systems," P.B. Richards; J. R. Fletcher, J.C. Bond; C. Delfosse - Millitary Medicine, Vol. 144 #4, April 1979

SPACE TECHNOLOGY TOOLS IN SUPPORT OF DISASTER RELIEF AND PREVENTION

Gloria W. Heath

Presented at "Space Technology in Disaster Relief Symposium" - attended by representatives of governmental and non-governmental organizations engaged in disaster relief and the producers of equipment they utilize.

University of North Carolina, Chapel Hill, NC, 27 October 1981

I. INTRODUCTION

An annual Symposium, of which I am organizing chairman, addresses the ways in which use of space-borne systems could assist those who, like you, bear responsibilities for alleviating the human costs in the wake of natural and man-made disasters. Our Symposium forms part of the week-long sessions of the International Astronautical Congress, attended by representatives of some 40 nations. The usefulness of our Symposium is that it encourages designers of space systems to investigate innovative applications of their technologies to the challenges faced by those actively engaged in disaster relief and prevention.

I would like first to review with you the already-demonstrated capabilities of space systems to assist in the conduct of disaster relief and prevention tasks and then to preview the capabilities that are on the horizon, for the important reason that user-engagement in planned demonstrations of these proposed systems can hasten realization of the benefits they promise. In this review, the essential question that you must ask is : What is needed in order to make these benefits available for disaster relief and prevention and what natural alliance of interests may there be to bring these promising possibilities to fruition?

As is well-known to you, the most critical challenge exists during the first few hours of an emergency or disaster. This is the period of time in which the greatest impact can be made in saving lives, in alleviating suffering, and in minimizing material losses. This is the period in which, despite the chaos that characterizes emergency or disaster occurrences, relief and rescue requirements must be assessed and deployment of suitable and sufficient resources initiated. This is the period of maximum communications channel demand but minimum or no-channel availability - owing to the effects or the cause of the disaster.

The contribution made by space-borne systems is essentially that they can provide accurate information rapidly, clearly, and dependably from the disaster site to those in command locations outside the affected area who are in the best position to initiate appropriate responses. The space-borne systems that have beneficial application to disasters are not limited to communications systems alone. Other significant contributors are the meteorological satellites that view worldwide weather systems and collect data from which weather and climate changes can be inferred; and the remote sensing satellites that observe earth's surface and detect changes that warrant attention.

II. COMMUNICATIONS SATELLITES: Disaster Applications

The early investigations centered on development of a small, air-transportable, earth station which could be delivered swiftly to a disaster site and would enable communications via satellite to be established with relief centers outside the affected area despite the destruction of normal channels of communications (disabling of power and telephone lines). Even during the development period, such COMSAT experimental earth stations were made available for use during actual disaster occurrences - such as the several Johnstown floods in Pennsylvania, USA. The European SYMPHONIE satellite and portable earth stations were utilized by the United Nations Peace-keeping Missions in the Middle East. More recently, a communications jeep fitted to be a control center was flown in to the Mount Saint Helens area following the eruption of May 18, 1980. Not only was the swift re-establishment of needed communications impressive but the quality of the communications far surpassed in clarity the usual (non-satellite) emergency communications heretofore utilized.

Preventive Health Care and Education

The earlier experiments via satellite focused on stationing a satellite over a remote area not reached by conventional communications and beaming to it educational programs and medical counseling. Such a program in Alaska was called by eskimo inhabitants "Doctor Call;" an educational program was conducted in remote villages of India - Project SITE - that brought information in native dialects on hygiene, agriculture, and population control.

Emergency Medical Service Project Conducted via ATS-3 Satellite

This demonstration combined the use of a communications satellite with existing terrestrial communication systems. It was directed toward the development of an Emergency Medical Response plan for thinly-populated areas such as the Alabama, Mississippi, Louisiana region of the USA. Characteristic of the region was the lack of adequately staffed hospitals - these lying at a distance from the remote communities. Particularly in Louisiana, a portion of the persons at risk were located on off-shore oil rigs. The Southern Regional Medical Consortium, which conducted the demonstration, consisted of paramedics, ambulances, helicopters and pre-arranged, expert medical staffing at the medical centers distant from the demonstration area. Communications, consisting of two-way voice, telemetry of vital signs and electrocardiagrams, were conducted from ambulance roof-top antennas and from portable antennas flown in to oil rigs by helicopter. A "Telecare" unit was developed to fit into a briefcase. In all cases, "affordable" cost guided the developmental objective.

The paramedic responding to the distress call could go beyond the usual treatments he was permitted to give and could perform sophisticated, life-saving treatments including drug therapy, fluid replacement, electroconversion, tracheal intubation - because the physician at the diagnostic center could bear the legal responsibility for the assistance given.

Meteor Burst Technology

An application that links both remote sensing and signal transmission by a "space-borne" system - but not a communications satellite - is that of meteor burst technology to relay from remote sites to central control laboratories the amount of snow cover, for example. The amount of snow can be determined by its weight on a platform - in the remote area. This measurement can be "gathered" when a call for the information is sent out from the center. The numerous emplacements of such platforms gives the center good knowledge of anticipated water runoff, which is essential for prediction of agricultural productivity and irrigation requirements.[1] An advantage of utilizing meteor bursts - which occur frequently - for "bouncing" signals is the independence from the need for a satellite overflight. Equipment involved is a means of detecting occurrence of a meteor burst and a means of triggering the transmission at the remote site so as to utilize the meteor burst, which is of very short duration - and a coding system that denotes the location of the measurement transmitted.

With innovative development, the technology might be applied to relaying other classes of measurements indicative of incipient "disaster." It might be applied also to providing for a distress-alert to be issued from a remote population - or individuals - to a central monitoring location which could relay information about the distress-alert to those in a position to render assistance.

Independence from communications satellites opens consideration of this means of communication in case of nuclear disaster.

III. REMOTE SENSING SATELLITES: Disaster Applications

Orbiting remote sensing satellites have earth's surface under observation continually for purposes of intelligent resource management. They can swiftly detect incipient and actual changes that warrant attention and serve as warnings of adverse conditions.

Locust Infestation Detection and Control

An example of use of remote sensing satellites for disaster response purposes is the surveying project conducted by the Food and Agricultural Organization of the United Nations - in connection with infestation of locusts in the Horn of Africa.

Killing the insects is not the underlying problem; the more pressing need is early detection. The migratory desert locust is dependent on rainfall to supply soil moisture for egg hatching and for supply of green vegetation on which the locust feeds during the nymphal and non-migratory stages. The quest is to find the locusts before they assemble in large

1. Such a SNOTEL program was developed for the Soil Conservation Service, U.S. Department of Agriculture.

numbers and start migrating. Previous to FAO's application of satellite surveying of the region, each country in the locust dispersion area would send out ground survey teams, without coordination, to cover extensive desert areas.

Replacing random searching for such areas, by limited numbers of ground personnel, with satellite overflight of the region resulted in a dramatic reduction in costs and an equally dramatic increase in effectiveness of control measures.

The FAO maintains a Remote Sensing Unit so that it may render effective assistance in enhancing agricultural production, in issuing warning of drought, desertification, and incipient crop disease. Also under study by this Unit of FAO is the interrelationship between natural disaster impact on a region and other events important to socio-economic development - information that, heretofore, has been sparse and fragmentary because conditions could not easily be recorded and maintained on an archival basis.

Flood and Flood Plain Management

One of the most feasible, effective, and direct uses of remote sensing that has been already demonstrated and utilized in numerous occurrences is the early detection of anticipated flood conditions and, more importantly over the long-range, in flood plain management. The data provided by remote sensing satellites illuminate the following crucial questions:

- what are the major sources of rising water that need to be under observation ?
- what is the nature of the soil and terrain beyond river banks ?
- what is the inundation pattern when flooding occurs ?
- what are the results of the flooding in respect to dwellings, to property, to dams ?

In respect to flood plain management, several innovative investigations are reported in the next section under METEOROLOGICAL ENVIRONMENTAL SATELLITES.

Future Disaster Applications

Proposals for future disaster applications of remote sensing include surveying disaster sites to assess the extent of damage and the resources needed for relief; and how to determine agricultural "recovery" of soil that has been inundated by salt water flooding, such as has occurred in Bangladesh.

IV. METEOROLOGICAL ENVIRONMENTAL SATELLITES: Disaster Applications

Observational satellites utilized for meteorological and environmental monitoring also utilize "remote sensing" - images of weather systems and measurements taken of ocean surfaces and at various levels of the atmosphere that provide a basis for weather and climate prediction and for detection of severe storms and the like. Representative of this are the disaster-related applications of the European Space Agency's METEOSAT. The program provides for the satellite to produce images of earth every half-hour; the images are disseminated by satellite to User Ground Stations in conjunction with the collection and distribution of environmental data gathered from platforms on earth's surface. From these images, for example, it is possible to estimate the intensity of hurricanes that are tracked; the absence of rain clouds over long periods of time that is indicative of incipient drought; infrared measurements of surface temperatures (the diurnal rate of temperature change) that can be linked to the soil moisture content (agricultural application).

Floods and Flood Plain Management: Data Collection

Present (non-satellite transmitted) methods of receiving "alerts" that flood conditions are approaching rely on routine collection of data, usually on a monthly basis; the information is transmitted by telephone lines or radio to the flood forecasting center. Drawbacks to such systems are that the flooding may well disable the telephone lines that are needed for transmission of the data; rivers that are being monitored often lie in valleys and, therefore, a radio alert system would require construction of repeater towers - which have a high installation cost, and maintenance costs that not only are high but which increase with passage of time.

In contrast, satellite transmission of a flood-condition alert offers a cost-effective alternative to telemetry transmission and high reliability. (It can be appreciated that if an alert is received too late, the warning has been useless. If warning is given too often and no flooding occurs, forecast warning issuance is of decreasing value.) The flood control center can, with assurance, issue a warning to those at risk from the flood waters. If the flooding occurs in small catchment areas, the time between the onset of the threatening rainstorm and the flood peak is short (one to two hours); in these situations, an alarm should be triggered at the location itself. In the case of large catchments, readings of maximum rainfall value or maximum river-level value are transmitted to the flood forecasting center where assessment is made by experts; appropriate warning is transmitted to populations at risk and the flood-response teams that are required to take action. The obverse situation is also of use in forecasting of disaster conditions - that of drought occasioned by lengthy periods of minimum rainfall or minimum river-level values. In considering either large catchments or small catchments, the initial alert is crucial to the system.

Recommendation has been made to governments involved in installing flood warning or forecasting systems that they consider the advantages of using space-borne systems, provided the geostationary satellite space-segment is available to them.

Earth Crustal Movements: Data collection

If earth crustal movements could be monitored on a continuous basis and assessed expertly, accurate and timely warning might be given of earthquakes, volcanic eruptions, land slides, and storm surges. Present systems, that do not rely on data collection by satellite, require emplacement of monitoring equipment to be inspected often (once or twice a week) for checking instrument performance and supply of chart paper. The monitoring sites may be in remote, inaccessible regions that are not served by telephone lines. Yet the measurements need interpretation which involves complex analysis by investigators who have available to them modern geodynamics-center equipment- centers that are located far from the measurement sites. In contrast, satellite systems can overcome these disadvantages and offer far higher cost-effectiveness.

Proposed Meteorological Satellite and Data Collection Applications

Proposal has been made of a novel means of issuing warning in native dialects to remote populations on islands in the Pacific that are subject to frequent tsunami occurrences. These are sea waves of enormous power, widespread in the Pacific Basin, that are generated by seismic causes (earthquake activity). The seismic sea waves originate off the coast of Chile and travel at an average speed of 440 miles per hour. The destructiveness is high, affecting a dozen nations. The difficulty is not in detecting them but in communicating quickly and reliably with the island populations - who are not served by conventional communications at all.

The proposal is to utilize a warning center already established in Honolulu that is part of a network of data collection platforms located in the Pacific Basin that collect data, store it, and transmit it when interrogated by overflying satellite. The Honolulu Center determines when and to whom warning should be communicated; this is relayed to the satellite command and acquisition center, which inserts the appropriate warning messages into the interrogation commands to be sent by the overflying satellite. The platforms in the Pacific Basin would be equipped to receive the inserted warning message and could issue a warning transmission that could be received by a particular island, triggering an alarm and a voice broadcast of the warning in the particular native dialect spoken by its inhabitants. To bring such a system into being would require the use of the two GOES-USA satellites already in place and a third satellite, the Japanese GMS,also already in place - in order to cover the Pacific Basin region at risk. It has been estimated that the equipment on the data collection platforms that transmit the warning messages could be provided at approximately $1,000 each - similar equipment has been developed for the issuance of marine warnings from ocean buoys.

V. DISASTER RESPONSE PERFORMANCE

A further aspect of disaster response is the use that can be made of disaster plans and the validity of the plans in actual disaster occurrences.

In the wake of Hurricane David in 1979 which struck the Caribbean area and mainland USA with equal force, more than 1,000 people died in the Caribbean area, victims of floods which swept away entire villages. By contrast, there was a total of 25 deaths in the United States. What made the difference ? The existence in the United States of well-tested systems that assured receipt of warning, involved realistic planning for evacuation, and led to public awareness and response.

A current space technology tool useful in establishing policies, plans and well-honed procedures that enable disaster relief teams to act efficiently when disaster threatens is a Disaster Planning Methodology - a computer simulation model that can predict the outcome of decision-making while planning is still in the formative stages and later can refine it during the actual stages of the disaster. Use of the planning model can improve swiftness of response, enable more efficient use to be made of communications, transportation, relief materiel, and provide more effective medical care.

Analysis supported by NASA is underway at the Naval Research Laboratory to determine what specific information is needed for all classes of disasters in order to determine the appropriate response for each type of natural and man-made disaster. Refinements of the computer model are expected to include prediction programs - for example, the probability that a tropical cyclone will pass within a particular area. This would be a key factor in bringing resources to bear where they are needed, making maximum use of limited manpower and resources.

A companion investigation provides a means of rapid visualization of earth-surface features to assess disaster scenes following occurrence of the disaster and aid in such tasks as planning evacuation and access routes by means of small, simple, low-cost equipment.

VI. CONCLUSION

In conclusion, I have reviewed with you this evening advances reported at our Symposia over the course of the last several years. Demonstrated and well-proven capabilities include rapid institution of communications facilities among relief teams inside the affected area. A detailed proposal for an integrated Land Mobile Space and Terrestrial Communications System has been developed that would bring improved emergency medical response to sparsely populated areas. The methodologies that can give early warning of dangers to be faced in food production are well established. Parallel methodologies can be employed to give warning of natural disasters earlier and reaching populations at risk more reliably. Of equal import is the deveopment of a methodology to guide disaster planning that accurately reflects the outcome of decisions embodied in the plan, identifies inadequacies and targets which ones of a number of possible disaster plans should be put into effect in a specific instance in a specific location.

It is the case that a number of space technology tools - although reflecting proven capabilities - utilize Experimental Satellites, the availability of which in future is not assured.

The institutional problems - as is so frustratingly evident in disaster response efforts - present the challenge. I am optimistic that the private Voluntary Organizations[1] can be very effective in wearing down these barriers. I believe the path will be through the perception that attaining national developmental objectives is linked imperatively with reducing losses in the wake of disaster occurrences - which need not be borne.

As very first steps, the disaster-response Voluntary Organizations could put forward in their high-level liaisons with Less Developed Countries

- wiser land-use planning in vulnerable areas
- arrangements for receipt of technical assistance for the employment of space tools in flood plain management
- on a regional basis, arrangements for bringing space capabilities to bear on highly focused investigations in which the users are deeply involved.

---

1 NOTE: Voluntary Organizations include disaster relief programs administered by religious and other humanitarian societies.

INDEX

This index covers papers presented at the International Space Safety and Rescue Symposia 1976 through 1981. These papers appear either in full or in summary or abstract form in this volume and/or in the microfiche supplements, Volumes 39-41, AAS Microfiche Series. The index consists of (1) a chronological, numerical index by year and by IAA number and (2) an author index. The papers of each author are identified by IAA number and the volume in which they appear.

NUMERICAL INDEX

VOLUME 54 SCIENCE AND TECHNOLOGY, SPACE SAFETY AND RESCUE 1979-81, 1983

VOLUME 40-1 AAS MICROFICHE SERIES, SPACE SAFETY AND RESCUE 1976, 1982

(Ninth International Space Safety and Rescue Symposium, International Astronautical Federation Congress, October 11-16, 1976, Anaheim, California)

---

* *These numbers may also be designated "IAF".*

Only abstracts appear in Volume 54, Science and Technology and the papers appear in full in Volume 40, AAS Microfiche Series.

NUMERICAL INDEX

VOLUME 54 SCIENCE AND TECHNOLOGY, SPACE SAFETY AND RESCUE 1979-81, 1983

VOLUME 40-2 AAS MICROFICHE SERIES, SPACE SAFETY AND RESCUE 1977, 1982

(Tenth International Space Safety and Rescue Symposium, International Astronautical Federation Congress, September 30 - October 1, 1977, Prague, Czchoslovakia)

Note: Only abstracts appear in Volume 54, Science and Technology and the papers, unless otherwise indicated, appear in full in Volume 40, AAS Microfiche Series.

** Only an abstract or summary was available for publication.*

IAA 77-A41 The Probability of Decompression Sickness as a Result of Going Out in a Pressure Suit from a Spacecraft with an Atmosphere Close to that on Earth, A.S. Barer, L.G. Golovkin, S.N. Filipenkov, I.N. Chernyakov, A.A. Sheikin

IAA 77-A42 Advanced Vehicle Concepts for Earth-Orbit Transportation and Rescue, R.C. Haefeli

IAA 77-A43 Survey of Recovery Capabilities, G.W. Heath

IAA 77-A43A Worldwide Disaster and Rescue Response Employing Space-Borne Systems, I.H.Ph. Diederiks-Verschoor

IAA 77-A70-A72 Not Available

IAA 77-A73 A Space-Based Public Service Platform for Terrestrial Rescue Operations, R. Fleisig, J. Bernstein, D.C. Cramblit

IAA 77-A74 Not Available

IAA 77-A75 Distress Detection, Location, and Communications Using Advanced Space Technology, W.E. Sivertson, Jr.

IAA 77-A76-A80 Not Available

IAA 77-A81 Worldwide Disaster and Rescue Response, Discussion Perspective, G.W. Heath

NUMERICAL INDEX

VOLUME 54 SCIENCE AND TECHNOLOGY, SPACE SAFETY AND RESCUE 1979-81, 1983

VOLUME 40-3 AAS MICROFICHE SERIES, SPACE SAFETY AND RESCUE 1978, 1982

(Eleventh International Space Safety and Rescue Symposium, International Astronautical Federation Congress, October 2-3, 1978, Dubrovnik, Yugoslavia)

Note: Only abstracts appear in Volume 54, Science and Technology and the papers, unless otherwise indicated, appear in full in Volume 40, AAS Microfiche Series.

* *Only an abstract was available for publication.*

NUMERICAL INDEX

VOLUME 54 SCIENCE AND TECHNOLOGY, SPACE SAFETY AND RESCUE 1979-81, 1983

VOLUME 39 AAS MICROFICHE SERIES, SPACE SAFETY AND RESCUE 1979, 1982

(Twelfth International Space Safety and Rescue Symposium, International Astronautical Federation Congress, September 16-22, 1979, Munich, Germany)

IAA 79-A18* Annual Survey of Spaceflight Safety Systems: 9th Supplement, Survey Period: July 1978 - June 1979 (AAS 79-325), N.E. Brown

IAA 79-A19 Manned Remote Work Station - Safety and Rescue Considerations (AAS 79-326), C.A. Nathan

IAA 79-A20† Mission Assurance for Spacelab Utilization (AAS 79-327), H. Schürmanns

IAA 79-A21† Safety Problems - Identification and Monitoring of Hazards for Manned Reusable Spacecraft (AAS 79-328), J. Leupen

IAA 79-A22 Considerations Associated with the Introduction of Female Crewmembers in Spacecraft and Space Stations (AAS 79-329) J.W. Brown

IAA 79-A23 Not Available

IAA 79-A24 Psychological Training - An Important Factor of Increasing Space Flight Safety (AAS 79-330), G.T. Beregovoy, I.V. Davydov, N.V. Krylova, I.B. Solovyeva

IAA 79-A25 Not Assigned

IAA 79-A26** Perspectives on Worldwide Disaster and Rescue Response Employing Space-Borne Systems, A Continuing Survey Paper: 1979 Supplement, G.W. Heath

IAA 79-A27 Not Assigned

Note: Unless otherwise indicated papers appear in full in both volumes.

* *AAS numbers were also assigned to some of these papers.*

† *Abstract in Volume 54, paper in full in Volume 39 AAS Microfiche Series.*

** *Paper appears in Volume 54 only.*

IAA 79-A28† The French Local User Terminal for the SARSAT Demonstration and Evaluation Phase (AAS 79-331), H. Castetbert, C. Lerr, M. Midroit, M. Winterholer

IAA 79-A29** The Application of Communications Satellite Technology to Emergency Medical Services, J. Freibaum, C.D. Burge

IAA 79-A30 Satellite Communications for Disaster Relief Operations (AAS 79-332), J.N. Sivo

IAA 79-A31† Worldwide Disaster and Rescue Response Employing Space Systems (AAS 79-333), E.E. Anderson

IAA 79-A32** Some Consideration of Satellite Technology Applications for Disaster Matters - Looking to the Future, H.G.S. Murthy, D. Felske

IAA 79-A33† COSPAS Project - A Satellite-Aided Experimental System for SAR Applications (AAS 79-334), Y.G. Zurabov, L.S. Pcheliakov, V.A. Bogdanov, I.S. Bronitsky

IAA 79-A34 Employment of Large Structure Communications Satellites for Emergency Calls (AAS 79-335), G. Landauer, E. Messerschmid

NUMERICAL INDEX

VOLUME 54 SCIENCE AND TECHNOLOGY, SPACE SAFETY AND RESCUE 1979-81, 1983

VOLUME 41-1 AAS MICROFICHE SERIES, SPACE SAFETY AND RESCUE 1980, 1982

(Thirteenth International Space Safety and Rescue Symposium, International Astronautical Federation Congress, September 25-26, 1980, Tokyo, Japan)

Note: Unless otherwise indicated, abstracts appear in Volume 54 and the papers in full in Volume 41, AAS Microfiche Series.
* *Paper appears in full in both volumes.*

IAA 80-26 Maritime Distress Radio Call Systems Employing Space-Borne Technology, W. Goebel, G. Landauer, E. Messerschmid

IAA 80-27 Development and Testing of a Satellite-Aided Maritime Distress Alerting System, J.J. Fee, Y. Kaminsky, W. Scales, R.E. Todd

IAA 80-28 Possibility of Creating Combined Search and Rescue Satellite-Aided System, Y.G. Zurabov, V.A. Bogdanov, I.S. Bronitsky

NUMERICAL INDEX

VOLUME 54 SCIENCE AND TECHNOLOGY, SPACE SAFETY AND RESCUE 1979-81, 1983

VOLUME 41-2 AAS MICROFICHE SERIES, SPACE SAFETY AND RESCUE 1981, 1982

(Fourteenth International Space Safety and Rescue Symposium, International Astronautical Federation Congress, September 11-12, 1981, Rome, Italy)

Note: Unless otherwise indicated papers appear in full in both volumes.

IAA 81-265* Space Technology Contributions to Emergency and Disaster Management, P.B. Richards

IAA 81-266 Satellite Alert Warning - Catalyst for an International Disaster Response Legal Regime? J.T. Stewart, Jr.

IAA 81-267/268* Maritime Distress Alerting System, U. Hammerschmidt, K.P. Galligan

IAA 81-269 Analysis of the November 23, 1980 Earthquake as a Design Basis for Satellite Emergency Communication, G. Barretta, S. Rossignoli, P. Daly, H. Fromm, D. Tits

IAA 81-270 Not Available

IAA 81-271 A New Shore-to-Ship-Alert Channel Via Maritime Satellites (SHOSAC), W. Goebel

* *Abstract in Volume 54, complete paper in Volume 41, AAS Microfiche Series.*

AUTHOR INDEX

Note: Papers appear in full unless indicated by the word "Abstract" or "Summary". A semicolon separates one item from another. "Mic" refers to AAS Microfiche Series; "S&T" refers to Science and Technology Series.